EVOLUTION

EVOLUTION
The History of an Idea

Revised edition

PETER J. BOWLER

UNIVERSITY OF CALIFORNIA PRESS
Berkeley Los Angeles London

University of California Press
Berkeley and Los Angeles, California

University of California Press, Ltd.
London, England

Library of Congress Cataloging in Publication Data

Bowler, Peter J.
Evolution, the history of an idea.

Bibliography: p.
Includes index.
1. Evolution—History. I. Title
QH361.B69 1983 575'.009 83-5909
ISBN 0-520-06385-6 (alk. paper)
ISBN 0-520-06386-4 (pbk.)

Printed in the United States of America

1 2 3 4 5 6 7 8 9

For Caroline and Ian

Contents

Preface to Revised Edition

After putting so much effort into preparing the original edition of this book, I was pleased to see it received enthusiastically by reviewers. Other scholars and teachers in the field evidently agreed with my own assessment of the need for a survey of this kind and felt that *Evolution: The History of an Idea* was successful in meeting that need. Continued sales indicate that it has gained a significant place as a textbook for courses in this area of the history of science. Clearly, if the book is to continue its function as an introduction to current thinking on the emergence of evolutionism, it must be brought up to date at regular intervals. This is one of the most active fields in the history of science, and the last few years have seen some major (and a host of minor) developments. I have thus prepared a significantly revised edition that should greatly increase the book's value to teachers and to anyone seeking an overview of the subject. Nearly one hundred works have been added to the bibliography, with corresponding references included in the text. There are dozens of minor additions to the text itself which provide brief outlines of new ideas and interpretations. There are also several more substantial additions, including sections on the origin of genetics, theories of human origins, and the interaction between evolutionism and the human sciences. The final chapter on the modern debates has been extensively revised. And there are indications of my current thoughts on the possibility of constructing a rather different approach to the subject, as outlined at the end of chapter 1. I have also taken this opportunity to correct a few minor errors, some of which were pointed out by reviewers.

PETER J. BOWLER

The Queen's University of Belfast
January 1988

Preface

Because the structure of this book is rather unconventional, most of this preface will be taken up with an explanation of why it has been written this way. The book is a survey of the history of evolutionism, presented not as an academic monograph but as an introduction accessible to someone with no background in either biology or history. It is not, however, a popularization in the sense of turning history into bedtime reading. It is aimed at those with a serious interest in the theory of evolution and its implications who have not tackled the subject in any detail before. Primarily, in fact, it is intended as a textbook for university courses in the history of science, although I hope that it will prove of value to professional biologists and historians looking for a compact guide to the field. Because the potential readership is so diverse, specialists from one side or the other will have to bear with the text when it describes what seems to them a trivially obvious point—it may not be so obvious to someone on the other side of the fence.

A number of books surveying the history of evolutionism are already in print, some of them of excellent quality. But most are a couple of decades old, and none seems to have been written as a simple guide to how historians are tackling the issues involved. Let me set out what I believe to be the essential characteristics of a book written for nonspecialists, including university students.

First, the book must be organized to allow a systematic introduction to the issues, taking nothing for granted. This is particularly the case in the history of science, where many of those who become interested do not have scientific backgrounds. A nonscientist may not understand, for instance, the principles of the Linnaean nomenclature—and unless he is told exactly what is involved, this will remain a puzzle to him whenever he encounters it. In addition, the historical issues must be presented so that the nonspecialist (in this case the scientist himself) can appreciate the different ways of looking at them. I have not hesitated to express my opinion on disputed issues, but I have tried to describe the alternatives and suggest where further information about them can be found.

Second, an introductory book must be comprehensive. It must cover all topics that could be of interest to anyone studying the field. This is especially true when the book is to be used for introductory level university courses, where something like the traditional idea of a textbook is required. So this is *not* a book on the Darwinian revolution alone. It is a history of "evolution," using that term in its widest possible sense to denote any theory postulating a natural process for the development of life on earth. Much important work was done in geology and natural history before the *Origin of Species* appeared, and one cannot properly understand the impact of Darwin's theory without some grounding in the earlier developments. Conversely, the post-Darwinian situation must be dealt with as well. Trying to teach nonscientists about the impact of evolutionism without introducing them to the events leading up to the "modern synthesis" is ridiculous. At all stages, the relationship between the scientific theories and the culture in which they appeared needs to be discussed.

The penalty for trying to provide comprehensive coverage of so wide a field is that the amount of space available for each topic is limited. This is why I have abandoned the normal academic practice of giving extensive quotations, footnotes, and so forth, to flesh out the narrative. I also have pared down to a minimum information on the lives and backgrounds of the scientists involved. The main purpose of this book is to introduce the ideas themselves, in all their complexity, in as straightforward a manner as possible. There are many books that provide such background material on the individual personalities, including the *Dictionary of Scientific Biography* (Gillispie, ed., 1970–80). I do not believe, in any case, that students at the introductory level are interested in such background or in the academic conventions. For them, the important thing is to get over the basic ideas as clearly as possible. In a teaching situation, the lecturer himself can supplement the textbook with extra material in those areas where he wishes to specialize.

This brings me to a third point: the book must serve as a guide to further reading. I anticipate that many professionals will find this book useful precisely because of its bibliography. For students preparing term papers, the bibliography will also prove invaluable. It might be objected that the material cited is too advanced for new students, but we have a responsibility to guide such students toward further reading. I hope that the references in the text will allow them to find the items in the bibliography most relevant to their needs. It would be easy for lecturers to provide their students with an additional guide to the bibliography. More advanced students approaching the subject for the first time also should find the bibliography useful. In the history of science, courses up to and including the graduate level are often taught to students with no previous background. Such students are placed in a difficult position, because they must absorb the basic points and then pass straight on to more advanced reading. I hope this book will ease their burden.

My original intention was to provide each chapter with its own bibliography, which might have seemed less intimidating to the uninitiated. This plan was abandoned because the amount of duplication would have significantly increased the cost of typesetting. I also had intended to list primary and secondary sources separately, but this proved impossible for a single bibliography, because the distinction becomes unworkable when dealing with twentieth-century science. Wherever possible, I have tried to cite translations of works published originally in a foreign language. For primary sources printed before 1900, I have concentrated on editions that have been recently reprinted and should be more widely available. The secondary sources concentrate on classics in the field, plus the specialized literature of the last few decades. I am grateful to my departmental secretary, Mrs. Hilary Joiner, who did the original typing for the bibliography.

As an introduction to the development of perhaps the most controversial of all scientific theories, this book is meant to help bridge the gap between the "two cultures" that still divide our society. The history of science is one of the few areas in which students from both the humanities and the sciences come face to face with each other, and can appreciate that there is a genuine relationship between what each is doing. Professional historians of science have a responsibility to ensure that suitable reading material is available to help in this reconciliation, without trivializing the issues through overpopularization. Yet despite constant calls for more to be written for the nonspecialist audience, most of us continue to produce our highly technical articles and monographs. This is my own effort to relate the issues I am familiar with at a level comprehensible to the nonspecialist. Whether or not the book is a success, the goal of providing this kind of literature is vital if the history of science is to serve its true function.

In its last two chapters, the book gives some background on the debates that currently are raging over the mechanism of evolution and the teaching of evolution in the schools. Scientists may be interested to see how some aspects of the modern debates raise issues that have been controversial throughout the growth of evolution theory. A few years ago, it would have been difficult to write a history of evolutionism except from the perspective that the modern form of Darwinism represents a triumphant climax to the process. Now, we see that the basic issues were not settled quite so conclusively. I do not expect scientists to derive any technical insights from reading about the work of their predecessors, but they may gain a better insight into the nature of the fundamental issues raised by what they are doing. On the question of creationism, I have expressed myself rather more forcefully than elsewhere in the book. The historian has as much right as anyone to comment with authority on a system that would, in effect, return us to a theoretical position last taken seriously by working naturalists in the early eighteenth century. Yet I have suggested also that scientists themselves may gain a better understanding of their own position when they acknowledge the complex status of scientific theories in general and evolution theory in particular.

My qualifications for writing this kind of book are twofold. First, I have spent a number of years teaching the history of evolutionism at various levels in universities in three different countries (Canada, Malaysia, and the United Kingdom). This, I hope, has given me some insight into the difficulties of presenting the essence of complex intellectual developments to students unfamiliar with the field. Second, I have published—originally by accident and later by design—research in most areas of the history of evolutionism from the eighteenth to the early twentieth century. Although there are others more competent than myself in any one area, few historians of science will have had direct experience in so wide a section of the time period that must be covered. The one area I am not directly familiar with is modern biology, including the material of the last two chapters. I have done my best to present a layman's introduction to modern Darwinism and its opponents. I hope that I have not oversimplified or misconstrued any important points, or misrepresented the views of those who are currently engaged in the debates.

My own intellectual debts are too numerous to record here in detail. It was Robert Young who first aroused my interest in the origins of modern evolutionism, but since then I have benefited from the thoughts of a host of scholars who are represented in the bibliography. Particular mention must be made of the two referees who read the original draft of the manuscript: John C. Greene and Malcolm Kottler. From their very different perspectives, they tore the whole thing apart and advised me on how to put it back together again in a much improved form. They also advised on a host of detailed points of information and interpretation. Needless to say, where opinions are expressed, they are my own, and any remaining mistakes are my responsibility.

PETER J. BOWLER

The Queen's University of Belfast
September 1982

1

The Idea of Evolution: Its Scope and Implications

For historians of science, the "Darwinian revolution" has always ranked alongside the "Copernican revolution" as one of those episodes in which a new scientific theory symbolizes a wholesale change in cultural values. In both cases, fundamental aspects of the traditional Christian world view were replaced by new interpretations of the universe. The medieval cosmology had pictured the earth as the center of a hierarchical cosmos stretching up through the perfect heavens to the abode of the Almighty. Because the physical universe was the stage on which the spiritual drama of man's creation, fall, and redemption was played out, his location upon the central body seemed only natural; and this fitted the commonsense notion that the heavens rotate around a stationary earth. Copernicus taught instead that the earth is merely the third planet orbiting around the sun. Although he wished thereby to display the true harmony of the universe, his followers realized that by breaking down the barrier between the earth and the heavens he had created a new, nonhierarchical universe operated by fixed laws of dynamics. Man was not the center of everything; he was simply the inhabitant of a single planet orbiting around what soon was recognized as an insignificant star, lost in the immensities of space. The heavens might still declare the glory of God, but it had become harder to believe that they did so for the benefit of man alone.

Yet one could still believe that man was spiritually far from insignificant. To the Christian, man remained unique in his ability to perceive the moral dilemma of his existence. He was still master of a world apparently designed to support him as lord of creation, the highest link in a "chain of being" that united all living things into a natural hierarchy. Surely no natural process could explain the existence of such an orderly system of life or of the spiritual faculties of man himself. The Book of Genesis assures us that everything was created by divine will in a period of only six days, with man as the final act of creation. The events of the Darwinian revolution—which actually began long before Darwin was born—undermined the traditional belief in man's innate superiority. As geologists investigated rocks, it began to appear that Genesis was not a good guide to the earth's history. There were vast periods of time before man appeared on the scene, when it was inhabited by strange creatures unlike any known today. Was it possible that natural forces alone had created this great sequence of living things ignored by the biblical account? And if so, must not man himself become merely the last of these natural products, no longer lord of creation but just a superior ape? Copernicus demoted man from his central position, but Darwin's theory required a complete reinterpretation of our spiritual role in creation.

Such changes were not achieved without arousing considerable resentment from those committed to the old world view. The trial of Galileo in 1632 revealed the extent to which the Catholic church was determined to resist Copernican astronomy. Yet in this case, opposition soon died down, and after 1700, no one seriously challenged the new cosmology. Opposition to Darwinism, however, shows no sign of dying down even today. There are still powerful forces at work in our society which compel people to believe in a God who is personally responsible for the creation of man. For those who accept a fundamentalist interpretation of Christianity, this means that the creation story of the Bible must be literally true—and hence that evolution must be opposed as the most extreme manifestation of subversive atheism. In 1925, a similar wave of fundamentalism led to the passing of Tennessee's notorious Butler Act that forbade the teaching of evolution in schools and to the "monkey trial" of John Thomas Scopes for contravening this regulation. The resulting publicity undermined this first attempt to oppose the spread of evolutionism, but now Darwinism is being challenged again by creationists who insist that their own position is just as valid even in the scientific sense and should be given equal time in schools. Whatever the merits of scientific creationism, the real purpose of this movement is to stem the tide of what is perceived as atheistic Darwinism. Clearly, the Darwinian revolution has not yet reached its final conclusion.

The enthusiasm of modern creationists can all too easily lead us to forget that the relationship between science and religion has not always been one of open warfare (Barbour, 1966, 1968; Peacocke, 1980; Russell, 1973). On the contrary, scientists often have been concerned to ensure that their ideas are consistent with some form of religious belief. It even can be argued that Christian values have played a vital role in the development of modern science (Jaki, 1978*a*). The belief that a wise Creator would build an orderly universe intelligible to man may have encouraged growth of the scientific concept of natural law. Christianity's essentially *historical* view of man's spiritual development may have provided the essential foundation for building evolutionary concepts. There certainly have been some religious thinkers who positively welcomed the theory of biological evolution. As long as the Bible story can be accepted as a revelation about the *purpose* of creation, rather than a detailed history of how it actually happened, evolution becomes the process whereby the divine purpose is achieved (Teilhard de Chardin, 1959). Some scientists are suspicious of the attempt to see a spiritual purpose in the biological process of evolution; but for many other thinkers, this is precisely what makes the idea of evolution so exciting.

Scientists who repudiate any attempt to find a religious meaning in evolution would argue that the theory is simply a product of the objective study of how nature works, to which no moral or spiritual values should be attached. Science may have been influenced by religious thought in its infancy, but over the last few centuries the rise of modern science has represented the emancipation of Western man from the bondage of ancient superstitions. Evolution is the inevitable product of the decision to treat even the origin of mankind as a subject open to rational investigation. Yet this image of science as a purely objective search for knowledge has its own moral and cultural values. Creationists would argue that it is precisely in its efforts to reduce man to the status of an animal under the control of only natural forces that modern science reveals the materialistic philosophy underlying its activities. Other critics would agree that in the area of "social Darwinism" we see the scientific theory as an outgrowth of the policy of competitive individualism that has dominated our society since the rise of capitalism (Montagu, 1952; Young, 1985). Evolution theory in particular and science in general thus can be regarded either as illuminations of the traditional concept of a divinely purposeful universe or as the source of an alternative philosophy by which some elements in our society have chosen to live. In either case, it is clear not only that the scientific theory itself can have wider implications but also that the supposed objectivity of the scientific method may be seen as part of a wider philosophy. Because of its immediate implications for human nature, evolution theory focuses those crucial questions about the relationship between knowledge and values that may underlie the prominent role science now plays in the world. A study of the history of evolutionism may help us to come to grips with these controversial issues.

Before we can undertake such a study, certain points must be clarified to give us a framework for understanding the rise of modern evolutionism. First, we must be able to recognize the full range of issues that distinguish the modern theory of evolution from the traditional religious view of nature. We also must be able to see how certain aspects of the traditional view could be preserved or even reinforced by some interpretations of the evolutionary approach. To do this, we need to appreciate the complexity of the basic idea of evolution, which can be expanded in a number of different directions, each with its own implications at a wider level. Finally, we must look more closely at the problems the historian faces as he tries to chart the rise of scientific evolutionism. In particular, these problems arise from the tension between the normal view of science as an objective search for knowledge and the suspicions of many critics that scientific theories are themselves value-laden contributions to philosophical and ideological debates.

THE OLD AND THE NEW WORLD VIEWS

The theory of biological evolution is really only part of a whole new approach to the study of the earth's past that has been developed over the last few centuries. Long before Darwin tackled the question of the origin of new species, geologists and cosmologists had begun to challenge the world view of medieval Christianity by postulating that the earth itself and even the universe as a whole have changed significantly over a vast period of time. Only within this new vision of an evolving *physical* universe did it become possible to imagine that living things also might be subject to natural change. The essence of the modern scientific viewpoint is that all features of the natural world, cosmological, geological, and biological, can be explained as the result of natural forces operating over long periods of time. Within this general program, we can distinguish several basic issues that form the challenge to the traditional world view.

The Expansion of the Time Scale

The traditional universe was extremely short-lived since the six days of creation were supposed to have occurred only a few thousand years ago. In the seventeenth century, Archbishop James Ussher tried to calculate the date of creation by working back through the biblical patriarchs to Adam and fixed the year as 4004 B.C. John Lightfoot, vice-chancellor of Cambridge University, declared that the final act by which man was created took place at nine o'clock on the morning of Sunday, October 23, 4004 B.C. Modern creationists do not fix the date quite so precisely but still insist that the earth was formed only a few thousand years ago. By the standards of modern science, these estimates are trivial: geologists and cosmologists now put the earth's age at between four and five billion years. This vast extension of the time scale took place gradually, as geologists learned more about the extent of the changes that have taken place on the earth's surface. Frequent efforts were made to limit the amount of extra time required by geological theory. But already by Darwin's time, no educated person doubted that the earth was at least some millions of years old. (The creationist challenge on this issue is strictly a twentieth-century phenomenon.) The whole issue was made particularly sensitive by the fact that paleontologists found no evidence for the existence of man except in the most recent geological past, thus reducing human history and prehistory to but a moment in the vast panorama of the earth's development.

The Concept of a Changing Universe

The traditional world view was essentially static. In the six days of creation, God formed the world just as we see it today, including the plants, animals, and man himself. Modern creationists recognize that some geological formations may have been laid down by Noah's flood but still believe that living things have remained unchanged since the creation. In the organic world, at least, the traditional view assumes that there *cannot* be any change because the forces of nature can only maintain the original forms created by God—they are not by themselves creative. This assumption was not derived purely from biblical authority but was backed up by the synthesis of Christianity with the philosophy inherited by the medieval world from the ancient Greeks. The views of Aristotle, in particular, were regarded as an important foundation of the belief that each species has a typical form maintained by the process of reproduction from one generation to the next. The hierarchy of natural forms stretching from the most primitive up to man—the "chain of being"—represented a complete and hence absolutely fixed plan of creation. By contrast, the modern view assumes that we live in a continually changing world, although the processes involved may be so slow that we do not notice them. Geology teaches us that the earth itself has passed through immense changes, while the fossil record reveals a series of extinct populations thought to be linked by evolution.

The Elimination of Design

The intention of creationism is not just to preserve a role for direct supernatural intervention in the origin of species but also to uphold the belief that each form of life has been designed by its Creator. In the classic form of the "argument from design" popular among naturalists until the early nineteenth century, the complexity of each specific form and its careful adaptation to the organism's way of life was held to be direct evidence of the Creator's wisdom and benevolence. Unaided nature never could have produced such structures; therefore, divine will had to be invoked as the only reasonable explanation of their existence. The belief that the structure or development of natural forms can only be explained by the *purpose* they are supposed to fulfill is known as "teleology." The whole thrust of modern evolutionism has been to eliminate the need for a supernatural purpose in accounting for the present structure of living things. Darwin conceived his mechanism of natural selection to show that everyday forces of nature can adapt each species to its ever-changing environment, without the need to suppose that the process is intended to achieve some predetermined goal. At an even more basic level, modern biologists also believe that natural processes can account for the origin of life from nonliving matter, by a process of "chemical evolution" leading to ever more complex physical structures that eventually take on the properties of life.

It is important to note that the argument from design can exist independently of the biblical creation story. Many nineteenth-century paleontologists accepted the supernatural origin of new species and invoked the argument from design, although they believed that production of new forms had occurred at various stages in the earth's history. At a rather subtler level, it is also possible to argue that the Creator *intended* the present structure of the earth to emerge from the original form in which He created the universe. In Descartes's interpretation of the "mechanical philosophy," the earth was formed by natural means from matter distributed in space—but because God created both the original distribution and the laws that govern the behavior of matter, He had clearly foreseen the end product and could thus be said to have designed the evolutionary process itself. Even Darwin accepted the concept that God had established the general laws by which life evolves, although he was forced to concede that the details of what happened were not the result of divine forethought. The real problem with natural selection, though, was that the "survival of the fittest" in a universal "struggle for existence" did not sound much like the kind of mechanism a benevolent God would choose to achieve His ends.

The Elimination of Miracles

The Genesis story of creation is clearly meant to uphold the belief that the Almighty not only designed all things in the universe but played a direct and personal role in supervising their formation. The biblical concept of miracle, however, does not confine the Creator's activity to the beginning—it allows Him to intervene from time to time throughout the continuing history of the world. Paralleling this, we have already noted that some early paleontologists were willing to admit divine intervention at the beginning of each geological period to account for the appearance of new species. Biological evolution, however, is intended to exclude any role for supernatural intervention in the world because it assumes that natural forces by themselves are sufficient to create new species. In the eyes of Darwin and his followers, it was only by accepting this policy of "naturalism" that the question of the origin of species could be opened up to scientific investigation. To appeal to the supernatural as soon as one reached the limits of existing natural explanations was to close off the route to any further research that might generate more satisfactory hypotheses. Miracles are by definition arbitrary violations of the normal laws of nature and as such cannot be studied by the methods of science. To admit their occurrence in order to explain the origin of certain structures in the world is to concede that a phenomenon lies forever beyond our comprehension—unless we accept the dictates of supernatural revelation.

The elimination of the supernatural was, however, no more straightforward than the elimination of design. The mechanical philosophy mentioned earlier eliminated the need for supernatural agencies except at the very beginning of the universe but retained design by supposing that the Creator intended the laws of nature to produce the results we observe. This view compared the universe to a gigantic piece of clockwork, built by the "clock-maker God" to run on inexorably toward its intended goal. There would be no need for God to concern Himself with His creation once it was formed—the religious philosophy known as "deism." Many religious thinkers, by contrast, believe that God must be involved with the universe at all times—the philosophy of "theism." It may even be supposed that the laws of nature continue to operate only because they are upheld by His will. In this case it will be less easy to make a clear distinction between laws of nature and miracles because both equally are manifestations of divine power, one operating continuously, the other at irregular intervals. Some nineteenth-century scientists tried to argue that there must be special "laws of creation" by which God continues to shape the development of life, laws that could anticipate future goals and work toward them because they embodied divine foresight. To Darwin, such a concept of law was worse than a miracle because it allowed the nonmechanical aspects of the supernatural to interfere continually with the regular operations of nature. For science to be possible, it was necessary to conceive the laws of nature so that they operated solely in a mechanical fashion, allowing the past (but not the future) to control the present by the normal rules of causality. To introduce God's foresight as the explanation of an evolutionary trend was just as much an abrogation of the scientist's duty to search for natural causes as was the more simpleminded appeal to miracles.

The Inclusion of Man within Nature

The Bible tells us that man was the last creation, formed in the image of God and given dominion over the rest of nature. To emphasize this superiority, Christianity assumed that of all the animals man alone was given a soul transcending the life of the physical body. Man thus stood above nature instead of being part of it. The theory of evolution emphasizes man's position as a member of the animal kingdom. He may possess characteristics developed far beyond those of any other species, but he is joined to the rest by a process of development in which no totally new element can have been introduced suddenly. This raises crucial questions for anyone who wishes to emphasize the role of man's higher faculties, whether or not they are seen as the product of a distinct spiritual element in human nature. Can our moral and ethical beliefs be explained away as the result of natural evolution, or do they show that in man, at least, nature has produced a being capable of transcending the struggle for existence? The unpleasant implications of social Darwinism were a direct consequence of the belief that all human characteristics depend on the processes of natural evolution. Similar issues again have been raised by introducing sociobiology to explain apparently altruistic behavior of animals in terms of natural selection—does this imply that man can be treated in the same way? Not only religious thinkers but also many social scientists oppose the attempt to explain human nature solely in terms of biology and insist that man has characteristics that lift him to a higher level of activity. The only possible way to accept the evolutionary link between man and the animals, yet preserve a spiritual component in human life, is to treat evolution itself as a spiritual as well as a material progression. Man then becomes only the latest product of a universal trend toward higher levels of activity, and the sense of purpose in nature that science has so carefully eliminated is reintroduced.

THE POSSIBILITIES OF CHANGE

The complexity of these broader issues will seem all the more obvious with a closer look at the idea of evolution itself. Only in the most general sense can we talk about *the* idea of evolution, as though it were a unified concept. At this level, "evolution" means no more than the belief that the existing structure of the world we live in has been formed by a long series of natural changes. As soon as we begin to unpack this basic statement, we realize that there are many different ways in which these changes can be imagined to have taken place. Many different concepts of natural change have, in fact, been explored by scientists at one time or another, with a consensus emerging only after considerable debate over alternatives. Geologists disagreed for centuries over the basic nature of events that have shaped the earth's surface, and the modern theory of continental drift has emerged only in the last few decades. Biologists too have postulated a whole range of different ideas about how life may have evolved. The Darwinian theory of natural selection is only one of the possibilities. Widely popular since its synthesis with Mendelian genetics in the 1930s, it now is being challenged once again by a number of working biologists. A basic framework is needed for trying to understand the many theories that have been suggested to account for the development of the earth and its inhabitants.

Even the word "evolution" is of little use to us here because it has been given many different meanings (Bowler, 1974). The Latin *evolutio* means "to unroll" and implies no more than unpacking a structure already present in a more compact form. The first biological use of the term "evolution" was to describe the growth of the embryo in the womb, which many people today still imagine to be a kind of small-scale model of the more general process of life's development on earth. Many early embryologists, however, believed that the growth of the embryo was no more than the expansion of a preformed miniature of the complete organism, already present in the fertilized ovum. This would be a process of a character quite unlike the popular image of progressive evolution, although it could quite aptly be described by the original Latin meaning. By 1800, this "preformation theory" had been discredited, and the evolution of the embryo was thought to be a goal-directed process by which a complex structure was built up out of unformed matter. This comes closer to the modern idea of evolution, but it is important to note that by using the growth of the embryo as a model, one is given the impression that living structures ascend a fixed pattern of development toward a predetermined goal. The earliest applications of the word "evolution" to the history of life on earth carry a similar implication because many nineteenth-century naturalists thought that the embryo recapitulates the ascent of life toward the pinnacle of creation: man.

The progressionist implication was retained in a rather different form by the philosopher Herbert Spencer, the person who did most to popularize the term "evolution" in its modern context. Spencer advocated a system of cosmic progress, which included a theory of the inevitable evolution of life toward higher forms. Darwin's theory came to be tagged "evolution," even though he seldom used the term himself; and most people still imagine that evolution is an essentially progressive process. Both Darwin and Spencer made an important step beyond the embryological concept because they believed the process was open-ended, rather than directed toward a single goal such as man. Spencer still insisted that evolution involved a necessary advance toward higher levels of organization, thus introducing a more sophisticated concept of progress. But Darwin was suspicious even of this, because he felt that the concept of biological progress was very difficult to define. The popular idea of evolution as progress is now seen to be inadequate on two counts. It is ambiguous, because we can define progress either as a movement toward a predetermined goal or in terms of ascending levels of general complexity. It is also misleading, because some interpretations of evolution involve only change, without implying any form of progress.

Clearly, even before we try to describe details of any particular theory about how evolution works, we must consider certain basic differences in the way that natural processes can be imagined to operate. In an attempt to clarify these basic differences, Stephen Gould (1977*a*) has suggested three fundamental distinctions for evaluating any theory of evolution. Each pair of alternatives allows us to define a particular attribute of the theory, and thus we can pinpoint the most crucial issues over which scientists have disagreed in the course of building the modern theory of evolution. We cannot, however, say that any particular combination of positions is the "correct" one: the latest scientific debates have reawakened basic issues that were once thought to have been settled for good. Even today, we are still debating the "eternal metaphors" of evolution theory.

Steady-State versus Development

To admit that the detailed character of the earth and its inhabitants may change is not enough to define a clear position. We still may ask whether or not the changes will be cumulative, adding up over a long period of time to give a totally new state of affairs. By assuming that the changes are small and vary in such a way as to cancel each other out, it still is possible to argue that the overall picture remains about the same in the long run. This would be a "steady-state" theory, a name popular in cosmology some years ago. A theory of "development" assumes that changes add up to give a state quite different from the one at the outset of the process. The essence of a development is that it involves a *direction* of change from a beginning to a quite different end point, not mere fluctuations about a mean.

The difference between the "big bang" and "steady-state" theories of cosmology illustrates this point. The big bang approach interprets observed expansion of the universe as the flying apart of matter caused by a gigantic explosion billions of years ago. Here is a true development: from the explosion of the "cosmic egg" there is a cumulative change as galaxies form and then move ever farther away from one another. The steady-state theory, by contrast, postulates the formation of new galaxies at a rate exactly compensating for expansion of old ones. An observer within the universe thus sees exactly the same density of matter *at any point in time*. The appearance and disappearance of galaxies cancel each other out, so that the average distribution always remains the same. The details change but not the overall picture. In this system it is meaningless to talk about the beginning of the universe, because the same pattern has been maintained throughout an indefinite period of time.

This cosmologic debate has its equivalent in geology. Most geological theories have been developmental in character: they assume that the earth had an origin at some point in time and that its original state was quite different from the one we now observe. A typical example would be the "cooling earth" theory, which supposes that the planet began as a hot, molten mass; gradually it cooled down and acquired a solid crust. Such a theory would predict a gradual decline in the level of volcanic activity through time. There have been attempts, however, to formulate a steady-state approach to the earth's history. The most notable case is Charles Lyell's "uniformitarian" geology, which greatly influenced Darwin. Lyell assumed that gradual disappearance of land through erosion has always been exactly compensated for by elevation of new land areas through earthquakes. Thus although individual continents come and go, the relative proportions of land and sea remain the same. Typical of the steady-state viewpoint, Lyell insisted that questions about the ultimate origin of the earth were outside the scope of geology.

In biology, any theory of evolution that insists on an element of progress toward "higher" forms of life (however defined) is an obvious example of the developmental viewpoint. But it is by no means essential to believe that the history of life on earth has been progressive. To back up his uniformitarian geology, Lyell insisted that even the highest forms of life have existed throughout geological time, with only cyclic changes in the character of successive populations corresponding to fluctuations in the environment. His opponents argued that the fossil record reveals a progression from the simplest forms of life to man, establishing the developmental interpretation that is still the popular image of evolutionism. Darwin accepted Lyell's geology and thus held that there has been no *outside* force tending to make life evolve in a particular direction (as there might be, for instance, if it had to adapt to a gradually cooling earth). Although suspicious of efforts to define biological progress, Darwin could not escape the common feeling that in some ways modern forms of life are more advanced than their earliest ancestors. The real function of the theory of natural selection, however, is to adapt species to changing conditions; and in the absence of any directional trend imposed by the environment, the result will be change without progress. Darwin compromised by making progress a kind of by-product of evolution, a long-range statistical trend that did not have to affect every line of development. Other forms of biological directionalism are possible, including an increase in the level of diversity to be found in successive populations.

Internal versus External Control of Evolution

The last paragraph hinted at another important question. Are the forces controlling biological evolution internal to life itself, so that development proceeds along a course determined by the inherent nature of the existing species, or does life evolve by responding to challenges from the external environment, with some degree of flexibility in how it meets each challenge? In the most obvious case of internally controlled evolution, the trend is imposed by a compulsion to develop toward a predetermined goal. Such a view naturally tends to be supported by those who believe that some form of purpose is built into the evolutionary process or that there is a pattern linking all the steps in creation into a unified whole. It is possible, however, to conceive of evolutionary trends predetermined quite naturally by inherent biological forces that may actually be harmful because they are not coordinated with the demands of the environment. Darwin, by contrast, assumed that there is no fixed direction of change, because each species responds as best it can to the challenges posed by an ever-changing environment. The resulting evolution occurs in whatever direction most conveniently solves the problem, unless the species is unable to change quickly enough and becomes extinct. Adaptation is thus the sole driving force of evolution, and to some extent the way each species solves the problems confronting it will be a matter of chance. There can be no regularity in evolution, no predetermined trend toward a fixed goal. Evolution becomes a totally open-ended process because there are no fixed pathways it must travel.

Continuity versus Discontinuity

This has often been treated as the most important distinction between various theories of geology and the development of life. A process of continuous change works gradually, over a long period of time, through adding together minor everyday differences. Discontinuity implies episodes of sudden change, when a relatively stable situation is overturned and replaced by something quite different. This was the old way of understanding the distinction between Lyell's uniformitarian geology and the opposing "catastrophism." Lyell insisted that the only forces that have affected the earth's surface in the past are those we still observe today, including volcanic activity and earthquakes (which are geologically quite trivial, however violent in human terms). Added together over vast periods of time, these small changes can have a major effect, such as destruction or elevation of a mountain range. The opposing catastrophists, however, insisted that the forces we observe today are not enough to account for such major events. Mountain ranges must have been thrown up by great upheavals far exceeding our modern earthquakes in scope. These catastrophes would serve as punctuation marks dividing the earth's history into distinct periods of geological stability. The cooling earth theory gave an explanation of why past events should have been much more violent than those of today, and some historians now argue that this element of directionalism was actually the more fundamental aspect of the catastrophist system.

Turning to biology, creationism can be seen as a theory of discontinuous change, because each new species is designed separately with no direct links to any previously existing form. If the creationist were to argue that a series of miracles was linked into some overall plan or pattern, this would tend to undermine the unique character of the supernatural events and thus blur the distinction. Darwin's theory, though, was uncompromisingly one of continuous change, because natural selection works by slow accumulation of minute variations among individuals of a population. There could be no sudden breaks in the process, no "steps" or "leaps." It is possible, however, to advocate a discontinuous theory of the natural origin of species; the best example of this is the "mutation theory" proposed by Hugo De Vries at the beginning of the twentieth century. Here genetic mutation produced new species instantaneously, because it gave rise to large numbers of individuals who differed radically from their parents. The modern genetic theory of natural selection still uses mutations as the raw material of variation, but it now has been shown that these changes are integrated into the population in such a way that variation appears to be continuous, just as Darwin had supposed. But a few modern biologists once again have raised the idea of discontinuous evolution, both within and outside a Darwinian framework. Thus the debate continues.

By introducing the idea of continuous evolution, Darwin forced naturalists to reconsider the traditional definition of the biological species. Originally, the species had been seen as a distinct, unchanging entity defined by key characteristics that automatically identified any individual belonging to it. Such a view had its origins as far back as the idealist philosophy of Plato, in which the species was thought to exist at a deeper level of reality than the individuals that composed it. The essence of the species was the idealized form or structure defining it, not the superficial characteristics of individuals in any one generation. This concept fitted well the belief that each species was designed by the Creator, who guaranteed its permanence by ensuring that individual organisms could not vary except within rigid limits. When Darwin suggested that evolution proceeds by accumulation of minute differences between individuals, he challenged this whole idealist concept of the species (Mayr, 1964). It is even possible to treat the rise of evolutionism as part of the more general process whereby the old typological concept of the species was undermined (Mayr, 1982). The alternative population approach treats the species as a group of interbreeding individuals, which may have significant differences among themselves. There is no ideal or permanent structure—if the average constitution of individuals making up the population changes, then by definition the species itself has changed.

THE HISTORIAN'S PROBLEMS

Because of the controversy surrounding the Darwinian revolution, there has never been any shortage of historical works trying to analyze its origins and implications. The year 1959 marked the centenary of the *Origin of Species*, and a number of works devoted to the history of evolutionism appeared at about this date. There were three general surveys, by Loren Eiseley (1958), John C. Greene (1959*a*), and Gertrude Himmelfarb (1959). Greene's book remains one of the most scholarly interpretations of the revolution; Eiseley's is more popular and rather opinionated; while Himmelfarb's seems to lack all sympathy with the Darwinian viewpoint. A somewhat uneven set of essays edited by Bentley Glass et al. (1959) concentrates on the pre-Darwinian period. Charles C. Gillispie (1951) demonstrates the extent to which British science was pervaded by religious sentiments in the decades before 1850. The principle of uniformity is explored by R. Hooykaas (1959) and Walter Cannon (1960*a*, *b*), while Francis C. Haber (1959) surveys the history of geology over a wider time period. Biographies of Darwin were published by W. I. Irvine (1955) and Sir Gavin de Beer (1963). Alvar Elegård (1958) provides a valuable survey of the Darwinian debate in the British periodical press. On the theme of social Darwinism, Hofstadter's book (new edition, 1959) was already a classic.

Thus, by the mid-1960s there was a wide range of books available, providing a comprehensive coverage of the rise of evolutionism. But this did not discourage historians from continuing to devote their time to this issue. Some revisions of the existing interpretation were made necessary simply by the emergence of a much wider body of primary information, for instance, the availability of Darwin's own papers, now at Cambridge University Library. To a large extent, though, modern studies have concentrated on reinterpreting the "orthodox" picture of existing facts that had become widely accepted by earlier historians of science. Many preconceptions and hidden assumptions have been challenged by these studies, leading to the growth of new insights into the forces at work within science. To a large extent, the issues debated by modern historians parallel a more general revolution in the way we think about science and its relations to the outside world. We are no longer quite so sure that science is a purely objective search for factual information, as the majority of scientists still like to pretend it is. Increasingly, the suspicion has grown that scientific knowledge is not just a matter of gathering new facts but is also influenced by the cultural and social environment within which scientists work. Historians have found the growth of so controversial a theory as evolution offers an excellent source of examples for exploring this suspicion.

The simplest possible model of science represents it as a straightforward process of factual discovery. However much the general public might imagine this to be a valid representation of the scientists' work, a moment's reflection shows it to be quite inadequate. Scientists are clearly interested not in individual facts but in the universal generalizations we call the "laws of nature," which would have to be abstracted from a collection of individual facts. The technique of extracting such generalizations is the basis of the old "method of induction," once hailed as the guiding principle of all true scientists. But it is no longer thought possible for knowledge to be gained by a simple abstraction from known facts—how, for instance, would one decide which facts were genuinely relevant to a particular phenomenon? Modern philosophers of science recognize that any investigation starts from a hypothesis proposed to explain how a phenomenon *might* operate; the hypothesis is then tested against the facts by observation and experiment. This is the so-called hypothetico-deductive method (Hempel, 1966). Theories are more general hypotheses that may not be testable directly but are used to guide and coordinate the proposal of hypothetical laws. If a hypothesis is successful in passing the tests to which we subject it, we might be tempted to regard it as an established truth about how nature works, but this is *not* a proper interpretation of knowledge gained in this way. It is always possible that a false hypothesis was lucky enough to pass the first, less rigorous tests and may then reveal its weakness by failing new tests in the future. It thus is necessary to regard all scientific knowledge as provisional in nature, accepted as a useful guide for the time being but open to potential falsification by further research.

If scientific knowledge is only provisional, why should it be given any higher status than other forms of knowledge? The conventional answer is that scientific knowledge is more valuable, because it is proposed and accepted in such a way that if it has any weaknesses, these will be recognized and corrected as soon as possible. A truly scientific proposition is formulated to maximize its testability—or, as any test may potentially refute it, its degree of "falsifiability" (Popper, 1959). Willingness to define its statements in a way that leaves them open to rigorous testing is precisely what distinguishes science from those pseudosciences that specialize in vague and slippery generalizations, which can never be falsified by any empirical test. It is recognized generally that the emergence of modern science corresponded to the growth of a profession of investigators who were dedicated to advancing only that kind of knowledge that can be tested. Some would argue, along with Popper, that scientists have consistently been guided in their choice of hypotheses by the criterion of which was the more falsifiable.

Although it can no longer be claimed that science reveals absolute truths about nature, most scientists have remained content with the philosophers' new approach, because it does not challenge the essential objectivity of the scientific method. In general, scientists believe themselves to be objective, because they submit their hypotheses to rigorous testing; and they like to think that their predecessors—the scientists of the last few centuries—were equally objective. This would imply that, once the scientific method had come into use, a cumulative process of discovery began and established the foundations we are still building on today. Clearly, there is a sense in which scientific research both broadens and deepens our understanding of nature. If it did not, to give an obvious example, the increasing sophistication of our technological control over nature would be impossible. While scientific investigations do extend existing theories to explain a wider range of phenomena, the notion of science as an essentially cumulative process runs into difficulties when we look at what are sometimes called the great "revolutions" in scientific thinking; these revolutions have shown previously successful theories to be inadequate and have replaced them with a totally new interpretation of how nature was supposed to work. Such revolutions obviously are not ruled out by the hypothetico-deductive method. If all hypotheses are accepted only on a provisional basis, because they have passed the tests they have been subjected to so far, then we are allowing for the possibility of a future refutation that will undermine the whole theory. In that case, a new hypothesis will have to be thought up to explain all the existing factual information *and* the new area of study that created problems for the old theory. As long as the new theory is more comprehensive than the old, we still can accept the cumulative nature of science, although we must concede that its theoretical foundations undergo major changes of direction from time to time.

In principle, as soon as an existing theory fails an experimental test, the whole scientific community should immediately abandon it and begin the search for a replacement. According to T. S. Kuhn (1962), however, revolutions are actually more complex affairs, because theories exert a much stronger influence over scientists than is allowed for by Popper's approach (see Lakatos and Musgrave, 1970). Successful theories establish themselves as the "paradigm" for scientific activity: they define not only acceptable techniques for tackling problems but also which problems are to be considered relevant subjects for scientific investigation. Furthermore, so great is the loyalty of the profession to the existing paradigm that many scientists will refuse to admit the significance of anomalous facts and will try to pretend that the old system still is functioning smoothly. Only when the number of anomalies becomes unbearable will a "crisis state" emerge, when younger, more radical scientists begin to cast around for a new theory. Eventually a new idea is found which will deal with the crucial problems left over by the old theory. It then will establish itself as the new paradigm, although it may be necessary to wait for the older generation of scientists to die off before the takeover is complete.

Kuhn's approach treats scientists very much as human beings. They develop an emotional and professional loyalty to the paradigm they were educated within, which tends to prolong the life of that paradigm long after it should have become obvious that changes were needed. Here is the first sign of a breakdown in the traditional notion of the scientist's complete objectivity. If Kuhn's interpretation is valid, the old guard that resists the oncoming revolution is being anything but objective—on the contrary, it is using every trick in the book to defend the theory to which so many of its practitioners have dedicated their lives. Still, Kuhn does imply that scientific objectivity prevails in the long run. Eventually the old paradigm is overthrown and replaced by a new one that deals successfully with outstanding problems. Paradigms become successful only because they are able to direct research programs that lead to genuine extensions in our range of knowledge. The scientific profession is organized around such programs, and the loyalty it gives to its paradigm becomes a hindrance to progress only during the brief period of crisis when the necessity for a more fundamental change first becomes apparent.

The classic example of a Kuhnian paradigm change is the Copernican revolution; and at first sight it seems reasonable to suppose that the Darwinian revolution should fit the same pattern. A world view based on the concept of a static universe designed by God was replaced—in the face of considerable opposition—by a totally different view of the universe as a naturally evolving system. The changeover does seem to have taken more time in the case of evolution theory, although this by itself is not a problem for the Kuhnian scheme. After all, the Copernican revolution took well over a century to complete. The Darwinian revolution was also more complex than the Copernican one, because in some cases the traditional viewpoint was able to make a temporary comeback. John C. Greene (1971) has argued that we see here not just replacement of one paradigm by another but an ongoing conflict between two paradigms in which the fortunes swayed this way and that until they were finally settled by the advent of Darwinism. Even this may be an oversimplification, though, because major changes took place within both creationist and evolutionary paradigms in the course of their development.

Perhaps the Darwinian revolution was more complex because it was really something more than a scientific revolution in Kuhn's sense (Ruse, 1970; Ghiselin, 1971; Mayr, 1972*a*). For Kuhn, a revolution is the changeover to a new paradigm within an established science. Before Copernicus, there was an earth-centered astronomy capable of explaining planetary motions with some success. If there was a paradigm in natural history before the Darwinian revolution, it was based on a static viewpoint that ignored the element of change so essential to the evolutionary approach. It was necessary to recognize first that the earth and its inhabitants could have a history, and this required not only the replacement of existing theories but also the establishment of a series of completely new sciences. Disciplines ranging from geology and paleontology to the science of experimental genetics had to be created before the modern form of Darwinism became possible. Small wonder that it took a long time for the framework of these new sciences to be established and that in the process many different kinds of ideas were tried out. The scientific debates that surrounded the emergence of evolution theory were not just the products of a revolution within a single science. Rather, they were the result of complex interactions between a whole series of emerging scientific disciplines, which had to be fitted together before the Darwinian revolution could be completed.

Although Kuhn's thesis partially undermines the objectivity of science, it ignores what some historians and sociologists consider the most pressing reason for questioning the accepted view of scientific knowledge. In the hypothetico-deductive method, new ideas about how nature might work are proposed and then tested; the objectivity of science lies in the testing process. But the formulation of a new hypothesis is not a mechanical response to factual problems confronting the scientist. It must go beyond the known facts and thus represents a leap of the imagination, an act as creative as that of any artist (Bronowski, 1975). If we acknowledge that any theory, however sophisticated, is only an approximation of the "real" structure of nature, then it is possible that different kinds of theories could function effectively as guides to research. It thus becomes necessary to ask why the scientist picks out the particular hypothesis to which he is intuitively drawn. Because he is a human being, living within a particular culture and society, it is difficult to believe that the ideas and values he has been taught will not play some role in stimulating his imagination in a certain direction. Existing theories are often perceived to have philosophical or ideological implications, so it does not seem unreasonable to suppose that (consciously or unconsciously) an awareness of these implications will shape the actual creation of each theory. If this is so, the search for scientific knowledge, like any other human activity, will have to be understood in the context of social values. Instead of being treated as something entirely separate that is construed to represent absolute truth, scientific knowledge will have to be treated in the context of the "sociology of knowledge" (Mulkay, 1979), and historians of science will have to take this into account (Barnes and Shapin, 1979; Shapin, 1982).

The suggestion that subjective factors play a role in science does not automatically imply that we must dismiss scientific knowledge as pure illusion. The method of continued testing ensures that no theory is accepted unless it can function effectively as a guide to factual discovery. Whatever its foundations, this guarantees that science is the only form of knowledge that is genuinely cumulative. The testing of hypotheses, at least, is truly objective (although there are a few cases of theories with strong ideological connections which have been supported by slipshod or even fraudulent evidence). For the most part, it is in the original conception of the theory that we must look for an external influence, some preconception from the scientist's environment that serves as a model for constructing his hypothesis about nature. The chief task of the historian of science, then, is to provide a balanced understanding of how theories are created within a social framework but are then exploited successfully as guides to the investigation of nature.

Historians always have realized that science interacts with other areas of culture, particularly religion, but all too often this has been seen as strictly a one-way relationship. Science, it was thought, accumulated its own form of factual knowledge, and if this undermined existing religious dogmas, then religion would be forced grudgingly to give way. This is the classic image of a "war" between science and religion (White, 1896)—a war that science must inevitably win. Exponents of such a viewpoint had to concede that scientists sometimes had tried to accommodate their theories to prevailing religious beliefs; but it was assumed that whenever this happened, religion caused a distortion of science that sent it off on the wrong track, until eventually a "pure" theory succeeded in overthrowing the perverted one. Once properly established, scientific knowledge might subsequently be shown to have a practical application to a social or philosophical question, but this had nothing to do with how that knowledge was obtained. Such an interpretation encouraged the division of historians of science into two camps: the "internalists," who studied the purely objective process of scientific discovery, and the "externalists," who looked at how society reacted to the knowledge provided by scientists.

If we now question the claim that scientific knowledge is completely objective, we see that it is no longer possible to treat the development of any theory apart from its social environment. *All* theories, whether "right" (i.e., still accepted today) or "wrong" (now abandoned) must have had some external input into their original formulation. The distinction between internal and external history of science becomes meaningless, because the development of science at all levels represents an interaction between objective and subjective factors. No area of science can grow in complete isolation, although in the case of the physical sciences, it may be more difficult to identify the role of external influences. In biology, the potential for an input from social or cultural factors is more obvious, since many theories have a direct bearing on our understanding of man himself. Evolution provides a classic example, where claims for such an influence were already being made long before the modern revolution in our attitude toward scientific knowledge. Even some nineteenth-century writers thought that the similarity between Darwin's "struggle for existence" and the competitive ethos of Victorian capitalism was too great to be coincidental, implying that the scientific theory was merely a projection onto nature of images derived from the social environment. More generally, evolution has spearheaded the invasion by science of those areas of knowledge once considered the exclusive preserve of theologians and moralists. The demand that all areas of knowledge, however sacred they may once have been, should be thrown open to science is in itself the reflection of a particular, secularized value system.

As general attitudes toward the nature of science have changed, so historians have begun to reinterpret various aspects of the Darwinian revolution. Before beginning a detailed historical study, it may be worthwhile to specify those themes within the traditional historiography that have been challenged by modern writers.

The "Precursors" of Darwin

In its least sophisticated form, the history of science degenerates into a search for the precursors or forerunners of some particular advance. Darwin's achievement has not escaped the attention of precursor-hunters. Naturalists as far back as Aristotle have been hailed as the "real" discoverers of evolution. More sensibly, a number of eighteenth-century writers including Buffon and Lamarck have been depicted as prototype evolutionists, who *nearly* put together an outline of the modern theory. Various motivations can be ascribed to those who engage in this kind of history. In a few cases, there is a deliberate attempt to play down the real significance of Darwin himself, by claiming that he was only building on a foundation already established by others. More commonly, though, precursor-hunting is a technique employed by scientists who cannot believe that truths so obvious to us today can have remained undiscovered for so long. If Darwinian evolution theory is the correct solution to so many biological problems, surely *someone* before Darwin must have glimpsed at least part of the truth. Those who approach history through the old image of science as the gradual accumulation of factual knowledge simply cannot accept the possibility of a genuine scientific revolution; they do not believe that there would have been a time when modern ideas were literally unthinkable because they could not be formulated within the intellectual climate of the time. Instead, they search the literature of the earlier period looking for brief glimpses of the truth, emerging in an embryonic way from traditional misconceptions about nature.

Precursor-hunting is particularly misleading in terms of eighteenth-century naturalists whose writings do show a genuine effort to grapple with the problem of biological change. It is all too easy to go through their works picking out passages here and there that—when put together apart from their original context—look something like an anticipation of modern evolutionism. The real task of the historian, however, is not to pick out just those items that may correspond to modern ideas. Rather, it is to reconstruct what scientists of an earlier period really thought about the problems that *they* found interesting. This means that their writings must be read in their entirety to reconstruct the context within which they discussed various issues. If points that we now recognize as crucial were buried in a mass of writing with its main concern elsewhere, they should not be given undue significance. The efforts of pre-Darwinian naturalists are interesting not because they tried to solve Darwin's own problems and failed but because they found so many different ways to explore the implications of a changing universe. A true understanding of eighteenth-century "evolutionism" can only be derived from a proper appreciation of the cultural values prevalent at the time. The natural history of this period fits into the general background of Enlightenment philosophy, and in some respects this background did not permit the creation of ideas exactly resembling those of today (Roger, 1963; Foucault, 1970; Bowler, 1974*a*).

The Geological Background

Many factors contributed to the change in the intellectual climate which made nineteenth-century Darwinism possible. Within science itself, there were major developments in geology and paleontology that formed an essential framework for Darwin's thoughts, and it is here that historians have focused a good deal of attention. Because Darwin acknowledged Lyell as an important influence, it has generally been assumed that uniformitarian geology was a step toward modern evolutionism, while the opposing catastrophism served only to obstruct the development of science. The need for greater care in assessing Lyell's system was first pointed out by Hooykaas (1959), and more recent work has followed up this theme (Rudwick, 1971, 1972; Bowler, 1976*a*). What Darwin got from Lyell was the idea of continuous change—but Lyell himself associated this with a steady-state viewpoint that is the very opposite of progressive evolution. The catastrophists kept the notion of *development* alive and used it to make the first reasonably comprehensive outline of the fossil record. Yet their contribution to the way we still think about the history of life has seldom been acknowledged, because most of them thought that successive populations were introduced by divine creation.

Two lessons can be learned as we try to produce a more balanced interpretation. First is the necessity of demystifying the legends that have sprung up around certain scientific developments. Lyell has—quite rightly—been hailed as an originator of a new and in some respects more realistic approach to geology and natural history. In focusing on his positive achievements, some historians have been led to oversimplify the complex situation in early-nineteenth-century science. Because catastrophism is generally assumed to be "wrong" as we understand things today, the supporters of this theory have not been taken seriously as scientists—despite the fact that many of them played a major role in establishing the modern sequence of geological periods. We must be prepared to admit that, despite the more obvious superiority of Lyell's views, our modern viewpoint is in fact derived from a synthesis of both sides in the nineteenth-century debate. By recognizing the contributions made by the catastrophists, we can see that the concept of discontinuous change may have had a positive role to play at certain periods in the history of science, even though we have our suspicions about it today. We also realize how dangerous it is to label any theory as absolutely right or wrong, a point emphasized by the willingness of some modern geologists once again to consider catastrophic events to explain the extinction of dinosaurs.

The second lesson concerns the role played by religion in nineteenth-century science. It is clear that the catastrophists had theological preferences for a theory based on sudden change and the supernatural creation of species. Their ideas have too often been dismissed as nothing more than an artificial construct, designed solely to maintain traditional religion. It is difficult to see, though, how a theory set up for this purpose alone could have served so well as a framework for certain kinds of geological research. Lyell's own alternative philosophy was the foundation for his theory. We must be aware that religious debates constituted a framework within which all scientists at the time had to function. All theories had some religious component *and* were useful in certain scientific situations. To pick out one theory and claim that it has somehow escaped external influence (and hence has become more "scientific") is to create an entirely artificial picture of the true situation. Once again, the context in which the science developed would shape the very nature of the theories suggested.

The Origins of Darwinism

Much work continues on the origins of Darwin's own theory of evolution by natural selection (see Oldroyd, 1984). On a host of technical matters, the study of Darwin's papers is now revealing how careful we need to be in our assessment of his autobiographical account of the discovery. The most controversial question, however, is still that of the extent to which he may have been influenced by nonscientific factors. The internalist approach always has stressed the objective nature of his approach, picturing his theory as the most plausible explanation he could find for a series of technical questions (de Beer, 1963; Ghiselin, 1969). At the opposite extreme are the modern exponents of the claim that Darwinism must represent an extension of the social philosophy prevalent at the time (e.g., Young, 1971*b*, 1973). The impact of Malthus's population principle—acknowledged to be important by

Darwin himself—remains a critical focus for the question, because it provides a clear link by which social attitudes could have been imported into the science. We are now much more strongly aware of how deeply the young Darwin had immersed himself in the general literature of that immediate period. Some historians think they still can trace elements of the old theological viewpoint in the early versions of natural selection, while others stress Darwin's awareness of the materialistic implications of what he was doing. In either case, the notion of a philosophically naive Darwin setting out on an objective study of the phenomena no longer can be accepted. To understand the origins of his theory, we need to look at both the general influences on his thinking and the ways in which he was able to apply his insights to particular scientific questions.

Within the technical study of Darwin's work, the most exciting insights have been suggested by scholars who have begun to emphasize the importance of his speculations on "generation," that is, sexual reproduction and growth (Hodge, 1985). We can now see that Darwin's materialism was shaped by his study of how reproduction can play a creative role through the production of new variations. In contrast to earlier interpretations, this approach discerns a continuity between Darwin's thought and the materialist speculations of the previous century. Going beyond Darwin's original discovery, Hodge also argues that our perception of how evolutionism developed should be altered to include an integral role for continued speculations about growth and reproduction. The artificial distinction between the biogeographical side of Darwinism and debates on the mechanism of reproduction should be abandoned. Evolution theory was deeply influenced by the prevailing fascination with the process of individual growth throughout the nineteenth century.

The Reception of Darwinism

These new ideas about the nature of Darwinism have implications for our assessment of the theory's reception and influence. Much recent historical work (some of it my own; see Bowler, 1983, 1986) has emphasized the extent to which late-nineteenth-century evolutionism was often non-Darwinian in character, that is, it did not rest on the parts of Darwin's thinking that modern biologists find most interesting. Even some of Darwin's staunchest supporters, including T. H. Huxley, made little use of the selection theory. "Darwinism" (meaning little more than the basic idea of evolution) became popular despite the fact that there were many objections urged against natural selection. This means that we need to ask a new kind of question about how the *Origin of Species* came to symbolize a new era in science. The triumph of evolutionism was as much a social as an intellectual process as it depended on the conversion of scientists who had major reservations about Darwin's new initiative.

Equally serious implications emerge when we switch our attention from the objections raised against natural selection to the non-Darwinian theories proposed as alternatives. If Darwin's radical insights catalyzed the transition to evolutionism but were ignored by many "post-Darwinian" thinkers, are we justified in treating the emergence of the selection theory as the key event in the theory's history (Bowler, 1988)? There is much evidence to suggest that a "developmental" model of evolution already gaining some currency in the pre-Darwinian era merely increased its influence after the *Origin* had stimulated a more general acceptance of transmutation. The non-Darwinian theories of the later nineteenth century were often progressionist and implicitly teleological: they assumed that evolution was intended to develop toward a particular goal, just as an embryo grows toward maturity. The analogy between evolution and growth allowed many of Darwin's contemporaries to ignore the more radical aspects of his theory—precisely those aspects that modern biologists find most useful. Only after Mendelian genetics had destroyed the plausibility of this analogy did it become possible for a truly materialistic view of evolution to prevail. From this perspective, Darwin's original theory was a catalyst rather than a true participant in the emergence of nineteenth-century evolutionism. Modern Darwinism owes at least as much to the geneticists' destruction of the developmental tradition as it does to Darwin's original discovery of the selection mechanism.

This reassessment of Darwin's position will be mentioned at appropriate points in the following chapters. I suspect that the world is not yet ready for a survey of evolutionism in which Darwin does not play a pivotal role. Nevertheless, in my view, our current fascination with Darwin's discovery of natural selection is at least in part an artifact of modern biology's commitment to the synthesis of selectionism and genetics. Historians have been encouraged to treat the origin of his theory as the central theme in the emergence of evolutionism because *we* now think that natural selection is the most important theoretical breakthrough. This has concealed the extent to which non-Darwinian ideas continued to shape the thought of late-nineteenth-century evolutionists. It has also promoted a more or less linear model of the theory's history, in which the modern synthesis appears at the end of a "main line" of conceptual development initiated by Darwin.

There are obvious reasons for modern biologists to wish to dismiss non-Darwinian evolutionary ideas as mere side branches of no real interest today. But cultural historians need to adopt a more sympathetic approach to the past. Popular attitudes to evolutionism are still influenced by an image of the theory and its implications established in the nineteenth century, and it is thus of vital importance for historians to understand how that image was created—including, if necessary, its non-Darwinian components. Yet many cultural and social historians have meekly accepted the scientists' view that non-Darwinian evolutionary ideas are not worth studying. Even writers who distrust the implications of modern materialism find it convenient to use Darwin as a symbol of how science has shaped modern values. Thus, the biologists and their critics have cooperated to retain a historiography in which Darwinism plays the central role in the emergence of modern evolutionism. No one can doubt that Darwin's theory had a major impact on modern science and thought, but we must bear in mind that historical research is beginning to suggest that this impact was much subtler than the traditional picture allows.

2

Early Theories of the Earth

Although Copernicus had published his sun-centered astronomy in 1543, only in the early seventeenth century did the new cosmology begin to appear as a serious challenge to traditional ways of thought about nature. Galileo and Kepler, in very different ways, began to explore the implications of the new idea and to provide it with the subsidiary hypotheses that would turn it into a plausible cosmology. It was Descartes, though, who first linked the new astronomy and the new physics into a complete world view based on mechanical principles. The solar system, indeed the whole universe, became a vast machine, a system of matter in motion governed only by the laws of mechanics. It became increasingly clear that the sun itself was just another star, a small element within the vast structure of the universe. Bernard de Fontenelle's *A Plurality of Worlds* (1688) emphasized how insignificant man's home now appeared within the physical universe and raised the disturbing prospect that other worlds might also be inhabited. Fontenelle realized that on the vast scale of the universe's activity, the span of human life was also reduced to insignificance. The stars themselves might change but so slowly that we do not notice the changes and have assumed the universe to be static. Once it was conceded that physical forces might produce such changes, the sheer magnitude of events seemed to demand an extension of the time scale far beyond the few thousand years that were supposed to have elapsed since the creation. By the middle of the following century, writers such as Kant (1755; translation, 1969) were speculating about processes of cosmic evolution that made the formation of the solar system just a minor episode within the vast cycle of nature's development.

The new cosmology provided an obvious framework for asking questions about the origin of the earth. If planetary systems were formed by physical processes, then the earth itself must have originated in a similarly natural way. This introduced the possibility that terrestrial changes were responsible for shaping the present structure of the surface. But the model provided by the cosmologic speculations was not the only inspiration for the emergence of a concept of geological change. The scientific revolution also flourished on the basis of a renewed faith in the significance of observation, the empirical method of gaining information that Francis Bacon believed would transform human knowledge. As those who followed this program turned their attention to the earth, they found a wide variety of objects whose structure raised obvious problems. Many "fossils" (the name originally meant anything dug up) seemed to resemble the external form of living bodies, now turned to stone. How had these objects been formed within the rocks, and why were some rocks laid down in regular layers or strata? Why were there signs of vulcanism in areas where no such activity had been recorded in human history? In attempting to answer these questions, naturalists of the decades before and after 1700 were forced to entertain the prospect of physical forces that had actually shaped part of the earth's crust over a long period of time.

Theory and observation thus both tended to undermine the static view of creation. But what kinds of processes might operate to shape the earth's surface? A simple solution to many problems was to assume that the great flood recorded in the Bible was responsible for depositing layers of fossil-bearing rocks. Other changes could be accommodated in the time-honored belief that the earth was akin to a living organism, subject to periods of growth and decay. The erosion of rocks, for instance, might be part of this gradual process of degeneration (Davies, 1969). The mechanistic attitudes of the new science would not sustain such vitalistic analogies, however, and naturalists increasingly began to fit their observations into a system based only on physical changes. Although debates continued to rage over the exact nature of the processes, a body of information and techniques was gradually built up to provide the conceptual structure on which all theories would have to be formulated. The eighteenth century was not just a period of speculation; on the contrary, it laid the foundations on which the heroic age of geology would later build (Porter, 1977).

The move toward an extended view of the earth's history was facilitated by broader intellectual trends that questioned the traditional account of human origins (Rossi, 1984). Some scholars now began to argue that the Bible did not account for the history of the whole human race and that some civilizations are older than the Judeo-Christian culture. This, in turn, made it easier to believe that the world itself had undergone a more extended period of development. By 1700, these new ideas were already creating difficulties for the Christian world view. Yet at the beginning, few believed that the old order would be completely overthrown. It was hoped that science would merely fill in the details of how God created the world for man, without disturbing the underlying pattern. The Christian religion would be preserved as an integral part of the intellectual structure that most seventeenth-century scientists hoped to construct, although they might disagree among themselves over the best way to effect the reconciliation. The Bible inevitably provided them with an outline of the Creator's purpose but left them with the problem of how to deal with the actual details of the creation story in the Book of Genesis. If the Bible was a divine revelation, did this mean that all its details must be true? Galileo already had confronted this issue in connection with those passages that seem to imply that the sun goes around the earth. While acknowledging the church's authority on spiritual matters, he had insisted that the language of the Gospels—meant for the comprehension of a primitive society—should not be allowed to put a strait-jacket on all subsequent thought about nature. Geologists, too, had to accept the possibility that some passages in the creation story were not to be taken literally. Protestants, of course, were even more concerned with the meaning of God's word, but each had the freedom to develop his own interpretation on this perplexing issue. If properly understood, revelation could not be contradicted by nature, and techniques were gradually developed for reconciling the Genesis story with the demands of the new science.

The mechanical philosophy also created difficulties. It was possible to suppose that God created the present structure of the universe exactly as we see it today, endowing the laws of nature with power to maintain the original pattern. But Descartes's approach suggested that the present structure had evolved mechanically out of an earlier state. As long as one believed that the original form of the universe was created by God, the idea could be accepted that any later developments occurred in accordance with His wishes. Many of Descartes's followers, nevertheless, began to lose sight of this point as they concentrated more on the mechanical processes themselves. It was difficult to imagine that God really superintended or even cared about the actual details of what happened in the world, because He had delegated all responsibility to the laws of nature. At best, one might suppose that He had planned the general outline of what happened, but to believe that every minute detail of nature's activity was intended by divine forethought seemed unreasonable. Thus emerged the clock-maker God of the deists and an increasing suspicion that such a God had no relevance for human affairs.

The new scientific ideas were only part of a more deep-seated revolution that eventually would sweep away much of the old theological viewpoint (Hazard, 1953; Wade, 1971). For all the concerns that Descartes and Newton might express about the link between science and religion, to later generations their achievements merely symbolized the power of reason to challenge ancient preconceptions. The philosophy of the eighteenth-century

"Enlightenment" was based on faith that reason could rework the human situation (Cassirer, 1951; Gay, 1966, 1969). The triumphs of the new physical sciences were just forerunners of a complete renewal of our beliefs about man and nature. In such a climate of thought, it was inevitable that constraints still imposed by religion on late-seventeenth-century science should be thrown off as shackles of an outdated bondage. What had begun as a search for new insights into how God had created the world ended with the proclamation that man's ability to understand the workings of nature would make any concern for the Creator superfluous.

THE MECHANICAL COSMOGONIES

Galileo's plea for freedom of thought was made on behalf of both the Copernican system and the mathematical approach to physics, which had rendered the motion of the earth plausible. The science of mechanics rapidly began to take on a wider significance, especially after it was incorporated into Descartes's system of thought. The so-called mechanical philosophy emerged, dedicated to the belief that all phenomena can be explained as a result of matter in motion. This emphasis on a universal physics opened up possibilities that simply could not have occurred to the medieval mind. In the traditional world view, there was no room for the idea that the earth could have been formed from matter originally scattered throughout the universe. The earth-centered cosmology was based on Aristotle's hierarchical distinction between the earth and the heavens, and because the two areas were held to be fundamentally different, it was impossible to conceive of one giving rise to the other (Kelly, 1969). The structure of the universe either must have existed eternally, as Aristotle himself supposed, or must have been established in its present state by the Almighty. But the mechanical philosophy held that matter throughout the universe was the same, hence there could be no absolute distinction between the earth and the heavens. It thus became possible to suppose that the earth or the whole solar system could have been formed from matter originally distributed through space in some other way. The "creation" of the earth could become a purely physical process, and the resulting theories of the earth's origin became one of the foundations of geological thought.

Fontenelle drove home his point about the insignificance of the earth by noting that the stars themselves are not permanent—occasionally one dies out, while another may be kindled elsewhere. Here he merely was following the lead already given by Descartes himself. The Cartesian philosophy of nature sought to explain the *origin* of all things in physical terms, without reference to supernatural creation. God does not design the individual structures of the universe; He merely established the basic laws of nature and made them responsible for all future developments. Descartes taught that the planets were carried around the sun in a vortex, or whirlpool, of transparent ether, and he insisted that it was possible for one vortex to run down and disappear while another was formed elsewhere. Thus, an individual sun and planetary system can be formed within the ceaseless motions of the physical universe.

It was a short step from a theory of the origin of a new vortex to an explanation of the origin of a single planet such as the earth. In his *Principles of Philosophy* of 1644, Descartes indeed suggested a mechanism whereby the earth could be formed from a star that cooled down to a ball of ash and then became trapped in the sun's vortex. To avoid criticism from the church, he admitted that his system only showed how the universe *could* have been formed mechanically, whereas we know from revelation that God actually created it directly. It is hardly surprising, however, that as the influence of his system grew, his followers insisted on treating the mechanical explanation seriously. Descartes thus founded one of the most significant trends affecting the early history of geology. Many subsequent theories of the earth were conceived deliberately as attempts to extend the mechanical philosophy into a complete cosmogony, or physical history of the universe. Eventually the details of Descartes's physics were replaced by those of Newton's, but as Aram Vartanian points out (1953), the basic Cartesian program continued to be the source of much Enlightenment materialist speculation. (On these early theories, see Haber, 1959; Greene, 1959*a*; Roger, 1974; Jaki, 1978*b*; and Laudan, 1987. See also the histories of geology by Adams, 1938; Geikie, 1897; and von Zittel, 1901.)

The new trend began as an attempt to reinterpret the Genesis story of creation in physical, rather than miraculous, terms—indeed, some of the earliest theories were little more than vaguely plausible mechanisms proposed to match the biblical events. This is obvious in the first attempt to elaborate the Cartesian system into a complete account of creation: Thomas Burnet's *Sacred Theory of the Earth* (1691; Gould, 1987). Burnet exploited a curious idea suggested in Descartes's own account: as the earth cooled, a layer of solid matter would be formed *above* the waters surrounding the planet's core. Burnet argued that this outer shell would be perfectly smooth, so that the original "creation" of the earth gave an ideal habitation for man in his initial state of grace. When man eventually turned away from God, he was punished by the catastrophe of the deluge. But this was now a purely natural event, caused by the collapse of the outer shell into the waters beneath. Only irregular fragments of the original surface would be left standing out of the water, forming the mountainous terrain of our present landmasses. Man's punishment thus became permanent—apart from the tragedy of the flood itself, the inconvenient and ugly mountains of today remain as evidence that this is a ruined planet, fit only to be inhabited by sinful creatures such as ourselves.

Burnet's attempt to provide a physical explanation of the events in Genesis encountered much criticism from those who wished to retain the original sense of the story. But his work also suffered because of its dependence on the Cartesian vortex theory, soon to be replaced by Newton's system of gravity. Newton himself tried to prevent his theories from being adapted to the search for mechanical origins, although in private even he could not help speculating about the formation of the earth (letter to Burnet reproduced in Brewster, 1855). Vartanian (1953) argues that despite Newton's public repudiation of such speculations, his system of physics was absorbed rapidly into the basic Cartesian program. Most eighteenth-century views about the origin of the earth were based on Newtonian, not Cartesian, physics—yet in another sense they continued the search for mechanical origins that Descartes had begun. Certainly, Newton's public warnings about the dangers of such speculations largely were ignored, and his own private thoughts about the earth's formation soon were mirrored in print by his followers.

The first of these Newtonian cosmogonies was the *New Theory of the Earth* (1696) by William Whiston, Newton's successor at Cambridge (Force, 1984). Like Burnet, Whiston adapted his physical system to the sequence of events in Genesis, providing a Newtonian alternative to the Cartesian "sacred theory." He suggested that the earth could have been created from a comet—a cloud of dust particles—that condensed to form a solid body under the influence of gravity. The deluge was caused when another comet swept by the earth, depositing large quantities of water onto its surface. This explanation of the flood already had been advanced by Edmund Halley, although his discussion was not printed until some time later (Halley, 1724–25). Whiston elaborated the idea at length, making the passing comet responsible not only for the flood but also for twisting the earth out of its originally circular orbit around the sun. As in Burnet's theory, the creation and the flood became purely natural events, the deluge corresponding to a permanent worsening of the earth's status corresponding to man's fall from grace.

Although Burnet and Whiston were anxious to provide a new foundation for the Genesis story, the implications of their work were potentially dangerous for religion. As the sciences grew in popularity, later workers would ignore both the letter and the spirit of the biblical story to concentrate on the physical processes themselves. The attempt to reconcile Genesis and geology proved no more than a temporary expedient, which soon broke down as scientists demanded freedom to follow out the logic of their own systems. A particularly serious problem arose over the question of the earth's age.

Burnet and Whiston had remained within the very limited amount of time normally accepted by Christians, retaining the belief that the creation of the earth and of mankind were virtually simultaneous. Burnet, at least, realized that erosion by wind, rain, and rivers was potentially capable of wearing down mountains, if it could act over a long period of time; but in his theory, mountains were formed by a single event and could not be rebuilt thereafter. The fact that they are still here therefore proves that they were formed only a short time ago. Burnet paid little attention to the empirical study of rocks and fossils and did not recognize the artificiality of his views. Others soon would begin such studies, however, that would challenge the plausibility of the whole time scale derived from Genesis.

Another dangerous implication of the new theories was their tendency to undermine the traditional belief in divine providence. Christians always had believed that God took a continuing interest in the world after the creation, even to the extent of miraculously interfering with the laws of nature He established. The deluge was interpreted as a moral event—man's punishment for turning away from God—and it was supposed that the floodwaters were a direct, that is, miraculous, effect of His wrath. In the new theories, the deluge was merely a physical event, produced inevitably by the mechanical operations of nature. Burnet had argued that an omniscient God could foresee man's moral history and could design the physical world so that it would produce a catastrophe at the time appropriate for punishment; but this claim represented the first step on the way to deism, to belief in a God who did not need to care about the universe once He had created it. Newton himself had retained a role for miracles, precisely because he believed in a God who *cares* about the world (Alexander, 1956). But the deists of the Enlightenment insisted on rejecting miracles as part of their wholesale campaign against Christianity (Torrey, 1930); and in such a climate of opinion, it is not surprising that naturalists began to lose interest in the moral significance of the physical events they studied.

These materialist implications already were apparent in the last Cartesian theory of the earth. This was Benoît de Maillet's *Telliamed,* published in 1748 but probably written between 1692 and 1718 (translation, 1968; see Carozzi, 1969). De Maillet pictured a universe made up of Cartesian vortices, but his real interest was the formation of the earth within the solar system. He made no effort to reconcile his system with Genesis and for that reason thought it best to present it as the work of an Indian philosopher whose name was his own, read backward. De Maillet assumed that the earth was originally covered to a great depth by water. Mountains were formed as currents shaped the bed of the ancient ocean, and as the water level gradually decreased, their tops eventually were exposed to form the first dry land. Marine erosion at the shoreline gradually wore down these primitive mountains, while the resulting debris settled on the seabed to form sedimentary rocks. As the sea retreated even farther, these rocks were exposed to form newer mountains. A gradual lowering of the sea level was thus the principal directional element in de Maillet's concept of the earth's history. Even now, he argued, the oceans still are retreating by a very slight amount each year.

The retreating ocean theory gained wide popularity in the eighteenth century, but none of its supporters went further than de Maillet in their refusal to compromise with the Genesis story. There was nothing corresponding to a recent deluge in *Telliamed,* and the whole history of the earth was treated as a process requiring vast amounts of time before human civilization appeared. De Maillet even suggested a natural origin for living things. His willingness to break so decisively with the traditional time scale cannot, however, be explained solely in terms of Cartesian mechanical philosophy. Descartes, Burnet, and Whiston all had produced physical mechanisms to explain the earth's origin without bothering to acquire anything more than an everyday familiarity with its present state. But de Maillet's views on rock formation seem to have derived from a study of the secondary rocks themselves, and it was this empirical foundation that provided the basis for his bold extension of the time scale beyond traditional limits.

THE EMPIRICAL TRADITION

The rocks of the earth's surface were a natural target for those who adopted Bacon's philosophy of scientific empiricism. But as soon as serious attention was paid to certain kinds of rock, some unusual features became apparent. The "sedimentary rocks" laid down in parallel layers, or strata, suggest that they had actually been formed from material deposited underwater. This was confirmed by the discovery of fossil sea creatures embedded in stratified rocks of even the most mountainous regions (Haber, 1959; Rudwick, 1972). Efforts were made at first to argue that fossils are not the remains of once-living things, but close observation soon convinced most naturalists that such a position was untenable. The work of Nicholas Steno (1669; translation, 1916) and John Woodward (1695) helped to establish the fact that fossils were indeed the remains of living things petrified within rock strata. This implied that areas that are now dry land must once have lain beneath the sea, while the material forming the stratified rocks was deposited. To explain how the land emerged from the sea after these rocks had been formed was the central problem of most eighteenth-century theories of the earth.

Fossils provided a challenge to the traditional view in yet another way. In some cases they appeared to represent living things of a kind no longer known on the earth today. Was it possible that some of the originally created life forms had become extinct? Deeply religious naturalists such as John Ray found it impossible to believe that a wise and benevolent Creator could have left His creatures to die out so casually. At first, Ray offered the hope that unknown species still would be discovered alive in some remote part of the globe (1692). Later in his career he became so concerned over this issue that he began to doubt the organic origin of fossils altogether; he followed his friend Edward Lhwyd in suggesting that they were purely mineral structures that had somehow "grown" within rocks (Ray, 1713). In this he departed from the consensus now emerging among most of his contemporaries, but Ray's fears pinpointed what was to become a highly sensitive issue as Enlightenment naturalists began to challenge the scriptural viewpoint more openly.

Once it was recognized that the fossils were genuine, the only hope of reconciling them with the biblical story was to assume that all fossil-bearing rocks were formed from debris swept across the earth's face by the Noachian deluge. This solution was offered in Woodward's *Essay toward a Natural History of the Earth* (1695). Despite the importance of Woodward's description of the fossils themselves, his theory was unsatisfactory. It was difficult to see how such an extensive sequence of strata could have been formed by a single catastrophic event. To produce the complex state of the surface that we see today, the earth must have undergone a whole series of developments. Given this more dynamic view of the past, the most obvious question was: how have sedimentary rocks been exposed as dry land? There are really only two possible solutions to this—either the water level must have gone down or the land itself must have been raised above the ocean. The first possibility is implicit in Woodward's suggestion that rocks were laid down during the deluge, but the idea would have to be drastically modified to incorporate a long series of developments more consistent with the complex state of the strata. The result was the retreating ocean theory first advocated in de Maillet's *Telliamed*. It assumed that when the earth was first formed, its whole surface was covered to a great depth by water, some of which has evaporated slowly into space. The sedimentary rocks would have been laid down while the whole earth was submerged and then exposed as dry land when the water level declined.

The alternative was to assume that the amount of water has remained constant and that earth movements have been capable of raising up new areas of dry land from what was once the seabed. This was suggested in Steno's *Prodromus* of 1669, and subterranean gases were cited as cause of the upward elevation. Steno also allowed flowing water to cause erosion of underground caverns into which overlying rocks could collapse. The power of earthquakes to elevate land also was suggested in Robert Hooke's "Discourse on Earthquakes" (published 1705) and in John Ray's *Miscellaneous Discourses* (1692; on Ray's work, see Raven, 1942).

Whichever theory was accepted, a second problem emerged in determining the amount of time needed to accomplish these immense changes. Steno, Hooke, and Ray all found it impossible to shake off the traditional view that the earth was but a few thousand years old. In this case, to achieve the observed results, past changes would have to have been on a scale far greater than anything experienced in recent times. Hooke and Ray, in particular, had to assume that earthquakes in the past were far more violent than those we experience today, because modern earth movements have only a trivial effect in producing an overall elevation of land. Hooke cited the sinking of Atlantis and other ancient myths as evidence that great effects indeed were produced very quickly in the distant past. The biblical deluge, of course, could have been another of these ancient catastrophes. Thus, a view of the earth's history that would later be called catastrophism emerged from the reluctance of many early naturalists to extend the biblical time scale. To compress all necessary changes into a short length of time, it was assumed that past events were on a catastrophic scale far beyond anything observable today. In 1691, the philosopher G. W. Leibniz published his *Protagaea*, explaining this decline in level of activity on the supposition that the earth once had been a very hot body—an early version of what we shall call the cooling earth theory. For Hooke, the falling off in the intensity of earthquakes had been a sign that the earth was entering a period of senility, but Leibniz's suggestion marks the beginning of a new and more materialistic approach to the problem.

GEOLOGY IN THE ENLIGHTENMENT

Growing lack of concern for the Genesis story was typical of the new intellectual outlook of the Enlightenment. The new breed of materialist philosophers was not prepared to accept imposition of arbitrary limits on the power of human reason, least of all by Christianity. Conservative thinkers still could produce works in which the flood was used to solve all geologic problems, but those who studied the rocks themselves demanded the freedom to follow the logic of their thoughts wherever it might lead. Traditional influences still had a certain amount of sway, enough to force a retraction from Buffon when his first attempt at a theory of the earth challenged Genesis too openly. But no one took the retraction seriously, and some years later Buffon was able to issue a revised version of his theory that still made only superficial concessions to orthodoxy. As the century progressed, the spread of deism and outright atheism encouraged even more naturalists to pay only lip service to the creation and the deluge.

The triumph of Newtonian science established a new basis for hypotheses of the earth's origin. Whiston's early attempt had concentrated on the earth alone, but to many it now seemed obvious that a particular planet could only be dealt with as part of the whole solar system. The simplest account of the system's origin derived the planets from the sun itself. This was proposed in the first volume of the comte de Buffon's *Histoire naturelle* of 1749. Buffon postulated that a comet had dived into the sun, throwing some of the intensely hot matter off into space. The sun's gravity would prevent this material from escaping while it cooled down to form the planets. Unfortunately, this theory did not square with the comparatively low eccentricity of the planetary orbits, nor did it account for the origin of the sun itself. Both of these problems were overcome in the "nebular hypothesis" first suggested by the philosopher Immanuel Kant in 1755 (translation, 1969). Kant assumed that the whole solar system had begun as a cloud of dust particles that condensed under the force of its own gravity and gradually acquired a tendency to rotate. Small amounts condensed into solid bodies circling around the main concentration, which ignited to form the sun. Be-

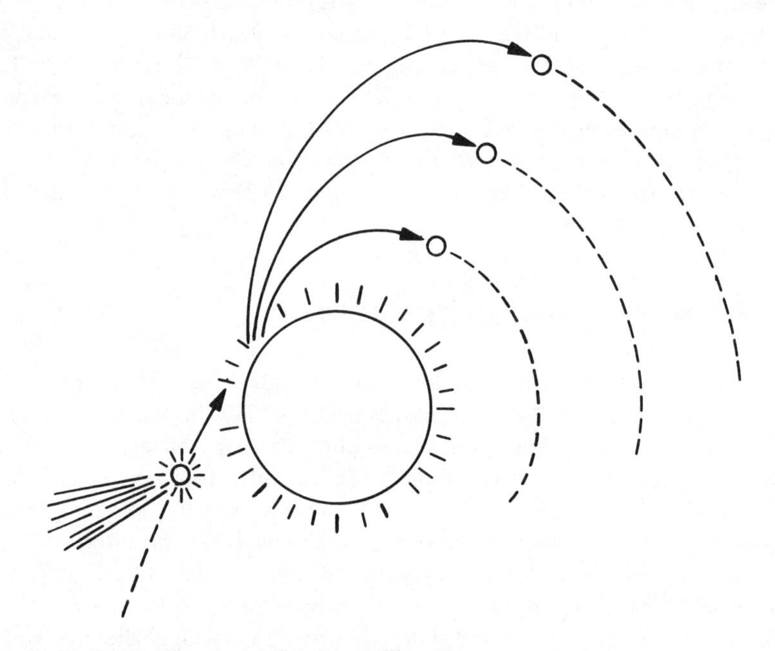

Fig. 1. Buffon's Theory of Origin of Planets.

A comet crashes into the sun's surface, throwing off globules of incandescent material into space. These are retained in orbit by the sun's gravitational pull and eventually solidify to give the planets. A similar theory in which the planets were pulled from the sun by the gravitational attraction of a passing star was popular earlier in the twentieth century. The great difficulty of such theories is to explain how the planets could settle down into orbits with the low eccentricities observed.

cause Kant's book was never issued, his idea received little attention, but a similar theory was later proposed by the astronomer Pierre-Simon Laplace in his *System of the World* (1796; translation, 1830). The theory seemed to be supported by the telescopic observations of William Herschel (Hoskin, 1964). Through his greatly improved instruments, Herschel was able to observe that some of the nebulae or hazy patches seen in the night sky were composed of gas or dust clouds apparently condensing into a central star, illustrating the nebular hypothesis in action.

The nebular hypothesis assumed that all stars condense in the same way and hence that the majority of them have planets circling around them. Gradual formation or "evolution" of a planetary system thus became a perfectly natural phenomenon. Such a belief did not necessarily remove the possibility that the whole process was designed by God—after all, He could have created the original gas clouds in such a way that they must eventually form planets of a certain kind according to His laws of nature. But Laplace felt that there was little point in preserving such a vestige of the old way of thinking. If the universe is a physical system that happens to have developed in a certain way over a vast period of time, little is gained by assuming that details of the process were worked out in advance by a Creator. The ultimate end of the search for a totally mechanical cosmogony was thus a gradual reduction of the role played by the Deity, until His connection with the universe was so indirect that it appeared negligible. Newton's worst fears were realized: elimination of the supernatural did lead inevitably toward deism and atheism.

By the time Laplace revived the nebular hypothesis, the spirit of the Enlightenment was already wilting under the effects of the French revolution. It was Buffon's less comprehensive theory that was developed as the basis of a complete geological system (*Histoire naturelle, I,* and *Les époques de la nature,* 1778; ed. Roger, 1962). In an effort to avoid theological censure, Buffon divided his history of the earth into six epochs, which could be said to correspond to the six days of creation—a trick used by many later geologists. His real intention was to provide a complete Newtonian cosmogony explaining the whole history of the earth and its inhabitants on materialist principles. His theory of the earth's origin from the sun led him to found one of the classic "directional" accounts of its subsequent history. Because the earth started hot, it must have cooled down steadily in the course of time, and the cooling defined the key stages in its development. The initial molten state was the first epoch, followed by a second in which the cooling allowed an outer crust to solidify. Following de Maillet, Buffon thought it necessary to invoke a vast ancient ocean to explain the formation of sedimentary rocks. In his third epoch, he thus supposed that great amounts of water vapor condensed and rained down on the surface to give a universal ocean. Parts of the seabed were exposed finally as the ocean re-

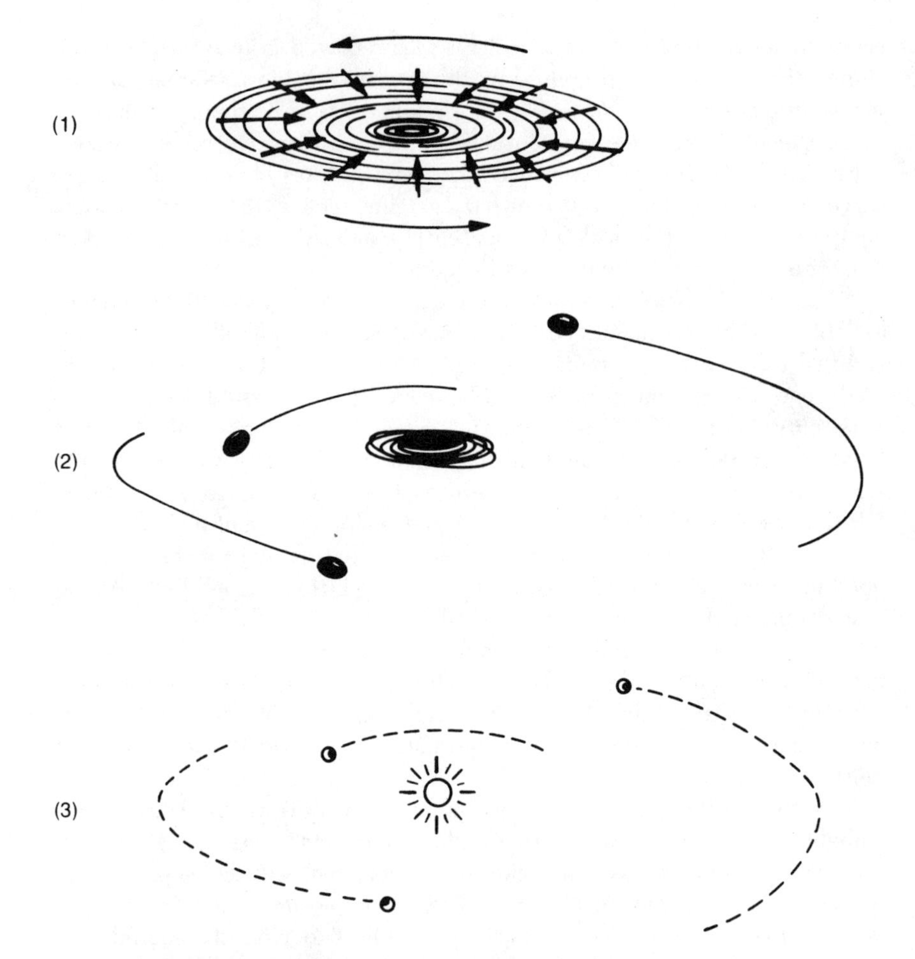

Fig. 2. Nebular Hypothesis of Origin of Solar System.

(1) A nebula—a great cloud of slowly rotating dust and gas—begins to collapse under its own gravitational attraction.

(2) Most of the material has begun to condense into a central body, but lesser amounts have formed rings rotating around the center and now themselves are beginning to condense.

(3) The central body has collapsed so far that enough energy has been released to make it incandescent, forming the sun. The smaller bodies have solidified to give the planets, which continue to orbit around the sun.

treated during the fourth epoch. The temperature was still much hotter than that of today, so that in the fifth epoch even the northern regions were inhabited by tropical creatures. Only in the last period did the earth come into its present state and the first men appear.

When compared with the schools of thought into which later geologists divided themselves, Buffon's theory is a curious mixture. On the one hand, it was based on the cooling earth hypothesis, which would appear to connect it with the school of "Vulcanism," the belief that subterranean heat is the chief agent of geological change. On the other hand, Buffon did not exploit this aspect of his theory. He held that volcanoes are caused by burning coal deposits, not by the residual heat left over from the earth's original molten condition. Nor did he try to argue that greater internal temperature of earlier epochs would have produced more violent earthquakes than those of today. Thus he did not anticipate the basis of what would later be called catastrophism and turned instead to the retreating ocean theory, in effect, abandoning Vulcanism for Neptunism.

What happened once the dry land was exposed by the diminishing waters? Buffon knew that the earth was very ancient—he had estimated 70,000 years in print but privately admitted this to be far too low a figure. The land surface must have been exposed for a considerable amount of time, during which the forces of erosion were at work. The surface as we see it may have been altered completely by the long-term effects of forces that we know are still in operation. In suggesting this possibility, Buffon partly anticipated the view later to be known as "uniformitarianism," the belief that the only agents to be taken into account in explaining the present state of the earth are those we still can observe in action today. But his commitment to his theory of the earth's origin prevented him from arguing that the reworking of the surface by observable causes might be so complete as to eliminate all traces of its original state. In particular, he accepted granite as a truly primitive rock, formed by the original cooling of the molten earth and unaffected by any later changes.

NEPTUNISM AND VULCANISM

The late eighteenth century saw a number of advances made in the study of rocks and of the processes that still continue to shape the earth's surface. To an increasing extent, these studies were made independently of any detailed theory of the planet's origin. Roger has suggested (Buffon, 1962; introduction) that by the time *Les époques de la nature* appeared, its speculative approach already had been rendered out-of-date by a renewed interest in pure observation. Questions still continued to be asked about the forces that shaped rocks, however, even if there was less interest in the earth's ultimate origin. Eventually this led to a polarization of geology into two camps, each stressing one of the two major forces that offered hope of explaining the facts.

The essential problem faced by all geologic theories was: how can sedimentary rocks formed by deposition from water now stand on dry land? Only two possible solutions had emerged, one relying on an absolute decrease in ocean levels, the other on elevation of land by earthquakes. The belief that all sedimentary rocks were deposited on the floor of a vast ancient ocean that has since disappeared became known as the Neptunist theory, after the Roman god of the sea. The alternative of invoking the power of earth movements was eventually connected with the belief that the interior of the earth exists in a state of high temperature and pressure and thus became known as Vulcanism, after the god of fire.

While de Maillet and Buffon both had adopted the Neptunist position and had tried to justify their belief in the universality of the ancient ocean in terms of their theories of the earth's origin, the Neptunism of the later eighteenth century abandoned such speculative attempts to justify the basic postulate in favor of an empirical study of the rocks themselves. It was held self-evident that sedimentary rocks were formed underwater and hence that the ocean once must have covered all the land. By abandoning speculation about the original state of things, the decks were cleared for a truly scientific study of the rocks themselves. This attitude was particularly prevalent in Germany and was encouraged by a concern for the practical values of geology. Both Johann Lehman and Abraham Gottlob Werner were teachers at mining schools, fully aware of the importance of mineralogy and stratigraphy for the location of ores.

In 1756, Lehman published an analysis of the structure of mountains that distinguished between primary formations, deposited from the original ocean before the appearance of life, secondary mountains of fossil-bearing rocks laid down during the biblical deluge, and recent or tertiary structures formed in the postdiluvial period. Lehman's simple use of the deluge suggests that he, at least, was not in sympathy with the more radical spirit of the Enlightenment. The same cannot be said of Werner, the leading figure in the development of Neptunism, who began teaching at the mining school of Freiburg in 1775. Werner was primarily a mineralogist, interested in classifying the substances that compose the earth's crust. Having worked out his system of classification, he made the natural assumption that those minerals found at the bottom of the sequence were formed first. From this, he went on to elaborate a complete Neptunist theory based on the order in which the various strata had been deposited from the vast ancient ocean.

Werner wrote only a few works explaining his system (translation, 1971), but he was an excellent teacher who drew students to Freiburg from the whole of Europe and inspired them to return home to apply the principles

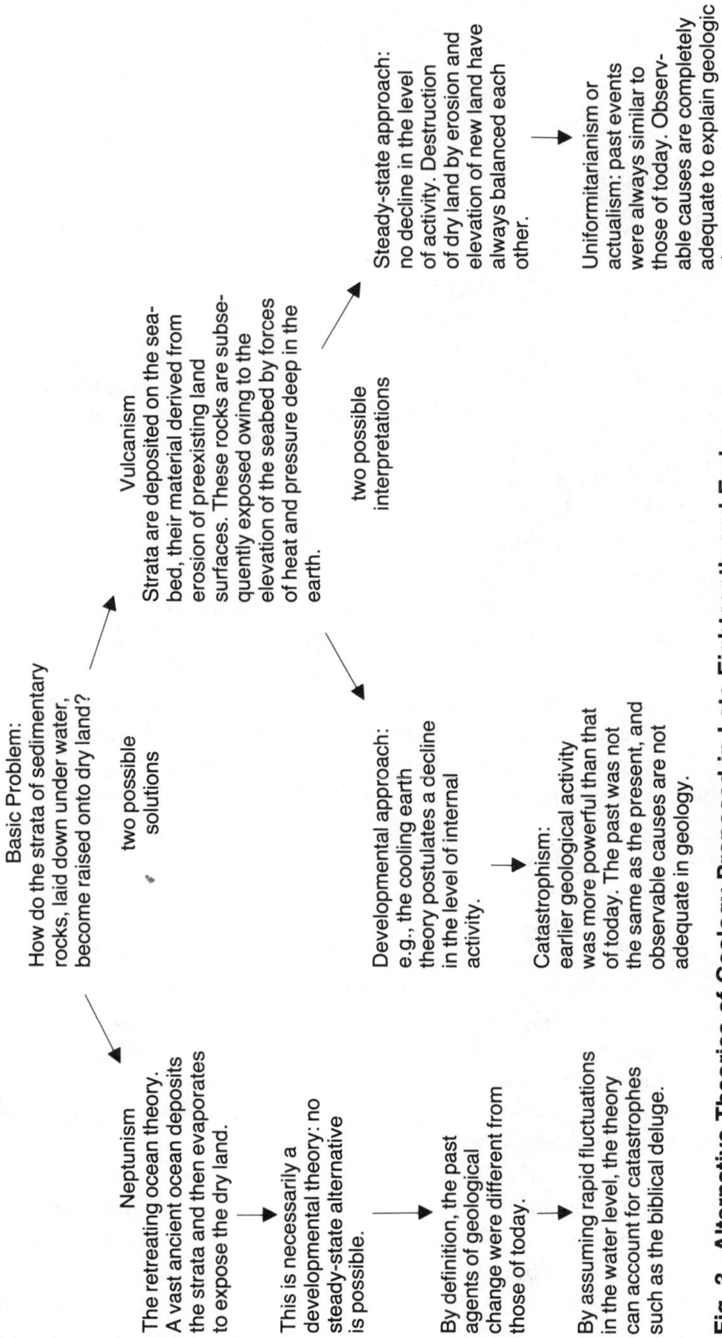

Basic Problem:
How do the strata of sedimentary rocks, laid down under water, become raised onto dry land?

two possible solutions

Neptunism
The retreating ocean theory. A vast ancient ocean deposits the strata and then evaporates to expose the dry land.

This is necessarily a developmental theory: no steady-state alternative is possible.

By definition, the past agents of geological change were different from those of today.

By assuming rapid fluctuations in the water level, the theory can account for catastrophes such as the biblical deluge.

Vulcanism
Strata are deposited on the sea-bed, their material derived from erosion of preexisting land surfaces. These rocks are subsequently exposed owing to the elevation of the seabed by forces of heat and pressure deep in the earth.

two possible interpretations

Developmental approach:
e.g., the cooling earth theory postulates a decline in the level of internal activity.

Catastrophism:
earlier geological activity was more powerful than that of today. The past was not the same as the present, and observable causes are not adequate in geology.

Steady-state approach:
no decline in the level of activity. Destruction of dry land by erosion and elevation of new land have always balanced each other.

Uniformitarianism or actualism: past events were always similar to those of today. Observable causes are completely adequate to explain geologic changes.

Fig. 3. Alternative Theories of Geology Proposed in Late Eighteenth and Early Nineteenth Centuries.

of Neptunism to the rocks of their native countries. Some historians have been puzzled by the popularity of the retreating ocean theory, apparently under the impression that it is such obvious nonsense that it is difficult to see how anyone could ever have taken it seriously (Geikie, 1897; Gillispie, 1951). Yet as Alexander Ospovat (1969) has shown, there were some good points in the theory's favor, and a more balanced account is now possible (e.g., Hallam, 1983). Because Werner had tied it to an extremely useful system of mineral classification, the theory gave geologists a conceptual framework for coming to grips with the vast amount of data they were gathering. Nor was the theory itself quite as simpleminded as some historians have supposed. It did not imply that the ancient ocean deposited a series of uniform layers of rock covering the earth like the skins of an onion. By accepting an uneven original surface and an irregular retreat of the waters, Werner could explain why different rocks were formed in each locality. Nor did he expect the rocks to lie always in neat horizontal strata. The evidence for extensive vulcanism was the only major stumbling block, which would eventually undermine the theory. In the meantime, Werner's Neptunism played a valuable role by allowing geologists to organize a vast mass of data that otherwise might have generated only confusion.

Later opponents of Neptunism invariably asked where all the water came from and how it disappeared. Werner sidestepped these awkward questions by declaring that they were outside the scope of scientific geology (or geognosy, as he called it). There is no evidence that earthquakes can raise mountains; therefore, we *must* assume that the whole earth once was covered with water. Our studies of existing rocks can tell us nothing about how the planet itself was formed, and the true scientist will ignore such speculative issues in favor of more detailed investigations of what we can observe: the sequence of stratified rocks.

Werner had no interest in scriptural accounts of the creation and the deluge, nor did he believe that the earth was only a few thousand years old. His later followers, however, had to deal with the new and more conservative climate of opinion the emerged toward the end of the century. Especially in Britain, fear of the French revolution bred a distrust of the atheistic philosophy of the Enlightenment. Geologists once again had to be concerned about the possibility of reconciling their systems with Genesis. Here Neptunism offered possibilities that had not been anticipated by Werner himself. It postulated a definite beginning for the earth's history which could be identified with the creation, while a resurgence of the ancient waters might explain the deluge. Thus, in the hands of later Neptunists such as Jean André Deluc, Werner's theory came to be associated with the revival of scriptural geology (Gillispie, 1951).

Werner and his students made two critical assumptions. The first was that volcanic activity has been a relatively trivial phenomenon throughout

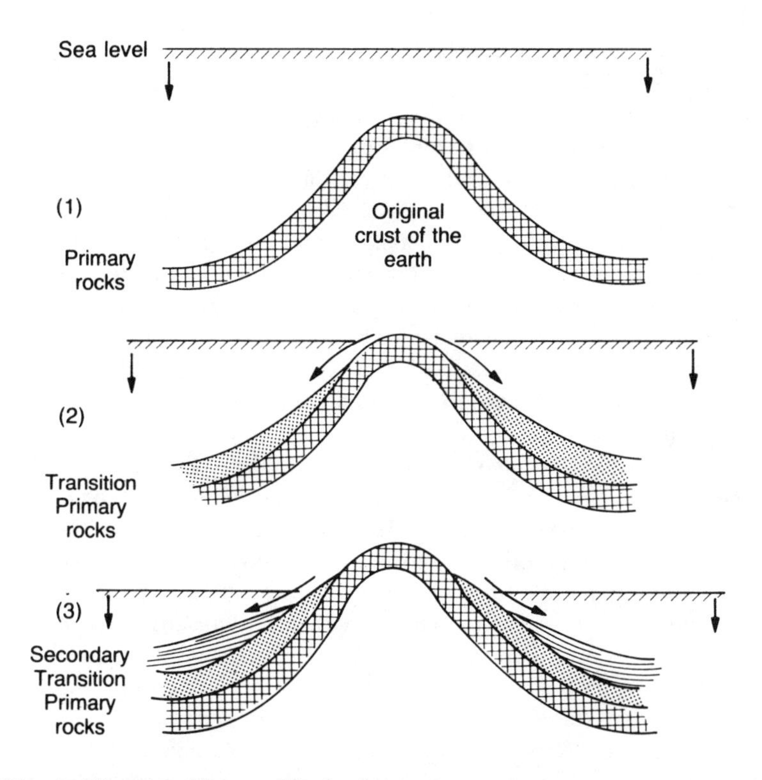

Fig. 4. Werner's Theory: Neptunism.

(1) Originally, a vast ocean covers the earth, holding great quantities of material in suspension or solution. As the sea level begins to decline, Primary rocks are deposited by crystallization onto the seabed. These rocks are universal formations that cover the whole of the original surface of the earth.

(2) Some time later, the decline in sea level exposes the first dry land. From this time on, there can be no universal formations. Transition rocks are laid down, partly from material crystallizing out of the ocean but also from sediments derived from erosion of the land surface.

(3) The ocean has retreated even farther, exposing larger areas of land. Transition rocks are now exposed, but the mountains will consist of Primary rocks. Extensive erosion sweeps large quantities of sediment into the ocean, where it is deposited to form the strata of Secondary or Flötz rocks. From time to time, there are great storms and even temporary resurgences of the sea level, and both create irregularities in the deposition of these rocks. Eventually, a further decline in sea level will expose some of the Secondary rocks, and further erosion will begin to form the material deposited as the most recent Alluvial formations. Only in the present has the further decline in the sea level exposed some of the Alluvial rocks, so these are only found in the most low-lying areas.

the earth's history. Volcanoes were regarded as the purely local effects of the underground burning of coal beds. Thus, it was not possible for large areas of rock to have an igneous origin—virtually all of the earth's crust must have been deposited directly from water. Even granite and basalt were supposed to have crystallized from water, despite the fact that such rocks are virtually insoluble. It was a central point within the Wernerian theory that rocks with crystalline structures must have come from solution in water, because the cooling of molten rock (such as volcanic lava) does not allow crystallization to occur. Evidence for the igneous origin of granite began to grow toward the end of the century, however, and even some of Werner's followers refused to accept their master's teaching on this point. If granite is an igneous rock that has crystallized as it cools, how many other kinds of rock making up the present surface of the earth might have a volcanic, rather than an aqueous, origin?

The Neptunists' second assumption was that there is no force capable of elevating the earth's surface. Modern earthquakes do not seem to provide such a force, because once the initial disturbance has died down there is little overall change in the position of the rocks. Hooke had suggested that in the past there might have been violent earthquakes that were indeed capable of elevating new landmasses. This provided an alternative explanation of how sedimentary rocks were raised to form dry land, but it required one to believe that very powerful forces have been at work deep in the earth. Eventually, evidence suggesting that volcanic activity had been much more extensive in the past provided the foundation for such a belief, thereby associating earthquakes and volcanoes as manifestations of the same basic phenomenon. Thus, the school of Vulcanism was born.

In 1740, Anton-Lazzaro Moro advanced the only *completely* Vulcanist theory ever to be suggested seriously. He regarded the strata of sedimentary rocks as layers of ash laid down during extensive volcanic eruptions. A few kinds of rock do indeed have such an origin, but Moro was unable to account for the fact that fossilized remains of marine creatures are found in the strata of many sedimentary rocks. The development of an effective school of Vulcanism involved adoption of a less extreme view: most stratified rocks are laid down underwater but are then elevated by forces of volcanic origin. First of all, however, it had to be shown that the center of the earth is indeed very hot. This was eventually demonstrated through a recognition of the extent to which igneous (i.e., once molten) rocks are distributed in the earth's crust. If such rocks are more frequently encountered than the Neptunists claimed, then volcanoes could no longer be dismissed as purely local phenomena and would have to be accepted as indications of major forces deep in the earth.

In the mid-eighteenth century, some naturalists began to suspect that vulcanism once had affected large parts of the earth. Jean Etienne Guettard reported (1752) that many mountains in central France exhibit the cone shape of a volcano, although there are no records of volcanic activity there during the period of human history. Nicholas Desmarest maintained that sheets of columnar basalt in areas such as the Giant's Causeway in Ireland had solidified from a molten state and were thus of volcanic origin. Although Desmarest was prepared to allow great antiquity to the earth, he still could not believe that the vulcanism of past ages was a force that had shaped the whole crust of the earth (Taylor, 1969). Only at the end of the century did the Scottish geologist James Hutton create the first really comprehensive Vulcanist theory. In it, he synthesized the various lines of evidence suggesting that the interior of the earth is very hot and then used this heat as the basis of his mechanism for mountain building. To distinguish the theory of the earth's central heat from a mere emphasis on vulcanism, it is often known as "Plutonism."

The first account of Hutton's theory appeared in the *Transactions* of the Royal Society of Edinburgh for 1788, followed by the two-volume *Theory of the Earth* in 1795 (Bailey, 1967; Gerstner, 1968, 1971; Dott, 1969; Gould, 1987). Hutton's own writings were rather cumbersome, but a little later John Playfair published his more popular *Illustrations of the Huttonian Theory* (1802). The theory did not challenge the basic assumption that the sedimentary rocks had been laid down underwater, but it rejected the claim that all material for these rocks once had been suspended in a vast primeval ocean. Hutton maintained that landmasses are constantly being eroded by wind, rain, and rivers and that the resulting debris is carried out to sea and deposited on the ocean floor. He now assumed that the earth's central heat would penetrate and harden the sediment to form rock strata. The new rock may then subsequently be elevated to form dry land by the force of earthquakes, driven once again by heat and pressure of the planet's core. Volcanoes were explained on the assumption that molten rock from the interior might occasionally find its way to the surface. Hutton proved that granite and basalt are both igneous rocks by discovering areas where they had intruded into the sedimentary strata. In this case, however, the molten rock had not reached the surface but had cooled slowly deep in the ground. Under these circumstances, they were able to acquire the crystalline character that Werner had taken as evidence of deposition from water. Hutton's views on the nature of granite and basalt were supported by a series of experiments performed with molten rock by James Hall.

If one now asked a person who had never read Hutton's works to reconstruct the probable foundations of his theory, his most logical conclusion would be to connect the idea of the earth's central heat with Buffon's cooling earth theory. This would account for the fact that the surface is now much cooler than the interior and would allow one to bring in Hooke's point that ancient earthquakes must have been more powerful than those of today. If internal temperature were higher and the crust thinner in the past, the disturbances would obviously have been more violent. This is *not* the logic of Hutton's scheme, however. Rather, it is the basis of the catastrophist theory that became popular some decades later. By one of those paradoxes that make it quite clear that the history of science is not a neat sequence of discoveries, Hutton came to his theory from a very different direction. He did not believe that the earth's central heat is gradually diminishing, nor did he think that past earthquakes were more violent than those of today. Instead he tied his Vulcanism into a steady-state world view, thereby founding a system that would be revived somewhat later by Charles Lyell under the name of uniformitarianism.

Hutton justified his steady-state approach on methodological grounds. Indeed, it has been suggested that his theory was largely an extension of his radical empiricist philosophy (O'Rourke, 1978). The scientific geologist, he argued, should do his utmost to explain the structure of the earth through operation of causes *that he can now observe in action*. This is the principle of the uniformity of nature, or more properly the method of "actualism" (Hooykaas, 1970). Postulating catastrophes on a scale we no longer observe is unscientific, provided that known causes operating at observed intensities can explain the phenomena. Because the slow changes we now observe could add up to produce major effects only over vast periods of time, Hutton was led to demand a vast antiquity for the earth. Given time, erosion by normal causes can produce the deepest valley or flatten a continent altogether. And over the same vast period, a series of small earthquakes no larger than those of today could raise a new mountain range from the sea depths.

This position was a major innovation, and because modern geologists still tend to be suspicious of catastrophes, Hutton has been hailed as a founding father of the science. In many respects, he did anticipate the modern view—the only major change is recognition that continents are not raised from the deep seabed but are confined to plates of lighter rock resting on the denser material below. Yet one aspect of Hutton's system shows that we cannot treat him purely as a scientific geologist. The actualist method is threatened by one problem: how can we be sure that there is no trace left in the rocks of the period of the earth's initial formation, when presumably conditions would have been very different from those of today? The modern geologist indeed believes that he has access to some rocks dating from such a primitive period. But Hutton denied the possibility of our ever locating any truly "primitive" rocks. He held that if ever there was an original formation of the earth, it is outside the scope of our investigations; we can perceive only an indefinite cycle of similar events, with "no vestige of a beginning—no prospect of an end." Although Werner had refused to speculate about the earth's origin, his theory was built on the assumption that it had been

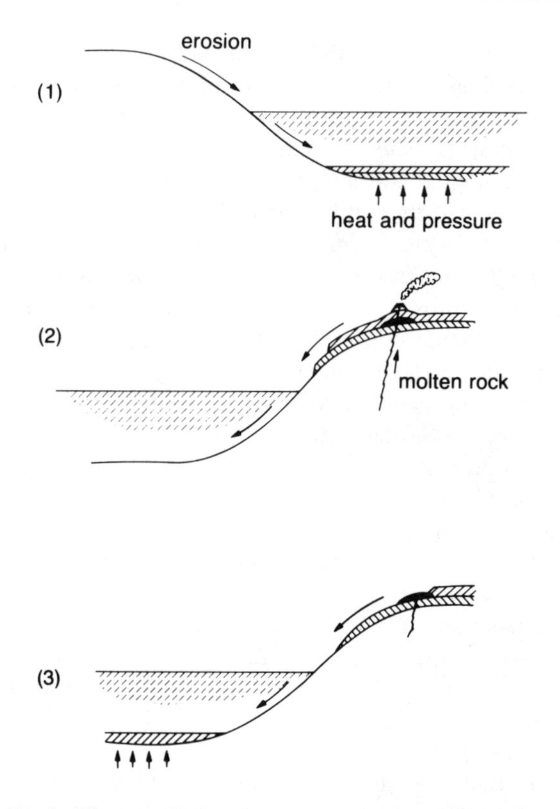

Fig. 5. Hutton's Theory: Vulcanism, or Uniformitarianism.

(1) Erosion by wind, water, and frost wears away at the land surface; rivers carry the debris out to sea where it is deposited on the bottom. Pressure and heat from deep in the earth "bake" the strata into sedimentary rock. Note that different layers of rock may be formed one on top of another, because the nature of the sediment will depend on what is being eroded from the land.

(2) Eventually, pressure acting upward on the old seabed has raised it by a long series of earthquakes until it now has become dry land, exposing the sedimentary rocks. These rocks have been distorted in the process, and cracks may have allowed molten rock from deep in the earth to rise upward. If this reaches the surface a volcano may be formed, but the igneous rock also may intrude into the strata of sedimentary rock where it will cool more slowly to give crystalline rocks, such as granite. Already erosion has begun to wear away the new land surface.

(3) Erosion has now worn away some of the sedimentary rock and exposed the granite. New layers of sedimentary rock are being formed on the seabed; eventually, they, in turn, may be raised upward by earth movements to form a new area of dry land. Note that because Hutton's is a steady-state system, the original land surface in (1) already would have been covered with sedimentary rock formed when it was below the sea.

formed under conditions radically different from those of today. To replace the retreating ocean theory with a system based only on observable causes, Hutton was forced to push the hypothetical origin completely beyond the bounds of consideration for scientific geology.

Hutton thus abandoned the image of "creation" in both its traditional and materialist senses. Yet he still believed that the earth was *designed* by a wise and benevolent Creator, even if we cannot fix a time when the cycle of its history began. The earth is not a chance agglomeration of matter; it is a carefully structured system capable of renewing itself perpetually—a true sign of the perfect workmanship of its Creator. The purpose of the whole exercise was to provide an eternal habitat for the enjoyment of living things, a purpose that reveals the strong theological foundation of Hutton's system. The concept of a *perpetually* self-renewing earth is not an obvious one, nor is it accepted by modern geology. However much Hutton anticipated the details of modern geology, we can only understand his formulation of such a system by recognizing that it was founded not just on the empiricist desire to use only observable causes but also on a particular assumption about the relationship between God and nature. Hutton's theory has a deist rather than a Christian basis: it depends on a God who designs the world but who does not impose a sequence of events on its moral or physical history. Such views were shared by other deists, including such a notable figure as Voltaire, and in more detail by Hutton's fellow Scot, George Hoggat Toulmin (Hooykaas, 1966; Porter, 1978).

Hutton's claim that the earth is indefinitely old exposed him to considerable suspicion in the new and more conservative atmosphere of late-eighteenth-century Britain (Gillispie, 1951). Werner's system still was acceptable to the new breed of scriptural geologist, because it did allow for a "creation" when the primeval ocean was first formed and could even be adapted to explain the deluge. Hutton's belief that the earth is virtually eternal, however, was seen as a denial of the creation and hence, quite unfairly, as a sign of atheism. Deluc (1790–91, 1798) and Richard Kirwan (1799) attacked the Huttonian system on both religious and geological grounds. Deluc, in particular, modified the Wernerian theory to support the biblical story, with six periods of deposition corresponding to the days of creation and a comparatively recent deluge. The Neptunist view also was advanced in Hutton's native Edinburgh by Robert Jameson (1808). But for the defense by Playfair, Hutton's Vulcanism might have been swamped completely by the opposition. Even after the collapse of Neptunism, catastrophist geologists of the early nineteenth century still repudiated the steady-state world view, at least in part on the grounds of orthodoxy. But there were some things Enlightenment geologists had established too firmly to ever be abandoned. The earth *was* very old (if not as old as Hutton supposed) and had undergone a vast sequence of changes. Nor should the rejection of Hutton's methodology be counted as a totally backward step: the essentially directionalist framework of nineteenth-century catastrophism would have its own role to play in the emergence of modern geology along with Lyell's revival of the uniformitarian system.

3

Evolution in the Enlightenment

The geological discoveries of the eighteenth century injected a new element into natural history. Until that time the creation myth had seemed adequate to account for the origin of the earth and its inhabitants. Now, however, if one could show that the earth was very old and subject to change, must this not also throw doubt on the story of the creation of life? Such a conclusion seemed inevitable as soon as fossils of extinct species were taken into account. If these were indeed the remains of once living things, then the naturalist had to recognize either that parts of the original creation had disappeared or that species can change through time. Both alternatives threatened the traditional view that God personally had designed a world that could preserve its structure as a habitat for mankind. Coupled with the growing distrust of Christianity so typical of the Enlightenment, this would ensure that the eighteenth century was the first period to take organic change seriously. (See Glass et al., 1959; Greene, 1959*a*; and Bowler, 1974*a*. In French, see Mornet, 1911; Rostand, 1932; Guyenot, 1941; Ostoya, 1951; Roger, 1963; and Callot, 1965. On Enlightenment thought in general, see Willey, 1940; Cassirer, 1951; Hazard, 1963; Gay, 1966–69; and Hampson, 1968.)

At first sight, the story of eighteenth-century evolutionism would appear to be a simple one. In the early decades of the century, the static, creationist view of nature was still pretty much intact. As materialist philosophers of the Enlightenment became bolder in their attacks on traditional religion, they were encouraged to seek explanations of how life originated on the earth that did not depend on supernatural intervention. The only plausible concepts available to them were those of spontaneous generation (the natural production of life from the combination of nonliving matter) and the transmutation of existing forms in response to environmental pressures. It was thus inevitable that theories containing an element of what we call evolution would be proposed. The logic of this interpretation almost demands that we see those naturalists who dealt at length with the possibility of organic change as the exponents of a genuinely modern evolutionism. Indeed, the only problem left is that of explaining why natural theology was able to make a comeback at the end of the century, checking what had seemed such a promising start for evolutionism.

In fact, the situation was a good deal more complex. There is no doubt that eighteenth-century naturalists and philosophers did try to grapple with the problem of change and that some of them came up with theories superficially resembling modern evolutionism. The reason for their ultimate lack of success lies in the extent to which their approach was shaped by its cultural environment. Eighteenth-century evolutionism was very much a product of Enlightenment thought, which was subtler than the simple interpretation given above would have us believe. Enlightenment materialism had its own unique character, and it did not break as decisively with the past as we might imagine. Some of the most constructive developments in natural history, furthermore, took place within an extended version of the *old* world view and were not easily reconciled with the search for a materialistic explanation of the origin of life. For these reasons, we should not be surprised that the evolutionary approach was abandoned at the end of the century, along with many other aspects of Enlightenment thought. A new start would be needed to build the genuine achievements of eighteenth-century science into a workable framework based on a rather different idea of natural change.

The most extreme exponent of this view of eighteenth-century evolutionism is Michel Foucault (1970), whose radical analysis rejects many themes normally debated by historians of science (Guedon, 1977). Foucault insists that the general interest in materialism is misplaced and does not help us to understand the "classical" natural history of the period. There was an attempt to classify natural objects according to their external characteristics; arrangements were set up in such a way that the network of possible relationships is defined by the method adopted for the classification process. Such a system is not open-ended: all possible forms are predictable from the nature of the system adopted by the naturalist to classify them. Even if it was appreciated that some elements in the pattern appeared on earth at a later time than others, this would only mean filling in slots already defined by the ordering process itself. Evolution, in other words, could never be anything more than the unfolding of a predetermined pattern. There was no possibility of a Darwinian, or open-ended view of evolutionary development until the naturalists of the early nineteenth century had overthrown this belief in a rationally structured order of things.

Foucault's approach seems intrinsically plausible when applied to those naturalists who worked within the tradition of natural theology, who would assume that their ordering process corresponded directly to the divine plan of creation. But we must also explore the far more controversial question of whether those naturalists who adopted a superficially materialist approach nevertheless remained enmeshed within a view of nature as a fundamentally orderly system. To this end, we need a framework of our own that will help us understand the many conflicting forces at work on the eighteenth-century mind. At the outset, we need to distinguish between those developments that took place more or less explicitly within the creationist system and those that were a product of an ostensible challenge to this approach. We must attempt to define key characteristics of the static creationism expounded in the late seventeenth century and then look at the extensions of this necessitated by technical problems of natural history and generally increasing awareness of the possibility of change. It would not be surprising to find that any discussion of the origin of new species within this tradition tended to assume that new elements of the divine plan appeared in a completely predetermined manner. Then we must look at the materialist challenge, trying to pinpoint its chief areas of concern in natural history. We shall see that some of the areas of greatest concern in the eighteenth century were subsequently ignored by naturalists of the Darwinian era. We shall also have to look for hints that would support Foucault's opinion that even the most radical materialists could not break away from the feeling that the laws of nature must function within a rational pattern.

DESIGN AND THE PROBLEM OF CHANGE

Naturalists of the late seventeenth century still hoped that their scientific work could be reconciled with Christianity. Retaining the concept of fixed species, each breeding true to its own type, they hoped that closer acquaintance with the variety of natural forms would reinforce the belief that each was the product of divine wisdom. Three components of this synthesis need to be described at some length because of their important consequences for the later history of evolutionism:

a) The application of the argument from design to explain the structure of individual species, generally in terms of adaptation for a particular way of life.

b) The belief that the overall pattern of relationships *between* species follows some easily recognized structure, such as a linear chain of being.

c) The extension of the Aristotelian notion of fixed species, falling naturally into groups of related forms, to practical problems of classifying the vast number of new discoveries being made as Europeans undertook a program of worldwide exploration.

The argument from design based on the adaptive purpose of each organic structure was a central feature of the school of natural theology that flourished about 1700 and survived well into the nineteenth century. John Ray's frequently reprinted *Wisdom of God in the Creation* (1691) was one of the most popular expositions of this view (Raven, 1942; Gillespie, 1987). William Derham's *Physico-Theology* (1713) was another popular work, while on the continent l'abbé Pluche's *Spectacle de la nature* developed the theme on a multivolume scale. Despite the challenge of the Enlightenment, similar views still were being voiced in William Paley's *Natural Theology* (1802) and were an influence on the young Charles Darwin.

The argument from design held that the order and complexity of the world—particularly as exhibited in the structure of living things—could not have been built up by nature itself and hence must have been imposed by an intelligent Designer. A frequent analogy was drawn with the relationship between a watch and the watchmaker. No one would believe that the pieces of metal that form the works of a watch assumed their complex shapes *naturally,* so we know that there must have been a watchmaker who designed and manufactured them. Because nature contains no power capable of building up the structure of a new species of animal or plant, the same argument for the existence of a Designer must apply. But more than illustrating the Creator's wisdom and power, the structure of living things also tells us of His benevolence. The form given to every organ is adapted to the function it performs in the animal's life-style, showing that the Designer cares for the welfare of His creatures. We may call this the "utilitarian" argument from design, because it assumes that every characteristic has a useful purpose. Man, of course, was the prime example of design, and the perfect structures of hand and eye seemed the most obvious illustrations of how our body has been created to serve us. At first there was a temptation to build the whole argument around man by suggesting that other species were designed to serve his ends. It is no coincidence that the horse, for instance, is so conveniently built for us to ride on. It was all too easy to take this anthropocentric view to ridiculous lengths, as when Pluche declared that ocean tides were designed to help ships in and out of ports.

The serious naturalists who developed the argument from design realized that it was impossible to see the whole of creation as formed for man's use. In the course of his studies, the naturalist found many species that were of no value to man, yet each in some way revealed itself as a product of the Creator's wisdom. Each was perfectly adapted to its own way of life, confirming the idea that divine benevolence extended to every part of creation. Note how well the claws and teeth of the predator, for instance, are adapted to seizing prey. Yet is there not a contradiction in supposing that a benevolent God created animals that can only live by killing others? The natural theologian was able to rationalize this by arguing that in the long run carnivorous animals *minimize* the amount of suffering in the world because they give the aged and sick members of prey species a clean, quick death. In addition, the idea of a "balance of nature" emerged as a forerunner of modern ecology (Egerton, 1973). Interactions between species were all thought to be harmoniously designed to ensure stability of the whole natural system. If some accidental cause allowed a certain species to increase its numbers unnaturally, its predators would themselves increase and soon restore the balance. Thus, the predator-prey relationship was incorporated into natural theology, and the implications of the struggle for existence were not recognized.

Natural theology was sophisticated enough to deal with many phenomena that might at first sight seem to be incompatible with divine benevolence, but some new discoveries were now creating problems. What about parasites such as intestinal worms—could they be reconciled with the argument (Farley, 1977)? This was one of few areas where some writers accepted the idea that parts of nature must have been intended to punish sinful man. More directly relevant to our topic was the problem of fossils, particularly those that appeared to be the remains of species no longer alive today. As a leading spokesman for natural theology, Ray found it difficult to accept the idea of extinction. If some of God's creatures can die out altogether, how can we continue to believe in the perfection of His workmanship? Sooner than destroy his vision of a perfectly formed creation, Ray chose in the end to deny the organic origin of fossils. As more fossils were discovered, however, this position became untenable, and the static view of creation had to be abandoned.

With the exception of ecological relationships, the utilitarian argument from design treated each species as an individual case of adaptation. Yet no one believed that a rational God would design His creation in a piecemeal fashion. Surely there must be some overall pattern linking the forms of the various species together into a harmonious plan of creation. The belief that such a pattern does indeed underlie the apparent diversity of life is an ancient one, going back at least as far as the Greeks. In its simplest form, the pattern was supposed to be a linear chain of being in which all species fall naturally into a single hierarchy stretching down from man to the simplest form of life (Lovejoy, 1936). This is an example of the kind of closed system of relationships envisioned by Foucault. Every link in the chain is predetermined by the need to provide a continuous series of relationships between the two extremes. Taken to its logical extreme, one could argue that once the outline of the pattern was understood, intermediate forms could be predicted even if they were not yet known to science.

The chain of being was particularly vulnerable to the problem of extinction, because in its original version the whole chain had to exist or the pattern of creation would be imperfect. Extinction of a single link would destroy the symmetry of the whole chain. The only way to accept an element of change would be to concede that the chain is not the complete pattern of nature as it exists now but represents a sequence that will only be fulfilled over a period of time. Perhaps only lower rungs of the ladder of creation existed during the earth's early history, after which life gradually mounted to higher levels. A "temporalized" version of the chain of being thus became the first truly progressive interpretation of the history of life. A theory based on this kind of progress toward a predetermined goal, however, is certainly very un-Darwinian in character. Even in its temporalized form, the chain of being was thus an obstacle to the development of the more open-ended idea of progress through diversity accepted today. The chain was gradually eliminated from serious biological discussions in the course of the eighteenth century, although it continued to be used as a poetic metaphor. It is important to note, though, that more sophisticated ideas of linear development continued to flourish as alternatives to Darwinism thoughout the later history of evolutionism.

According to Foucault, the revolution in our understanding of natural relationships that permitted the Darwinian concept of open-ended development did not begin until the very end of the eighteenth century. Then, he maintains, the new approach to classification pioneered by Georges Cuvier broke away from the notion of a fixed order of nature and allowed naturalists to appreciate that there is no limit to the diversity of life. Although there are certain underlying themes on which organic forms are based, adaptation of these themes to the demands of each individual life-style does not follow any preconceived pattern. Note that in Foucault's opinion, it is Cuvier—an archenemy of evolutionism—who nevertheless created the framework of the Darwinian world view. It is the change in the relationships among natural forms that is crucial, from an orderly to an open-ended system, and not the question of whether diverse forms are produced by design or by natural evolution. Foucault insists that all versions of classical eighteenth-century natural history adhered to the formalized view of natural relationships, even when they broke with the chain of being. The chain was the simplest imaginable order of creation, and the more sophisticated taxonomies merely went on to acknowledge a more complex structure for an equally closed system of relationships.

This creates a major problem for our understanding of the developments that occurred within eighteenth-century taxonomy (the theory of classification). Even before 1700, naturalists such as Ray had begun to ignore the chain of being as a guide to how species actually are related. They believed that it was necessary to start by surveying degrees of resemblance among various species, without any reference to the hierarchy of complexity. A workable system of classification would express these relationships in a convenient way and would allow any newly discovered form to be fitted in by establishing its resemblances to known forms. By the mid-eighteenth century, the problem of devising a satisfactory system had become acute and was solved by introducing a series of new techniques. Historians traditionally have interpreted these developments as a revolution that laid the foundations of modern biological taxonomy. The system of nomenclature established by Linnaeus is the direct ancestor of that still used today. As *we* understand the Linnaean system, it is certainly open-ended, permitting the easy inclusion of an ever-widening and quite unpredictable range of natural forms. Foucault would maintain, however, that this is not how Linnaeus himself envisaged the system. We have adapted his techniques to our modern viewpoint, forgetting that as originally formulated, they were meant to define a closed pattern of relationships that corresponded to the divine plan of creation.

Significantly, when Linnaeus at last conceded the necessity for an element of temporal change in nature, he chose to stress not adaptation to new conditions but hybridization as the source of new species. Hybrids between existing species posed a problem for the traditional view in which each natural form is a fixed unit within the divine plan. But Linnaeus was able to exploit the problem as a way out of the even greater difficulty posed by the concept of a developing universe. The hybrid is merely a new combination of existing characteristics and thus fits readily into the belief that natural change represents no more than the filling in of new slots in a preexisting order of creation. Although very different in structure, Linnaeus's theory of hybridization thus fits along with the temporalized version of the chain of being as an illustration of how the eighteenth-century view of change was confined by a system of closed relationships. But what of those naturalists who abandoned the faith in a divine order of nature for materialist philosophy? Surely they, at least, were able to formulate a more positive approach to change that is far more difficult to square with Foucault's interpretation? For the materialist, there was no supernatural agency to guarantee the stability of any particular combination of material particles, which should imply that the environment has unlimited ability to shape organic forms. But we cannot simply jump to the conclusion that Enlightenment materialists were in a position to foreshadow the modern theory of evolution. Their philosophy was applied to the study of natural history via a set of problems that had become crucial in the previous century. This situation directed their thoughts along channels that would seem incongruous to a modern materialist and may have prevented them from breaking completely with the old tradition of natural order.

Consider what was available to the eighteenth-century naturalist who enquired into the adequacy of natural laws to replace divine creation in the origin of life. For the most part, such laws had to be conceived within the mechanical philosophy that had emerged from the triumphs of the physical sciences in the seventeenth century. Descartes's philosophy had divided the world brutally into two: a mental plane on which the human mind lived and a physical plane revealed to the mind through the senses. The physical world was essentially mechanical; all changes were brought about by mere rearrangement of particles of matter. The body of an animal was nothing more that a physical structure governed solely by the laws of mechanics (a robot, we should call it today). Descartes refused to compromise his dualism by allowing special vital forces to control the living body. His "animal machine" doctrine became the starting point for much eighteenth-century thought about the laws of organic nature. Borelli's *De Motu Animalium* of 1680 successfully showed how the relationship between bones and muscles could be understood in mechanical terms. When applied to more fundamental biological questions, however, the animal machine doctrine created more problems than it solved.

The most crucial question was that of reproduction or "generation," traditionally the process that guaranteed the stability of species by ensuring that each individual was formed in the same mold. Here the earlier belief that living matter was governed by special vital forces seemed particularly appropriate. Aristotle had assumed that fertilization by the male supplied an organizing force that molded the passive matter supplied by the female to create the complex structure of the embryo. Could purely mechanical forces perform so delicate a task precisely enough to ensure maintenance of the basic form of the species? To experienced naturalists, it seemed inconceivable that an animal machine, a mere piece of clockwork, could manufacture another machine in its own likeness. The mechanist thus had to assume that the embryo was preformed within the fertilized ovum and then grew only by absorbing matter into its structure. William Harvey had held that all animals grow from eggs, and the ovum was sometimes seen as the "germ" or seed containing the complete new organism in miniature. But this still did not solve the problem, because the mother could not be supposed to have formed the germ within her own body. The germs that would grow into her children must have been present within her body *when she was born*. To preserve the logic of the argument, her own germ must have been contained within *her* mother and so on back through the generations to the first woman, Eve. This "preformation theory" (more properly, the theory of "preexisting germs") held that the whole human race was literally created by God in the beginning, enclosed one within the other, like a series of Russian dolls, waiting to be unpacked generation after generation (Cole, 1930; Needham, 1959; Adelmann, 1966; Gasking, 1967; Roger, 1963; Bowler, 1971).

That such an apparently ridiculous theory was discussed seriously for over a century seems incredible. That it *was* taken seriously is an indication of the influence exerted by two factors, the animal machine doctrine and the argument from design. At first, the mechanical philosophy in biology did not lead to atheism, because the only way to explain how animal machines were formed was to assume that they (or their germs) were created by God. Only gradually did a new level of confidence build up, allowing naturalists the hope of extending materialism to cover the origin of life. The first step was to show how the more sophisticated Newtonian model of physical law could account for normal formation of the embryo within the womb; after this it might become possible to extend the argument to include the first formation of living things on earth. Several important theories of the origin and development of life did indeed arise from such a deliberate challenge to the germ concept, and the fact that these theories set themselves up deliberately as alternatives to the concept of preëxisting germs reveals that the problem of mechanical generation of life was still central. The animal machine doctrine and the concept of germs defined a characteristically eighteenth-century framework for discussing the question of the origin of life. Subsequent collapse of this framework meant that later workers such as Darwin not only ignored these bizarre concepts but also could ignore the very question of the first appearance of life on the earth.

By replacing the germ concept with the belief that each living body is built up by physical forces, materialists had in principle removed any guarantee of the fixity of species. In their search for a causal explanation of how the world has been shaped into its present form, naturalists such as Buffon thus were able to investigate the possibility that environment might affect a species over a long period of time. Yet there were good reasons why experienced naturalists would be reluctant to abandon the idea of organic stability altogether. Their account of embryological development had to tackle the very serious problem of how matter could organize itself into the complex and purposeful structure of a living body and then preserve a similar structure over a long series of generations. If the permanence of species was not established by divine command, it was, nevertheless, a real fact that natural forms are not infinitely mutable. In addition, there were practical advantages for the taxonomist in assuming that species are fixed. The philosopher could very well speculate about the instability of nature, but such a position represented professional suicide for any naturalist who hoped to provide his readers with a workable system of classification. Buffon, clearly aware of these practical considerations, decided in the end to retain the fixity of organic forms by assuming that matter is only capable of arranging itself into a limited number of stable patterns, thereby defining the great families of the animal kingdom. Here we see how a materialist could be forced to compromise with traditional belief in a stable order underlying the superficial changes of the physical world.

Philosophers such as the baron d'Holbach, on the contrary, saw no reason to limit their speculations to accommodate the facts of biology. Some of the boldest expressions of an evolutionary viewpoint can be found in the writings of Enlightenment materialists, who used them in their attack on established religion. By undermining the Genesis story of creation and the argument from design, they hoped to bring down the intellectual structure that sustained the traditional code of morality. It made sense, then, to maintain as brazenly as possible that species had no fixed structure, that nature could create any conceivable form out of random combinations of atoms, and that only the most successful of these forms would survive. It would be wrong to see in these wild speculations a genuine anticipation of the Darwinian selection theory, yet in a more general sense they are as close as any Enlightenment thinker could come to a concept of totally open-ended change.

THE CHAIN OF BEING

The germ theory represented the conservative wing of mechanical philosophy, which attempted to preserve the traditional connection between God and nature. It thus is not surprising to find it allied with another ancient concept, that of the chain of being. The idea of the chain has a long history stretching back to the Greeks (Lovejoy, 1936; Bynum, 1975). It was an attempt to see nature as a highly structured system—which meant a system designed overall by God. Starting from the naturalist's intuitive sense that living things can be ranked in a hierarchy of complexity from the highest (man) down to the most primitive, it was assumed that a linear plan of creation linked the two extremes. Each species could be assigned a unique position in the chain, its closest relatives placed immediately above and below it, so that the highest and lowest points on the chain ultimately were linked via a series of regular intermediate steps. As originally understood, the chain was a static plan of natural arrangements, representing creation as it was first formed and as we still see it today. The theory of preexisting germs allowed one to see how God ensured maintenance of the structure of the plan, because He created the series of germs that enabled the reproduction of each species to take place. Eventually, two philosopher-naturalists who adopted this combination of the germ theory and the chain of being—Charles Bonnet and J. B. Robinet—came to see the chain as a plan of development through time. Successive elements of the chain would appear one after another in the course of the earth's history, the whole predetermined by the system of germs. This is what Lovejoy called temporalization of the chain—the injection of a time element into an originally static plan. Although this view still supposes that all developments are merely the unfolding of a fixed plan of creation, it does attempt to grapple with the notion of a changing universe.

Perhaps the most popular defense of the germ theory came from the Swiss naturalist-turned-philosopher Charles Bonnet (Whitman, 1894; Savioz, 1948; Glass, 1959*b*; Bowler, 1973; Anderson, 1982). Bonnet gained fame through his discovery of the parthenogenesis of aphids. The fact that females of this insect could reproduce themselves for several generations without fertilization by a male convinced Bonnet that miniature structures for a whole series of generations must be stored in the females. His *Considérations sur les corps organisés* of 1762 and *Contemplation de la nature* of 1764 developed a complete theory based on the germ concept: the germs were tiny miniatures containing the whole structure of the new organism, generations enclosed one within the other on the principle of *emboîtement* or encapsulement. Bonnet insisted that such a system was necessary within a mechanical world view: the laws of nature were not capable of arranging matter into the complex structures of living things and could only provide for filling out preexisting structures that must have been created by God.

Of course, there were problems, particularly with the facts of heredity. If the whole organism comes from a germinal structure contained within the mother, how can the offspring inherit any of the father's characteristics? Bonnet solved this problem by suggesting that male semen was needed to start the germ developing and thus could transmit some of the father's individual peculiarities. In the end, he concluded that the germ defines only the structure of the species, not of the individual: it contains only the basic characteristics ensuring that the organism will grow to be a man, a dog, a horse, and so forth. All individual characteristics are produced by the matter absorbed into the germ as it grows, first from male semen and then from the mother's womb. Bonnet also insisted that the germ was not an exact miniature of the adult organism, which one might expect to recognize through a microscope. It contained only an outline of the basic structure, which had to be filled out to become visible—it would be no more recognizable in its original state than a man-shaped balloon would be when deflated.

Bonnet was also an enthusiastic supporter of the chain of being (Anderson, 1976). He believed that by placing each species next to its closest relatives, one could build up a perfectly linear arrangement linking man with the lowest forms of life (see diagram). Such an arrangement represented the divine plan of creation, and the orderly nature of the plan had a number of important implications. It reemphasized the belief that extinction of any

Man
Monkeys
Quadrupeds (Mammals)
Bats
Ostrich
Birds
Aquatic Birds
Flying Fish
Fish — Sea Lions? Whales?
Eels
Sea Serpents
Reptiles
Slugs
Shellfish
Insects
Worms
Polyps (Hydras)
Sensitive Plants
Trees
Shrubs
Herbs
Lichens
Mold
Minerals
Earth
Water
Air
Etherial Matter

Fig. 6. The Chain of Being.

This is a simplified version of the chain in Bonnet's *Contemplation de la nature* of 1764. The chain is built up from a series of superficial resemblances, many of which appear ridiculous in the light of modern classification based on internal structure. Note how the necessity of making a continuous linear pattern leads to some unusual relationships. Although most later naturalists considered reptiles superior to fish, Bonnet ranks them the other way round, because this allows him to make the transition from fish to bird and then on to mammals. Even so, Bonnet was forced to admit the possibility of branches leading off the chain at some points, indicating that the idea of a totally linear arrangement was beginning to break down under its own weight.

species is an impossibility. If the chain were planned as a complete whole, God must ensure that it does not become unbalanced through wiping out parts of the overall pattern. The species were the "links" in the chain, so their structure must be absolutely fixed and permanent. The germ theory gave Bonnet a way of seeing how the Creator could have ensured this absolute stability; He formed the germs of each species and thereby had ensured personally that each member of every enclosed series would grow true to type.

If the Creator formed the whole series of germs on the same pattern, the species would not be able to change. Bonnet's system, however, was potentially flexible in that one also could suppose the Creator to have formed different kinds of germs destined to grow in different periods of history. Eventually, Bonnet indeed came to believe that the chain of being was not a static plan but had unfolded step by step through time to give a progression of life from the simplest forms at the bottom of the chain to the most complex at the top. The whole process was predesigned by the Creator through the different series of germs He formed in the beginning, each planned in such a way that it would become realized in flesh at a particular point in time.

This idea of universal progress was explored in Bonnet's *Palingénésie philosophique* of 1769. Rather curiously, it originated from Bonnet's interest in the Christian notion of the resurrection of the body, which he had speculated might be achieved through a second germ supplied to each soul by its Creator. He argued that perhaps animals have souls restricted within their more limited bodies and may themselves be resurrected in the future with a higher form of body. Then man will move on to a higher plane of existence, animals will become men, and plants will become animals. But why limit this resurrection to a single event in the future? Perhaps every soul has been reincarnated in the past through a series of physical bodies, all developed from germs originally provided by God. At each reincarnation the soul would appear in a more perfect body, and thus the history of life would be a gradual ascent up the chain of being from the simplest bodies in the distant past to the more perfect form enjoyed by man today. Bonnet concluded that from time to time a geological catastrophe would wipe out existing forms of life, but the germs responsible for future resurrections would survive and would be able to develop into a whole new population when conditions settled down.

Bonnet thus came to the idea of biological progress through the temporalization of the chain of being. Exact details of how the germs were supposed to be reincarnated were never worked out, and the whole system seems so obscure that it is difficult to see it as a step toward modern evolutionism. Some historians, in fact, have dismissed Bonnet's system as merely an attempt to sidestep the implications of development by making everything depend on a single original creation (Whitman, 1894; Glass, 1959*b*). It is certainly true that progress represents only the unfolding of a divine plan locked up in the series of germs. Yet there are later naturalists, widely accepted for playing roles in popularizing the idea of evolution, who also saw organic development as a process with a built-in goal. Bonnet was at least in tune with the growing desire to eliminate the need for miraculous interference with the ongoing processes of nature, even if he did compress all of God's design into a single creative act. He also was making a genuine response to growing interest in the earth's history, using geological catastrophes as the agent that prepared the way for each new population. Indeed,

Bonnet also showed that he was aware of the need for each population to be adapted to the conditions of the period in which it lived—a more forward-looking idea than simple ascent of a linear chain of being.

The move from a static to a progressive version of the chain of being was also made by the French philosopher Jean-Baptiste Robinet, between the first and fourth volumes of his *De la nature* (1761–66; see Murphy, 1976). There was, if anything, an even greater element of speculation in Robinet's work than in Bonnet's. He did not believe that the chain of being is divided into links representing distinct species. Instead, he explored another image of the chain as a *continuous* pattern, more like a rope. Species, he claimed, were illusions, because careful inspection would always reveal a complete spectrum of forms linking any two points on the chain. The division of this continuum into species is based merely on the comparative rarity of the intermediates and is favored by naturalists purely for convenience. Species do not exist in nature, only individuals corresponding to every possible point along the chain. A number of eighteenth-century thinkers were attracted to this view, which might at first sight be taken as an anticipation of evolution. In fact, though, the idea of continuity was part of the older tradition that had to be abandoned before anything resembling Darwinism could appear (Zirkle, 1959*a*). Darwin did not destroy the species concept; he merely reinterpreted it: in his system, species are still *distinct* (but not immutable) entities. Robinet's view could generate only a concept of change as progress along a preordained sequence of form, not the modern idea of branching evolution.

Robinet accepted the germ theory of reproduction but believed that germs are spread throughout nature, waiting for suitable environments in which to develop. In his fourth volume, he argued that germs corresponding to successive points on the chain of being must have grown to maturity one after another in the course of the earth's history. At first only the simplest germs were able to develop, but these served as parents for germs of the next highest degree, enabling life to progress along the chain toward man. There is considerable difference between the views of Bonnet and Robinet, but the fact that the two leading supporters of the germ theory should both have adapted it to the progressive interpretation of the chain of being shows that even conservative ideas were yielding to increased awareness that man is a comparative latecomer in an extremely ancient and ever-developing world.

THE NEW CLASSIFICATION

The chain of being was based on the belief that a system of classification built up from the closest similarities between species would naturally fall into a hierarchical, linear pattern. This view began to break down in the eighteenth century as exploration revealed more and more species that had to be fitted in. It now became clear to most naturalists that nature was too complex to be described in terms of a simple linear pattern. A more flexible method of representing organic relationships was needed.

First it was necessary to establish the basic unit of the classification system. It is possible to argue that nature consists only of individual organisms that do not fall into well-defined groups, as did those in Robinet's vision of a continuous chain of being. Yet casual observation suggests that for the most part individuals do belong to distinct groups, traditionally known as "species," that could serve as the basic unit of classification. Naturalists realized, however, that it was not always easy to establish which natural groups deserve to be accorded the status of true species. John Ray attempted to deal with this problem by arguing that trivial, local differences must not be allowed to break up the basic unity of the true species that God originally had created. It was ridiculous to add a new species to the list every time one discovered a form differing in some slight degree from those already known. Such differences could have been produced by the continued effect of local conditions on the original form. The Creator will have ensured, however, that these modifications cannot go so far as to obscure the character of the form He designed. Thus, the distinction emerged between *species* as real entities created by God and *varieties* formed within the species as a result of changed conditions (Ray, 1724; Raven, 1942; Sloan, 1972).

Having accepted the reality and fixity of species, one could tackle the difficult problem of expressing natural relationships between distinct forms. Ray made important steps in this direction, but it was the Swedish naturalist Carolus Linnaeus who founded the modern system of classification. Linnaeus wanted to sidestep the sterile debates over the mechanical philosophy and revolutionize biology from within, by working out the true relationships between the various forms of life. If the species were created by God, one could assume that a rational Creator would have formed the world according to a meaningful order that man himself could hope to understand. Linnaeus believed that he had been privileged to see the outline of the Creator's plan and his efforts to represent it would become the basis of a new biology. His technique was outlined in his *Systema Naturae* (1735)—a work that began as a slim pamphlet but grew over the decades into a multivolume classic, gaining its author worldwide fame (Hagberg, 1953; Blunt, 1971; Larson, 1971; Stafleu, 1971; Broberg, 1980; Frängsmyr, 1984).

The concept of design by God lay at the heart of Linnaeus's philosophy of nature. The very fact that we can classify species into an orderly system bespeaks the existence of a rational Creator. The relationships of similarity between species which today are seen as a sign of common evolutionary ancestry were for Linnaeus elements in the divine plan, real only in the sense that they existed in the mind of God. The world, however, is not just a formal pattern of relationships—it also has to work in practice. The Creator designed each species to fit its particular way of life, as emphasized by natural theology; but Linnaeus and his followers were more interested in what today we should call ecological relationships *between* species inhabiting a certain area. Each was dependent on the whole for its livelihood, and God had ensured the permanent stability of the system by designing a series of checks and balances that would maintain the population of each species at an appropriate level. Thus, the "balance of nature" was preserved and relationships that would later be seen as part of the struggle for existence were incorporated into natural theology (Bilberg, 1752; Egerton, 1973). For all these reasons, of course, it was impossible for a species either to change beyond rigidly defined limits or to become extinct, because this would upset not only the formal plan of creation but also the balance of nature.

Given the faith that there is a divine plan of creation, the naturalist must discover and represent the structure of that plan. The natural starting point will be to group together those species with obvious similarities into a higher-level grouping called the "genus" (pl. "genera"), then to group the genera according to more basic resemblances, and so on. But how do we work out these degrees of resemblance? It must be supposed that in the Creator's plan *every* relationship has a meaning; thus, a truly natural system of classification will have to take into account all the characteristics of every species. Linnaeus believed that such a natural system was the goal of his work, although at the beginning of his career, he was overwhelmed by the immense amount of information that would have to be processed in order to set up the natural system. As a preliminary step, he decided to establish an "artificial system" that would classify according to resemblances in a *single* characteristic. This would not give a perfect arrangement, but it would be a rough outline of the plan of creation that could be modified as further experience showed necessary. In his own field of botany, Linnaeus drew on Camerarius's discovery in 1694 of plant sexuality and made the reproductive organs the key characteristic on which his system would be based. This was not an arbitrary choice, because the reproductive organs were responsible for maintaining the structure of the species.

The success of Linnaeus's artificial system depended on the ease of assigning any species to its correct position. The plant kingdom was divided into a number of basic classes, which in turn were divided into orders. To determine the class and order of a particular plant, it was necessary only to count the stamens and pistils in its flower. Orders were divided into genera and species by closer inspection, taking into account the shape and proportion of the flowers. The same taxonomic ranks were also introduced into zoology, where Linnaeus distinguished six classes in the animal kingdom. Modern biologists find it necessary to use a larger number of ranks than Lin-

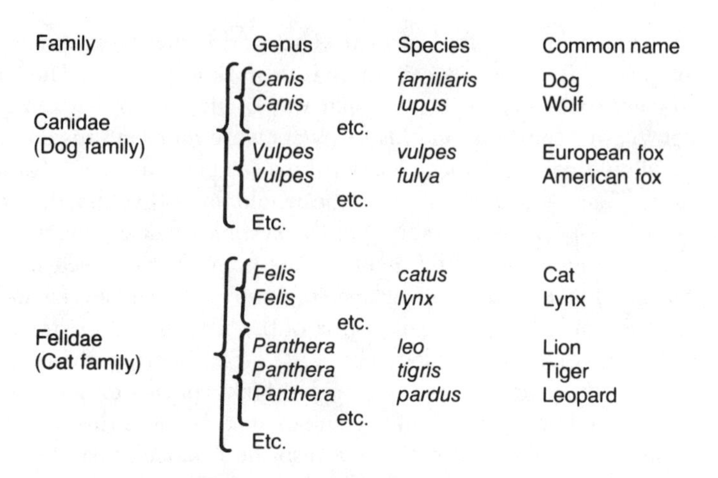

Fig. 7. Classification and the Binomial Nomenclature.

The illustration here uses species that will be familiar even to those readers with no biological training. Examples are shown from four genera belonging to two different families: note how the most closely similar species fit into the same genus, while genera with a fairly obvious similarity fit into the same family. The families Canidae and Felidae belong to the order Carnivora (flesh eaters) of the class Mammalia.

naeus allowed, so that the genera are first grouped into families before being combined into orders and classes. The diagram here shows an example of the modernized Linnaean system drawn from the animal kingdom, with the grouping of species into genera and genera into families. The diagram also illustrates another of Linnaeus's innovations, the "binomial nomenclature," which identifies each species by two Latinized names. The first name defines the genus, the second the particular species. By international convention, the starting points for all botanical nomenclature are Linnaeus's *Species Plantarum* (1753) and the fifth edition of his *Genera Plantarum* (1754). Zoological nomenclature originates from the tenth edition of the *Systema Naturae* (1758), in which the binomial nomenclature was first applied to all known animal species.

Linnaeus's system made no effort to represent the relationships between species as though they fell into a linear pattern or chain. Indeed, it does not imply a hierarchical ranking at all in the original sense of the term "hierarchy." Modern biologists treat the Linnaean system as hierarchical in a different way (the more basic taxa are "higher" in the sense that they include the lower ones), but the system destroys the kind of superior-inferior ranking implicit in the chain of being, which was based on an intuitive sense of the organisms' complexity of organization. Other naturalists aspiring toward a natural system continued to believe that a linear arrangement would be possible (e.g., Adanson, 1763). But Linnaeus was essentially a pragmatist: if it became obvious that the Creator's plan was not a linear pattern, then he was prepared to let the relationships build up in whatever seemed the most natural way. In the diagram, the fact that cats are placed above dogs is merely accidental—the families are simply two different kinds of animals, neither of which is in any obvious way "superior" to the other. A single species does not just have two closest relatives, one to be placed above it on a scale and one below. There may be a number of close relatives, although the nature of the relationship would be different in each case. Such a system of relationships cannot be expressed as a one-dimensional chain but will require at least two dimensions. Linnaeus in fact drew an analogy between these relationships and the way countries fit together on a map.

Even the close relationship between the species in a genus at first was supposed to be nothing more than a formalized part of God's plan, but the similarities used to establish this degree of association occasionally were so striking that there was a temptation for the naturalist to wonder whether the genus might be formed by the gradual splitting of a single original form into a number of closely related ones. While Linnaeus originally denied this most strenuously, he eventually came to accept the concept that such a multiplication of the species within a genus might occur through action of different environments. Without perhaps realizing the full consequences of his actions, he thereby threatened the traditional distinction between species and varieties. In some circumstances, local conditions could produce so great an effect that a variety eventually might become a distinct new species. The question, then, of how one could tell which species were original creations by God would become a very ticklish one.

Formation of a new species through the action of external conditions is certainly a kind of evolution, but there was another mechanism for multiplication of species on which Linnaeus placed more emphasis: hybridization (Roberts, 1929; Glass, 1959*b*). Species normally breed true to type, but hybrids between two different forms are not uncommon—the mule is the best known example, although it is sterile and hence cannot count as a new species intermediate between the horse and the ass. But Linnaeus and his students eventually came to believe that in the plant kingdom it was possible for two species to crossbreed and produce a hybrid that could reproduce itself successfully. Because the hybrid was distinct from both parents, it thus would constitute a new species. The first example discovered was a new form of toadflax, described by one of Linnaeus's students under the name *Peloria* (Rudberg, 1752; Hartmann, 1756). In his "Disquisitio de Sexu Plantarum" of 1756 (reprinted in Linnaeus, 1749–90, vol. 10), Linnaeus suggested that

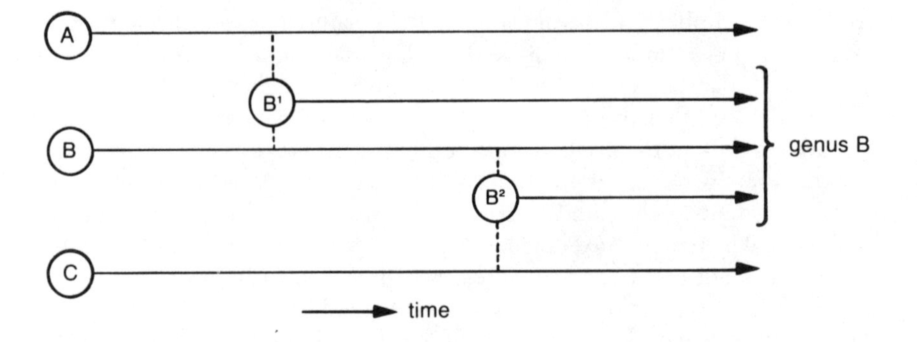

Fig. 8. Linnaeus's System of Hybridization.

A, B, and C are originally created species that continue to breed true throughout. At some point in time, a female of B is fertilized by pollen from A to give a hybrid species B^1, which then continues to breed true. Linnaeus believed that in such a cross the female parent would determine the basic form of the hybrid, with the male influencing only some superficial characteristics. In this case, the hybrid will be a new species within the same genus as B. A later cross of a female of B with a male of C produces a second hybrid species B^2, again a member of genus B. A and C are, of course, capable of producing their own hybrid variants when their females are involved in a cross with a male from another species. The order A-B-C is not meant to represent a chain of being, so that any original form may be capable of hybridizing with any number of others to give a whole cluster of new species within each genus.

in the original creation God formed only a single species as the foundation of each genus and left the multiplication of species within genera to a natural process of hybridization.

Other investigators dismissed Linnaeus's hybrids as trivial modifications or as infertile forms (Kölreuter, 1761–1766; Adanson, 1779). Only one is accepted as genuine today, although the possibility that new plant species may arise through hybridization generally is conceded. The fact that Linnaeus exploited this phenomenon as a means of increasing the number of species through time illustrates the wide range of options available to anyone in the eighteenth century beginning to grapple with the problem of organic change. He believed that the basic structure of nature still was designed by God through creation of the original genetic forms, but the purely natural process of hybridization was responsible for filling in details of the plan, without the requirement of divine intervention. At least a limited form of natural development was thus possible, and the system of classification might express real, rather than purely formal, relationships.

THE NEW THEORIES OF GENERATION

Whatever the practical success of Linnaeus's system, his efforts to turn naturalists away from the mechanical philosophy were not particularly successful. The fascination of trying to explain how living things worked in physical terms had taken too strong a hold. Descartes's original program had stressed the need to search for the mechanical origins of every structure in the universe, as with the theories proposed to account for the origin of the earth itself. It was inevitable that some more radical thinkers should have the ambition to extend the materialist program to include the generation of life, challenging the theory of preexisting germs created by God. The overthrow of Descartes's own physics in favor of Newton's was ultimately a positive benefit to the overall Cartesian program. Because Newton had postulated the mysterious force of gravity, which could act at a distance across empty space, the concept of what constituted a "mechanical" system was now far more sophisticated. The body need not be merely a piece of clockwork but could be governed by more complex (but still physical) forces. Once Newton's physics finally had stormed the bastions of French culture, the first efforts to explain generation through operations of natural forces were made.

If the living body was formed by natural forces instead of from a preexisting germ, a host of possibilities were opened. In Bonnet's theory, the series of germs created by God guaranteed fixity of the species, but if generation occurs through a material process, this guarantee is lost. If the structure of the parents' body is changed, may not such a change be passed on through the process of forming the new embryo? And if such "acquired characteristics" are inherited over many generations, might this not change the structure of the whole species? Beyond natural transmutation lay an even more exciting prospect: explaining the origin of life on earth. If a material process is responsible for normal generation, might we not assume that under certain special circumstances forces of nature could act directly on inorganic matter to create a living organism? The "spontaneous generation" of life from nonliving matter was the final goal of the materialists' program, because it excluded the Creator from all direct control over the world.

A step along the road to a materialist explanation of the origin of life was taken by de Maillet, whose *Telliamed* provided one of the first theories of the earth completely to dispense with the creation and the deluge. Although published in 1748, the book had been written much earlier when the simpler Cartesian physics made complete elimination of the germ theory impossible. Yet in keeping with his materialist outlook, de Maillet refused to invoke a miracle to start life off once the earth was formed. Instead, he adopted a version of the germ theory in which germs exist independently and are distributed throughout the universe. Normal generation occurs when a germ of the appropriate species finds its way into the womb of a female who can serve as its mother. Before there was life on earth, it may have been possible for the waters of the vast ancient ocean to provide an environment in which the first germs could mature. The original members of each species would be formed not through a miracle but through natural development of germs that had found their way onto the earth.

De Maillet held that germs would give rise directly to all of the species we now observe, without any progression from simple to complex. Each species, however, had to undergo one modification in the course of the earth's history. He believed that when life first appeared the whole earth was covered by water, hence the first members of every species would have to be aquatic. Each species would have to evolve into a terrestrial form once the first dry land appeared. Every modern species, he held, has an aquatic counterpart, which represents its original form. Sailors' stories of mermaids were taken as evidence for an aquatic form of mankind, while flying fish were supposed to be the ancestors of birds. However ridiculous such ideas might seem, they show that de Maillet was prepared to admit an effect of external conditions on the species. Yet he also insisted that the basic form of the species was recognizable throughout the transformation—the mermaid is still human. Thus, the kinds of germs available within the universe determine the basic form of the species that can appear on earth or on any other planet.

De Maillet had eliminated miracles and partly circumvented the argument from design by allowing germs to adapt to different conditions as they grew. He also tried to avoid postulating a supernatural origin for the germs themselves by claiming that they always have existed throughout the whole universe—which is itself eternal. A similar theory of the origin of life also was advocated by the extreme materialist philosopher La Mettrie. Yet even this modified version of the germ theory is not really satisfactory in a materialist setting. Because the structure of each living thing derives largely from its germ, one still can ask how the germ was formed or "designed." Making germs eternal along with the universe only evades the problem, because life still requires something more than the laws of nature to explain its origin, and this "something"—the germs—still might have a divine origin. A completely materialist position could be maintained only by eliminating germs altogether and making matter itself responsible for the origin of life.

The first major challenge to the germ theory came from a leading Newtonian: Pierre Louis Moreau de Maupertuis (Brunet, 1929; Glass, 1959*a*; Sandler, 1983). Maupertuis's ideas were presented in his popularly written *Venus physique* of 1745 (translation, 1968). In it, he raised a number of arguments against germ theory, including the fact that every observer from Aristotle to Harvey had confirmed the growth of the embryo by "epigenesis" (the sequential addition of parts), not by expansion of a preformed miniature. Maupertuis also studied heredity, especially in the case of polydactyly (full report in Maupertuis, 1768, vol. II, letter XIV). If a person were born with six fingers on one hand, was this abnormality designed by the Creator when he formed the germ of that person, or was the extra finger produced by an accident of growth? Assuming that God does not deliberately design malformations, how could the product of an accident be inherited through a whole series of generations? The only solution was to scrap the germ theory and accept the obvious fact that both parents contribute equally to their offspring. Maupertuis held that both parents produce a fluid semen that, when mixed in the mother's womb, develops into the embryo. (The concept of a female semen may seem odd today, but it has a long history stretching back at least to Aristotle.) The semen contains particles derived from each part of the parent's body, and these are drawn into their respective places in the embryonic structure by the force of Newtonian attraction.

At the end of *Venus physique*, Maupertuis applied his theory of generation to a problem with considerable significance for evolutionism: the origin of human races. Because semen derived from the parents' bodies was now responsible for the formation of their offspring, it was possible for a new characteristic appearing within an individual to be perpetuated through later generations. As in the case of the sixth finger, once the variation has appeared, it can become a permanent feature of the species. In the case of racial characteristics, how were they formed in the first place, and how were they concentrated into distinct groups of men? Maupertuis suggested two possibilities, corresponding roughly to natural selection and what would later come to be known as Lamarckism. Perhaps the Negro's dark skin appeared by some accident of the reproductive process and was then preserved by isolation as black individuals were driven into the less-hospitable tropics. Or perhaps the climate of the tropics actually had produced the dark color in people who came to live there, and later it became permanently fixed through heredity. Whichever speculation was accepted, the new theory of generation allowed one to circumvent the common belief that the structure of any species is permanently fixed. Totally new characteristics could appear and be preserved through the process of heredity.

These suggestions were expanded into a wider theory of the development of life in the *Système de la nature*, which Maupertuis published under an assumed name in 1751 (reprinted 1768, vol. II). He now attacked the deeper problem of the ultimate origin of life, adopting a completely materialistic approach. Perhaps when the earth originally was covered with water, some particles of matter might have been sufficiently active to arrange themselves into the first living structures without requiring a womb in which to develop. Spontaneous generation through natural causes thus replaces divine creation. Maupertuis also noted the possibility that the original forms of life might have divided themselves through natural transmutation into the different species we now observe. This extended the suggestion on the origin of human races—now whole new species could be formed through adding up small changes introduced into the process of heredity.

By raising the possibility of spontaneous generation, Maupertuis pinpointed a problem that already had begun to bother him in *Venus physique*. How are the particles in semen able to "remember" the place they are supposed to occupy in the growing embryo? More puzzling still, if living structures can be spontaneously generated from inorganic particles, even more than "memory" must be involved. Particles of matter must have some fundamental tendency to organize themselves, and Maupertuis now argued that this must involve something more than purely material forces. Particles of matter must have a kind of "will" or "awareness" to recognize the functions they are supposed to perform. Here we see the strangest outcome of the Enlightenment's trend toward materialism. In attempting to explain life as a product of matter alone, the philosopher is forced to attribute the properties of life to the particles of matter themselves!

Maupertuis's later speculations hardly can be counted as scientific, but a theory of generation very similar to his served as a starting point for one of the most comprehensive pre-Darwinian accounts of organic change, that of Georges Louis Leclerc, comte de Buffon. When Samuel Butler (1879) tried to discredit Darwin, he did so by arguing that all the essential points of evolutionary theory had been foreshadowed by Buffon. From time to time, this claim has been repeated by historians (e.g., Guyenot, 1941). No doubt there are passages in Buffon's voluminous writings which seem to anticipate aspects of the Darwinian theory, but in the end what emerges is a picture of nature based on quite different foundations.

A more relevant question concerns the extent to which Buffon's views belong to the atheistic wing of Enlightenment materialism. He certainly was committed to the search for a causal explanation of the present structure of the earth and its inhabitants. As Wohl (1960) has pointed out, Buffon was far from a simpleminded materialist, yet in practice his program often amounted to no more than a Newtonianized version of the search for mechanical origins. Roger (1963) has suggested that Buffon was an atheist who wished to dispense with the notion of a Creator who ordered the workings of the universe. Certainly, he was no Christian, or even a very typical deist, yet in the end he refused to accept the more radical materialists' vision of nature working completely at random. The origin of life is a material process but not a "trial and error" affair, because the physical world operates within very strict limits.

In 1739, Buffon was appointed superintendent of the Jardin du Roi in Paris (the modern Jardin des Plantes), and he began to plan the publication of a comprehensive survey of natural history (Hanks, 1966). The first three volumes of his *Histoire naturelle* appeared in 1749, and as the succeeding volumes poured from the press, they established their author's reputation as France's leading naturalist. If Linnaeus's system of classification triumphed because of its practical applications, Buffon provided a widely popular alternative for those who believed that man could discover the natural causes of all things.

Buffon's first volume launched an immediate attack on Linnaeus and all those who reduced natural history to the search for abstract relationships. Many historians have been puzzled, because at this point he even seems to deny the existence of species as distinct entities. Buffon argues that species and their groupings into genera, and so on, are merely figments of the taxonomists' imagination. In reality, there are only individuals, and we may sometimes find an intermediate form between two supposedly distinct species. Thus, the claim that nature presents a complete spectrum of forms appears to be used against setting up any system of classification. Yet in the other two volumes published in 1749, Buffon was already writing of species as fixed, distinct entities, a view he maintained with some modification for the rest of his life (Wilkie, 1956; Lovejoy, 1959*a*; Farber, 1972; Eddy, 1984). Phillip R. Sloan (1976) has pointed out, however, that Buffon's position is not really inconsistent. He certainly objected to the Linnaean program of creating abstract relationships between species which were assumed to exist only in the mind of the Creator, but he did not want to deny that there were any meaningful connections between individuals. Such links must be seen in terms of physical relationships existing through time: if we can show that a group of individuals belongs to a distinct population maintained by reproduction, then we shall have identified a true species.

The definition of the species as a group preserving itself through reproduction threw great emphasis on Buffon's theory of generation. He argued that life could be properly explained only if its production could be accounted for in material (i.e., Newtonian) terms. The germ theory was dismissed as providing no real explanation of generation, because it threw everything back to original creation by God. Following Maupertuis, Buffon adopted the view that the embryo is formed from a mixture of male and female semen in the womb. The semen consisted of "organic particles" derived from food superfluous to the organism's nutritional requirements. The crucial question was: how do these particles "know" how to arrange themselves into the complex structure of the embryo? Here Buffon introduced the concept of the "internal mold," an entity supposedly capable of directing the particles into place. The mold was a characteristic of the species, as though some fundamental constraint forces the particles to arrange themselves only into a particular structure. This, of course, would preserve the form of the species from one generation to the next, but Buffon never explains exactly what the mold is or how it uses purely physical forces to guide organic particles into place. The ambiguity of the concept illustrates the difficulty of creating a satisfactory theory of generation without something like the preexisting germ to provide necessary information for constructing the embryo.

However unsatisfactory, the mold concept allowed Buffon to begin his description of the animal kingdom within the framework of absolutely fixed species. In his article on the ass (Vol. IV, 1753), he deliberately raised the question of whether or not closely related species such as the horse and the ass might be descended from a common ancestor, only to answer it in the negative. His rejection has sometimes been dismissed as insincere, but it is based on arguments developed in such detail that it is difficult not to imagine Buffon taking them seriously. Over the next decade, however, he became convinced that related species do have a common ancestor. In the section "De la degénération des animaux" (Vol. XIV, 1766) he openly supported the position denied in 1754. The relationship between certain Linnaean species now was to be regarded as a real one, based on a process of divergence from an ancestral form over a period of time. Buffon's collaborator, the anatomist Daubenton, did not share these opinions, however (Farber, 1975).

All the closely related forms ranked within a Linnaean genus (or modern family) now were supposed to be derived from a single original population, which became divided into separate groups through migration to different parts of the world. Each group was then influenced by the climatic conditions of its own area and gradually changed its form. Buffon argued that the effect of external conditions was produced by the different qualities of the organic particles assimilated into the reproductive systems of the organisms. The end product of the migrations was a series of similar, but distinct, forms called species. Yet Buffon continued to insist on the absolute fixity of species, and it is clear that he now saw each family (not each Linnaean species) as based on a single internal mold. The so-called Linnaean species are *not* real species, only strongly marked varieties. In theory, each "species" could breed with any other member of the same family, although in practice, the crossing may not work because of accidental factors. Buffon had in fact conducted experiments attempting to produce hybrids between related species and claimed to have been successful in several cases (although modern authorities are suspicious). He believed that the mule is not a sterile freak but a potentially viable product of parent forms whose reproductive incompatibility is only superficial. By contrast, hybridization between members of different families will be impossible, because their reproduction is based on different internal molds.

In arguing that a single ancestral form might diverge into a number of "species," Buffon came close to the modern concept of evolution. His recognition that migration to different parts of the world might cause the divergence also marks a pioneering effort in the study of geographic distri-

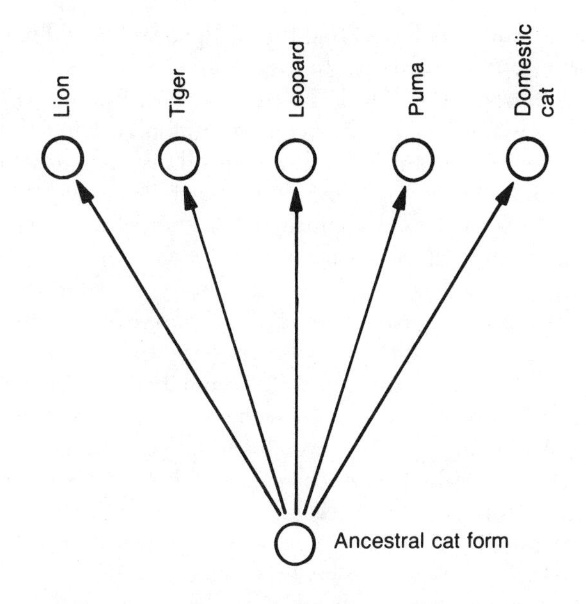

Fig. 9. Buffon's Theory of "Degeneration."

The various members of each modern family have been formed from a single ancestral type. Buffon supposed that the two hundred or so mammalian species known to him had been derived from thirty-eight original forms by the kind of divergence illustrated here for the cat family. He also supposed that as members of the pure original type migrated around the world they encountered different conditions, and their bodies responded by "degenerating" into the local races that Linnaeans (and at first Buffon himself) mistakenly classified as distinct species. These are not in fact true species because in theory they can interbreed. Nor are the local modifications permanent: the amount of degeneration adds up over a number of generations until it reaches the limits defined by the fundamental nature of the type. If the peculiar conditions responsible for the degeneration could somehow be removed, the local modification would disappear as rapidly as it was built up. The modern theory of evolution assumes that similarity of structure implies community of descent but imposes no such limits and does not suppose the changes to be reversible.

bution as a clue to the history of life. But his idea that the environment acts directly on the organism through organic particles is a long way from the theory of natural selection. There is also the question of the origin of ancestral forms from which members of each family have diverged. In the modern theory, these too must have evolved from some even more basic form, whereas Buffon insisted that the character of each family is permanently fixed by its internal mold. Yet families must have had a starting point in time, because Buffon believed that the earth at first would have been too hot to support life. His alternative to evolution on a large scale was, in fact, spontaneous generation (Wilkie, 1956; Roger, 1963; Bowler, 1973).

Spontaneous generation had been part of Buffon's original theory of generation. He believed that a mass of disorganized organic particles could arrange itself naturally to give the simplest forms of life, what we today would call microorganisms. This was confirmed by a series of experiments performed in association with John Turberville Needham in which boiled meat gravy was sealed into a flask and then observed to be swarming with microorganisms a few days later (Needham, 1748). These experiments were criticized by Lazarro Spallanzani (1769), who quite rightly argued that Needham had not sterilized his flasks well enough to eliminate the "germs" responsible for producing the microorganisms. The whole issue was a complex one, however, and the position adopted by Needham and Buffon was not so clearly in error as historians of biology once assumed (Roger, 1963; Farley, 1977; Roe, 1983, 1985). Suffice it to say that many naturalists at this time still believed in the possibility of matter spontaneously organizing itself into living structures. By extrapolating from his experiments, Buffon now could argue that in certain periods of the earth's history, special conditions might have allowed the natural spontaneous generation of even the higher forms of life.

In a supplementary volume entitled *Les époques de la nature* (1778), Buffon surveyed the whole of the earth's history, dividing it into seven epochs (a parallel of the seven days of creation designed to allay the church's fears). In the third epoch, Buffon believed that the earth had cooled down enough for the first organic matter to be formed, which would in turn spontaneously organize itself to give the first forms of life. These early forms were not, however, the ones we know today—they were a different set of species suited to the higher temperature. As the earth cooled, these early forms migrated toward the equator and then became extinct as the temperature dropped below their level of tolerance. In the fifth epoch, more organic matter was formed, and a second spontaneous generation gave rise to the ancestors of the species we know today. Since that time, these species have changed to a limited extent, modified by the conditions of different areas and gradually getting smaller as the earth cools. Still Buffon insisted, as in 1766, that the basic structure defining each family remained fixed throughout these superficial modifications.

Buffon's views on spontaneous generation must be interpreted with care, because they differed in one crucial respect from the way in which the same basic ideas were exploited by more radical materialists. For Buffon, the coming together of organic particles to give the first living things was *not* a purely random process, because whatever the circumstances, it would always yield exactly the forms we know. In fact, he argued that the same species would be produced on each planet of the solar system when it reached the appropriate temperature (supp. vol. II, 1775). Thus, the internal mold of the family is fixed not only during its lifetime here on the earth but through the whole history of the solar system. Apparently, the molds are a permanent feature of the universe, ensuring that wherever life is formed it will conform to certain basic structures. Living matter only can combine to form certain kinds of structure, just as chemical elements only can form certain stable compounds.

This insistence on the permanence of organic structures sets Buffon apart from more extreme materialists who argued that nature creates life through a trial and error process that shows no evidence of design by a benevolent Creator. Buffon himself had no sympathy with those who saw every detail of every species as illustrating the power of the Creator, so the permanent forms he envisaged cannot be called "designs" in the traditional sense. Yet he would not picture nature as a completely open-ended system, and his acceptance of permanent guidelines built into the very structure of nature might indicate a last vestige of the argument from design. Certainly, Buffon's position fits in with Foucault's claim that eighteenth-century naturalists could not see evolution as anything more than the unfolding of a preordained series of forms. For Buffon, natural evolution in response to environmental change was always constrained within the limits defining each family structure. At the same time, however, he saw these limits defined not by the method of classification (as Foucault suggests) but by some fundamental characteristic of the matter that constitutes the natural universe.

THE MATERIALISTS

There were more radical thinkers than Buffon who wished to eliminate anything that would limit the creative power of nature, atheists who challenged the argument from design and presented nature as the sole reality. Everything we see in the world must be a chance product of the ceaseless activity of material nature, including the various living species and man himself. Fixed species, to these thinkers, represented vestiges of the old creationist myth—if nature was truly creative in herself, she would be capable of generating *any* kind of structure. In some respects, the Enlightenment atheists came close to the philosophy of modern evolutionism, although their ideas were seldom developed with anything like the rigor of a scientific theory.

The materialist program was pioneered by Julien Offray de La Mettrie. In his *L'homme machine* of 1748 (reprinted 1960; Vartanian, 1950), La Mettrie ignored the ultimate origin of life to concentrate on showing that man himself must be treated as a purely material entity. The mind or soul is not a distinct spiritual element but merely a product of the physical body. Living matter is itself inherently capable of maintaining the processes of life without the intervention of mysterious vital forces or spirits. La Mettrie had been fascinated by the ability of the "polyp" or freshwater hydra to regenerate two complete organisms if the original were cut in half, a phenomenon recently discovered by Abraham Trembley (translation, 1973; see Baker, 1952). If such a simple piece of living matter contained within itself power to regenerate a whole organism one could assume that life was a basic property of matter itself. The whole traditional apparatus of souls and vital forces could be swept away and replaced by pure materialism. Man becomes a machine—provided we transfer some of the distinctive properties of life to the matter of which the machine is composed! The problem for Enlightenment materialists was that the physical sciences of the time gave them no indication of how inanimate matter could achieve or even maintain the complex structure of a living thing. Thus, there was a constant temptation toward "hylozoism"—the belief that matter itself is in some primitive way alive or aware. Such views frequently were seen as extensions of the earlier philosophies of Leibniz and Spinoza (Vernière, 1954; Barber, 1955; Roger, 1963; Yolton, 1983).

When La Mettrie finally turned to the question of the origin of life, he was unable to break away from the theory of preexisting germs and adopted a view similar to that of de Maillet (*Système de l'Epicure*; reprinted in La Mettrie, 1774). Two of the most accomplished materialists of the Enlightenment—Denis Diderot and the baron d'Holbach—succeeded in eliminating the germ theory from their thinking and replaced it with spontaneous generation. In addition, they both saw nature as a totally flexible system in which there could be no absolutely permanent structures. Of the two, Diderot is the more sympathetic figure, a complex personality torn by emotional distaste for atheism, which he found intellectually inescapable. As editor of the great *Encyclopédie*, Diderot challenged the establishment by publishing critical commentaries on all topics, however sensitive; he thus ranked with Voltaire as a major figure in the awakening of conscience that paved the way for revolution.

In his *Pensées philosophiques* of 1746, Diderot still wrote as a deist, accepting the argument from design and the theory of germs. But by the time he wrote his *Lettre sur les aveugles,* he had moved on to an extreme form of materialism (Vartanian, 1953; Crocker, 1959; Roger, 1963). The "letter on the blind" followed La Mettrie's policy of arguing that the mind is completely dependent on the body. Diderot examined the case of the blind mathematician, Nicholas Saunderson, to argue that such a person would live in a different mental world than ours. In conclusion, Diderot created a fictional impression of Saunderson's deathbed, with the blind man rejecting the platitudes of the attending clergyman. How, he asks, can there be benevolent design in a universe that can produce monstrosities such as himself, lacking the most vital organs? Diderot then has Saunderson articulate a vision of the origin of life that harks back to the materialists of antiquity. At the beginning of the earth's history, nature *experimented* with spontaneous generation of many forms of life, most of which must have lacked essential organs. All these monstrosities would have died out, but occasionally nature would by chance hit on a form capable of living and reproducing itself. Thus, the species that inhabit the earth were formed by a sort of trial and error process. This does not, however, resemble the theory of natural selection, because Diderot was postulating the elimination of forms spontaneously generated from brute matter, not an ongoing process within an established species.

In this original view, Diderot saw matter as creative only in the sense that it could lock together at random to give potentially living structures. Once successful forms were established by chance, they continued to the present, unchanged except for the occasional "mistake" in the form of a monstrosity. After reading some of Buffon's works, Diderot realized that if living things were produced by chance, instead of by design, there was no reason why they should preserve their form through successive generations. Significant change must be possible after the original spontaneous generation. In his *Rêve de d'Alembert* (translation, 1966), he again uses a fictional device, presenting his speculations as the ravings of his sleeping friend, the mathematician d'Alembert. Matter is spontaneously active, may even possess a primitive level of awareness, and can organize itself into the complex structures of living things. Needham's experiments were cited as evidence that the spontaneous generation of life is going on even now; and possibly even the largest organisms might be produced in this manner under certain circumstances. In response to their needs, animals can develop new organs, which can be inherited by their offspring and thus fixed in the species. Diderot was also fascinated by the production of monstrosities (Hill, 1968). Nature does not always breed true to type: it is constantly trying out bizarre new forms, some of which may be able to perpetuate themselves. By accepting chance production of new structures as part of the regular process of generation, Diderot broke completely with Buffon's vision of a universe constrained to produce only certain preordained forms. If nature is truly active and creative, the philosopher and naturalist should not try to impose arbitrary limits on her powers.

Diderot was far from the kind of hard-headed materialist so despised by some later thinkers. When supporters of the Romantic movement expressed distaste for materialism, they were more likely to be thinking of the *Système de la nature* of the baron d'Holbach. This work was published in 1770 under a false name and became known as the "Bible of atheism" (new ed., 1821; translation, 1868; Naville, 1967). Although he was a friend, Diderot seems to have distrusted d'Holbach's militant dismissal of all religions as frauds designed to uphold social repression. The *Système de la nature* attempted to provide a new utilitarian social philosophy, but d'Holbach saw that this would have to rest on a materialist conception of life.

D'Holbach refused to compromise his materialism by suggesting that matter itself had some primitive kind of awareness. It would have to be accepted as fact that when inert matter was organized into a complex structure, it took on the properties of life. If not itself alive, nevertheless, matter was still a far more complex affair than had been assumed in the old mechanical philosophy. D'Holbach followed Diderot in picturing the material universe as an essentially active system that could organize itself to generate living structures wherever the circumstances were appropriate. All things are in constant motion and fire is the most active of the elements. D'Holbach was a member of the school of chemical philosophy founded by Georg Ernst Stahl which gave the element of fire (phlogiston) an important role (Metzger, 1930). All forms of matter are, however, governed by affinities that draw the particles into combination. The spontaneous generation of life itself becomes little more than a kind of chemical reaction, and living structures are formed whenever the correct substances are brought together. D'Holbach used Needham's experiments to support this view but like Diderot assumed that under certain circumstances even the more complex forms of life could be generated in this way. D'Holbach also agreed that once living things had been formed, they would be constantly changing. Nothing in nature is permanent: it is constantly experimenting with new forms, as illustrated by the appearance of monstrosities.

Of all the eighteenth-century thinkers, Diderot and d'Holbach came closest to seeing the production of living things as a totally open-ended process. Because they wished to replace the Creator's design with the forces of nature herself, they tended to give unlimited scope to those forces. The basic activity of the universe was to keep everything in a state of flux, thus there could by no fixed species and no predetermined plan of development. Yet these thinkers did not formulate a comprehensive theory of evolution: their insights were confined to occasional hints. Perhaps this was inevitable in a group whose primary interest was philosophy rather than natural history—yet La Mettrie, Diderot, and d'Holbach all had paid very serious attention to the sciences at some points in their careers. A more fundamental reason for their failure to outline a theory of evolution can be seen in their fascination with the possibility of spontaneous generation. It was absolutely essential to the materialist program that life be supposed to arise directly from inorganic matter. But the materialists were never content to take Needham's experiments at their face value and always extended the notion to include the spontaneous generation of even the highest forms of life. Their real concern was the direct relationship between matter and life, which was satisfied far more easily by a consideration of spontaneous generation than by mere changes in living things once formed. Thus, although they recognized the possibility of evolution, the materialists did not explore it in detail, because in their eyes it remained only a secondary issue compared with the ultimate origin of life.

ERASMUS DARWIN AND LAMARCK

Many of the ideas discussed so far have been based on assumptions quite alien to those of modern evolutionism. Buffon is the only figure who has been seriously proposed as a forerunner of Darwin, and even his view of the nature of species turned out to be quite unlike anything accepted today. At the end of the eighteenth century, however, there are two figures whose ideas seem to come much closer to the modern concept of organic development: Erasmus Darwin and Lamarck. Both avoided the temptation to see even complex forms of life as derived from spontaneous generation and hence were forced to take more seriously the processes by which living things can actually change through time. Both have been hailed as founders of modern evolutionism. To a large extent, this acclaim was generated as a by-product of the opposition to Charles Darwin's theory of natural selection. As in the case of Buffon, it was Samuel Butler's efforts (1879) to discredit Darwin which led to an exaggerated view of the earlier workers' contributions. At the end of the nineteenth century, a school of "neo-Lamarckians" emerged, consciously opposed to Darwinism and determined to see Lamarck as founder of their alternative evolutionary mechanism (Packard, 1901). The neo-Lamarckians, however, concentrated only on those aspects of their hero's work that they could fit into a modern framework. Recent historians have looked instead at the "whole" Lamarck and have recognized that his own conception of his theory was quite different. Erasmus Darwin and especially Lamarck are important because they elaborated the most complex of the Enlightenment's efforts to deal with the problem of organic change; but we should not be misled by superficial similarities into assuming that they contributed directly to the Darwinian revolution.

Erasmus Darwin occupies a unique place in the history of evolutionism. He was a colorful personality and perhaps the only thinker we shall encounter who put forward some of his ideas in the form of poetry. Works such as the *Botanic Garden* (1791) and the *Temple of Nature* (1803) were popular in their day, although Darwin's couplets are not adapted to modern taste. Erasmus was also the grandfather of Charles Darwin, and because his (non-poetic) *Zoonomia* (1794–96) proposed a theory of evolution, he naturally has been a target for those who wish to show that it was only by gleaning insights from his precursors that the younger Darwin was able to formulate his theory of natural selection. It was Charles Darwin's endorsement of a biography of Erasmus (Krause, 1879) which sparked Charles's open feud with Samuel Butler (see Darwin, 1958). Modern enthusiasts have followed Butler in finding passages in Erasmus's works which seem to anticipate the selection theory (King-Hele, 1963; Darwin, 1968). But such anticipations always turn out to be superficial; for example, Erasmus's account of the "balance of nature" maintained *between* the species has been mistaken for the "struggle for existence" *within* each species used so effectively by his grandson.

Erasmus Darwin's views must be interpreted within their own context (Harrison, 1972). He was a deist who believed that God had designed living things to be *self-improving* through time. In their constant efforts to meet the challenges of the external world, they developed new organs through the mechanism that the Lamarckians would make famous as the "inheritance of acquired characteristics." The results of the individual's efforts are inherited by his offspring, so that by accumulation over many generations a whole new organ can be formed. Darwin seems to have assumed that the overall results of this effort to adapt to the environment would be a gradual progress of life toward higher states of organization. Curiously, he claimed to have developed his idea of transmutation not from natural history but from David Hartley's account (1749) of how the soul is affected by the habits of life.

Darwin was a physician rather than a naturalist, and the only consistent account of his theory is a single chapter in the *Zoonomia*. Jean Baptiste Pierre Antoine de Monet, chevalier de Lamarck, was a professional naturalist who wrote extensively on his own theory. Largely ignored in his own time, it is not surprising that he has received more attention from historians. To the neo-Lamarckians of the late nineteenth century, he was the founder of an evolutionary mechanism compatible with the knowledge of their own time, an alternative to natural selection. It was assumed that Lamarck's theory postulated the evolution of all living things from a common ancestor and that he had proposed the first plausible mechanism to explain how species adapt to their environment.

Closer study by modern historians has revealed, however, a very different picture of Lamarck's real intentions. Gillispie (1956, 1959) was the first to draw attention to the fundamental differences between Lamarck's view of nature and that of the post-Darwinian period. Gillispie presented Lamarck as a "Romantic" thinker, a view that has not been accepted by later writers. It may be true that the concept of an inherently active nature developed by Lamarck may bear some resemblance to the Romantic world view, but his real inspiration was Enlightenment materialism, which had also stressed the essentially creative power of nature. Lamarck adapted this to his own needs, giving an evolutionary theory that was quite unlike Darwinian natural selection. His views certainly changed during his career, and later expositions seem to bear a closer resemblance to the modern concept of evolution. But throughout his writings, there are strands of thought that represent a very different idea of how life develops (Hodge, 1971; Schiller, 1971*a*; Burkhardt, 1972, 1977; Mayr, 1972*b*; Barthelémy-Madaule, 1982; Sheets-Johnstone, 1982; Jordanova, 1984).

Strictly speaking, Lamarck cannot be counted as an eighteenth-century evolutionist, because he did not come to accept the possibility of transmutation until just after 1800. He is one of those unusual figures who make a major shift in outlook at a comparatively late age. The fact that he was over fifty when he abandoned his original commitment to the fixity of species allows us to see him as a product of the Enlightenment. His theory represents a combination of several themes characteristic of eighteenth-century attempts to deal with the origin and development of life. There is the materialist belief in spontaneous generation, made more reasonable by limiting it only to the simplest forms of life. There is the steady ascent of a scale of organization, which may owe something to the temporalized chain of being. And finally, there is a process by which living things can change in response to new conditions, something accepted in various forms by many Enlightenment thinkers. Lamarck's tragedy was that he put all these ideas together at a time when the Enlightenment's way of thought had gone out of fashion, and he paid the price of being dismissed as a crank by many of his contemporaries.

In his early botanical works, Lamarck showed an interest in the hierarchical arrangement of classes, although from the first he realized that plants and animals would form two parallel hierarchies, not a continuous chain of being (Daudin, 1926). At the same time, he was developing his unconventional theory of chemistry and a unique system of geology based on uniformitarian principles (translation, 1964). In 1794, Lamarck was appointed to the Muséum d'histoire naturelle, now reorganized from the old Jardin du Roi by the revolutionary government, and was given the task of classifying the invertebrates. He adapted so well to this that he is regarded as one of the founders of invertebrate taxonomy; but while developing this new skill, he abandoned his original commitment to the fixity of species. His theory of organic development was first outlined in 1802, then reorganized as the basis of his best-known work, the *Philosophie zoologique* of 1809 (translation, 1914; see also the introduction to the *Histoire naturelle des animaux sans vertèbres*, 1815–22).

Lamarck's theory of matter postulated an undetermined number of chemical elements capable of being combined into myriad compounds (thus he opposed Lavoisier's new chemistry based on fixed, simple compounds). Lamarck at first, however, did not believe that matter possessed any inherent power to build itself up into compounds. The main cause of chemical reactions was the active power of fire—but this was always *destructive*, tending to break down the molecules into simpler combinations. Compounds can only be built up by a nonmaterial force, the force of life. For Lamarck, life was a force imposed on the material universe, capable of shaping it constructively. He even held that all compounds in the earth's crust have been formed through action of living things (not as ridiculous as it might seem: chalk and limestone are composed of the shells of minute sea creatures). In his original view, this living force manifested itself in the form of a hierarchy of fixed distinct species.

Eventually, Lamarck converted to the materialist belief that life is the product of matter via spontaneous generation. Various suggestions have been made to explain his motives. Gillispie argued that Lamarck accepted the development of life by reversing the process of degradation, which he had postulated in all chemical reactions; but Hodge (1971) points out that Lamarck explained his own motives somewhat differently. Lamarck always had used a hierarchical arrangement in his classification, and this suggested to him that the series of forms might correspond to the historical sequence by which classes were produced. When his studies of invertebrates showed him that the simplest forms of life have no specialized organs at all, he perceived that such forms might be simple enough to be generated directly from unorganized matter. The key to this spontaneous generation was the active power of a subtle fluid corresponding to electricity and the nervous fluid. Under certain circumstances, this active fluid could act on gelatinous matter to form the simplest living organism, carving out channels it would circulate through to vivify the structure. Thus, living things were produced naturally by the powers of material nature.

Because only the simplest form of life could be produced directly, higher forms must have been derived from these simple ones by some kind of progressive development over many generations. For Lamarck, the species constituted a hierarchy of structures ranging up to the most complex, and this hierarchy represented the historical pattern along which life had advanced. The active powers of the nervous fluid carve out ever more complex channels, and each generation advances slightly beyond the level of its parents. It has been argued that Lamarck's acceptance of this idea of continuous progress does not derive from the temporalized chain of being as advocated by Bonnet (Schiller, 1971*b*). Certainly, Lamarck did not believe in a unilinear sequence of forms and allowed two main branches instead of a single chain. Yet Bonnet too had accepted the possibility of a branching chain, and it is not unreasonable to suppose that Lamarck translated the unilinear chain into the more sophisticated notion of a hierarchy of organization. The element of overall linearity remained, because Lamarck insisted that, in theory, the progressive trend would produce an unambiguous sequence of forms. The hierarchy was not just an abstract scale defining degrees of organization but a predetermined path by which life advanced. It

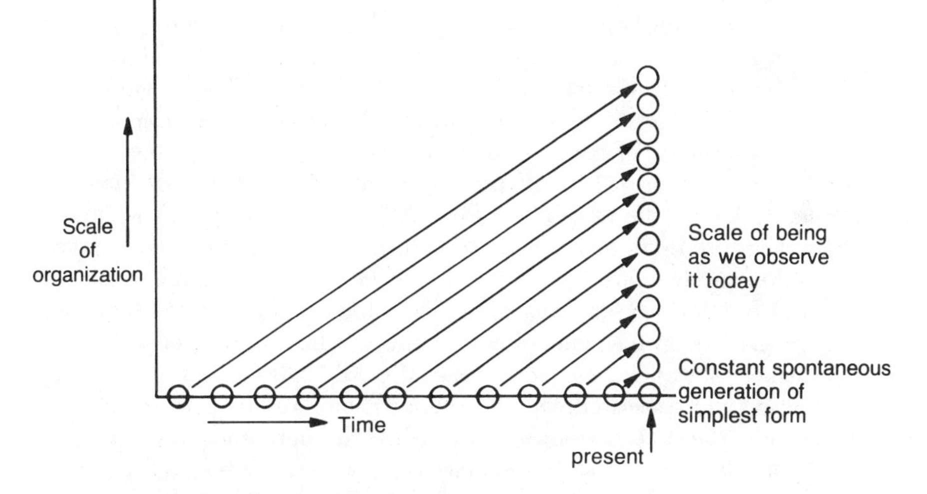

Fig. 10. Lamarck's Theory of Organic Progression.

Each point on the scale of being we observe today has been derived by progression from a separate act of spontaneous generation. The lower down the scale the organism is today, the more recently its first ancestor was produced. Thus, evolution is not a system of common descent but consists of separate lines progressing in parallel along the same hierarchy.

is possible even that Lamarck saw the path of development as mapped out in advance by the Creator, because he was certainly not an atheist.

If the active power of nature compels life to mount steadily up the chain of being, how, he asks, can we still see the complete hierarchy today? Why have all living things not raised themselves to the same level as man? The answer to this is crucial for defining the difference between Lamarck's theory and the modern one. He did *not* suppose all forms alive to have evolved from a common ancestry. On the contrary, he believed that the organisms at each level of the scale today have progressed to that point separately; organisms at different levels are derived from different acts of spontaneous generation at different points in time. The direct formation of the simplest forms of life has gone on continuously throughout the earth's history. Today's highest forms have progressed over many generations from the earliest appearance of the simplest forms. Organisms now halfway up the scale have progressed from acts of spontaneous generation in the more recent past, while the simplest organisms we observe today just have been formed. The evolution of life thus consists of a whole series of lines, each advancing separately along the scale of being, as shown in the diagram.

As a taxonomist, Lamarck knew that he could not, in fact, observe a simple linear arrangement of forms in the animal kingdom. The chain of being as it exists for him has a number of major branches and many gaps, illustrated in a diagram added to the *Philosophie zoologique*. Branching has occurred because the mechanism that forces each line of development along the scale of being is not the only one involved. Lamarck knew from his geology that the surface of the earth is subject to constant, if very slow, change, and fossils told him that living things also have changed through time. He refused to accept the possibility of extinction: the fossil species must have evolved into those of today, because nature is powerful enough to prevent any of her productions from being driven to extinction. There must be a mechanism by which life can adapt to changing conditions at the same time as it progresses up the chain of being. This mechanism was, of course, the inheritance of acquired characteristics—the only part of the theory still remembered and associated with its author's name. Yet for Lamarck himself, it was only a secondary factor that disturbed the pure line of progress.

Lamarck believed that the animal's needs determine the organs its body will develop. This did not mean that the animal could grow itself a new organ by willpower alone, however. The needs determine how the animal will use its body, and the effect of exercise, of use and disuse, causes some parts to develop while others wither away. The environment creates the animal's needs, which in turn determine how it will use its body. Those parts that are strongly exercised will attract more of the nervous fluid; this fluid will tend to carve out more complex passages in the tissue and increase the size of the organ. Disused organs will receive less fluid and will degenerate. Lamarck gave no detailed theory of inheritance but assumed that the characteristics acquired as the result of effort would be transmitted to the offspring, thereby enabling the effect to become cumulative. To give a famous example, the short-necked ancestors of the modern giraffe were at some point in their history forced to begin feeding from trees. All the individuals stretched their necks upward and, as a result, this part of the body grew in size. The next generation inherited the extra neck length and stretched it even farther, so that over a long period of time the giraffe gradually acquired the long neck we see today.

The inheritance of acquired characteristics can be seen as an alternative to Darwinian natural selection, as a means of explaining how living things slowly adapt to their environment. In almost every respect, however, the system into which Lamarck built this mechanism differs from that accepted in the post-Darwinian era. Darwin was not interested in the ultimate origin of life, whereas for Lamarck, spontaneous generation was an integral part of the materialist viewpoint. For Darwin and almost all later naturalists, evolution was a process of divergence by which all forms of life have branched out from a few ancestral forms due to the long-continued effects of geographic isolation and adaptation to new conditions. For Lamarck, a series of distinct lines of evolution moved independently along the same

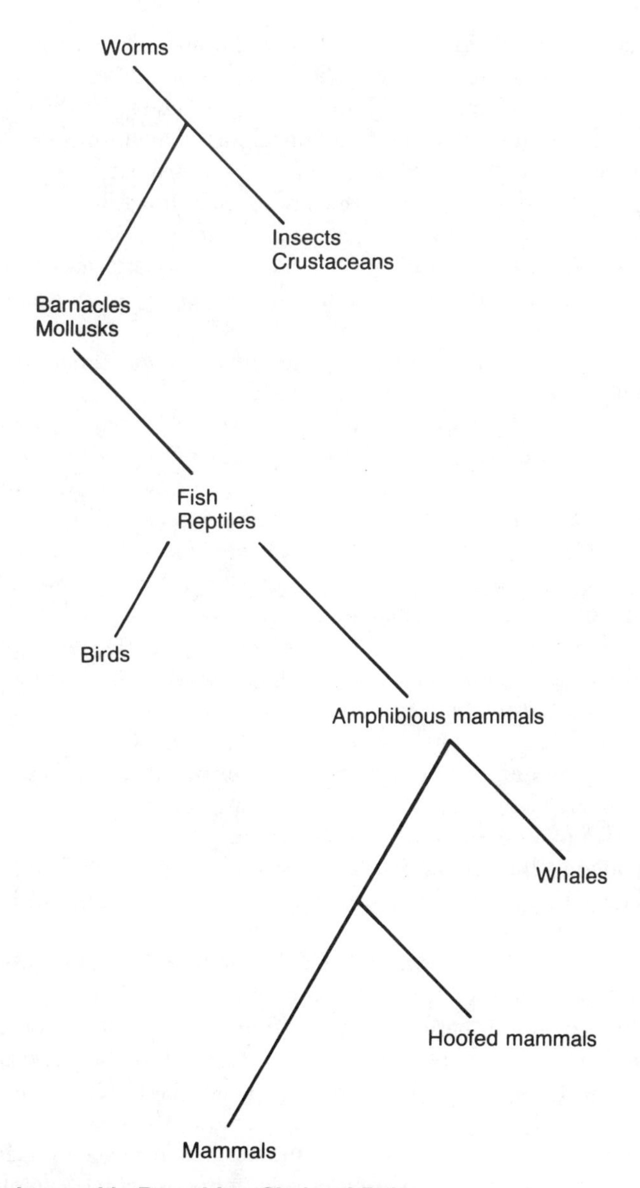

Fig. 11. Lamarck's Branching Chain of Being.

Adapted from a diagram in the *Philosophie zoologique*, this shows how Lamarck thought the theoretically linear scale of organization had been forced into various branches by the necessity of life adapting to changing conditions throughout the earth's history. Note that this is *not* meant to be a system of genealogical relationships: no one part of the chain is derived directly from any other. The diagram is merely a more realistic representation of the chain produced in the previous diagram.

scale of organization, progressing inevitably upward. The chain of being would be forced into branches by the effects of different conditions, but this was a secondary phenomenon, and Lamarck had no interest in the problems of geographic distribution that were so crucial for Darwin. Species still had a real existence in Darwin's theory, in the sense of distinct breeding populations, but Lamarck denied the existence of species altogether. He pictured both his evolutionary mechanisms, progressive and adaptive, as essentially continuous processes that would never produce sharp divisions in nature. Lamarck even predicted that the march of biological discovery would fill in the gaps between what were regarded as distinct "species." In almost every respect, his viewpoint harks back to the eighteenth century, not forward to the nineteenth.

Although in some respects alien to modern evolutionism, Lamarck's theory was worked out in more detail than any other Enlightenment account of the origin of life. He attracted no immediate followers, however, and died in obscurity. To a large extent, his eclipse was engineered by his greatest rival, Georges Cuvier (Burkhardt, 1970). Cuvier not only built up a rival system of biology but also rose to political power in France, an eminence that enabled him to ensure that Lamarck's views were dismissed as outdated speculations. Cuvier developed techniques of comparative anatomy that even Lamarck adopted, but Lamarck remained with the invertebrates, and Cuvier was appointed to a more prestigious post at the Museum classifying the vertebrates. Cuvier astounded the world by applying his techniques to the reconstruction of extinct species from their fossil remains, while Lamarck failed even to see the possibility of using the new paleontology to support the idea of an evolutionary progression.

Cuvier divided the animal kingdom into four basic "types" that could not be ranked into a hierarchical order as Lamarck and many of his predecessors had maintained. The real power of the Linnaean system based on parallel, rather than hierarchical, relationships now could be appreciated. Although the types were distinct, representing four viable plans on which an organic form could be based, each was infinitely flexible in the way its external modifications could be adapted to the demands of the environment. The relationships between species were now seen to be based on the fundamental similarities of their internal structures, rather than an ordered ranking of their external characters. Foucault's thesis is that this approach made possible the Darwinian view of natural development, in which a single original form diversifies through adaptive radiation, its basic characteristics remaining to serve as the link by which we classify the descendants. We have seen that this thesis needs to be qualified in a number of cases, yet it remains true that to a surprisingly large extent, eighteenth-century evolutionism was conceived as change taking place within an orderly framework.

Cuvier himself was no evolutionist, though—it was his classification that was modern, not his views on the mutability of species. The sense that he developed for understanding the way in which internal parts of the body fit together convinced him that species are fixed and distinct. The new comparative anatomy gave so detailed a picture of each animal's internal structure that it seemed impossible for such complex forms to be created by a natural process. Materialist speculations on the origin and mutability of life were dismissed as incompatible with the new depth of biological knowledge. The Enlightenment's more speculative approach was swept away by the increasing conservative tastes of Napoleonic France. In Britain, an even stronger reaction was creating a new interest in natural theology and making it impossible for Hutton's antibibical geology to be taken seriously. Lamarck founded no school because he developed his theory twenty years too late, when the materialist spirit he had turned to was being swept away by political and cultural revolution. Evolutionism would now have to make a new start, taking into account the new sciences and the new thought processes of the nineteenth century. In many respects, Darwinism would be shaped more by these ostensibly conservative developments than by the speculations of Enlightenment evolutionists.

4

Changing Views of Man and Nature

If most eighteenth-century ideas of evolution failed to anticipate later developments, new values and attitudes were helping to prepare the way for the Darwinian revolution in a less direct manner. Questions about the essential nature of man himself were being asked, and the answers would help to fix the framework of much nineteenth-century thought. To what extent does man's relationship to his closest animal cousins, the apes, imply that he has a position *within* rather than *above* nature? What is the origin of the different human races, and is this connected with the question of our relationship to the apes? What is the nature of the human mind, and to what extent can it be studied scientifically? Finally, what is the nature of human society? Does it have a genuine capacity for change or development through time? Eighteenth-century thinkers explored some radical new answers to these questions, but they were slow to adopt what we should call an evolutionary approach to the origin of man. Only with some difficulty did they build up the concept of human history as progress from primitive beginnings; nor did they make any effort to see that progress as the continuation of earlier developments in the history of life on earth.

In the intellectual ferment of the French revolutionary era, however, the idea of progress began to seem more plausible, and it was established as a standard theme for nineteenth-century thought. The basic idea could be explored in a number of different directions. The industrial progress anticipated by advocates of laissez-faire economics in Britain was a far cry from the almost mystical notion of the universe perfecting itself, a view expressed by the German idealists. It also must be emphasized that these different schools of thought managed to develop the idea of social progress *before* it became popular to account for the origin of man in terms of biological evolution. The progressive view of human society was a stimulus for, not a product of, the theory of organic evolution. Only later in the century, after the success of the Darwinian revolution in biology, were the two levels of progress synthesized into a comprehensive vision of universal development.

We shall tackle these issues in two stages, investigating first those eighteenth-century developments in biology and anthropology that bore on the question of man's relationship to the apes. This will reveal the extent to which the thinkers of the Enlightenment and the early nineteenth century failed to grasp the implications of human evolution. Then we shall go on to look at the philosophical and ideological debates that paved the way for the new framework of nineteenth-century thought. To give some coherence to a complex and extremely diverse process, these developments will be considered as three major national styles of thought: in France, Britain, and Germany.

MAN'S PLACE IN NATURE

The period around 1700 saw intense speculation on human origins, with some scholars beginning to argue that our distant ancestors had lived little better than brutes (Rossi, 1984). In these circumstances, it was inevitable that naturalists should begin to wonder about the relationship between humans and their closest animal relatives. As travelers went around the world, they encountered new races of men, some living under conditions so primitive that there was a temptation to question their true humanity. Specimens of the great apes were collected, confirming that such creatures represented man's closest relatives in the animal kingdom. How close was this relationship, and to what extent could the lowest forms of man be distinguished from the most intelligent ape? These questions began to raise the kinds of issues the materialists were debating but at a far more down-to-earth level. These were not abstract speculations about the nature of mind but practical questions the taxonomist must answer in order to complete his system of nature.

Man had always been assumed to differ from the animals in the possession of an immortal soul, but it now was becoming clear that his physical body, at least, was closely related to the apes (Greene, 1959*a*). Travelers' tales abounded of apes that showed almost human capabilities and even a preference for human females. The naturalist would have to investigate these stories and try to get more detailed information, and one of the most obvious approaches was to get better specimens of the great apes. An early anatomical description of the chimpanzee was provided by Edward Tyson in 1699. This helped to dispel the possibility that the apes might be almost human, yet the resemblances were inescapable. When Linnaeus came to classify the animal kingdom, he included man along with the apes in the order Anthropomorpha. When criticized for linking man with the brutes, he defended himself by asking any naturalist to show him a *physical* characteristic by which the two could be distinguished more clearly. Although prepared to admit that man was morally and intellectually superior to the apes, Linnaeus insisted that the naturalist must classify by physical resemblances alone, and at this level, the relationship was plain.

The more radical materialists made surprisingly little effort to exploit the implications of Linnaeus's classification of man. The possibility of arguing for a gradual evolution of man from animal ancestors does not seem to have excited them. It was Buffon who took the problem of the apes most seriously, in the fourteenth volume of his *Histoire naturelle* (1766). He accepted a physical resemblance between the apes and man, and because he was developing his theory of "degeneration" at this time, the possibility of an evolutionary relationship must have seemed very real. Yet Buffon shied away from this conclusion and argued that the resemblance was only physical; stories of intelligent actions by apes were fabrications, and in fact, the dog came closer to man in this respect. For Buffon, our mental faculties distinguish us completely from the animal kingdom, and these cannot be explained in material terms. The only materialist to suggest a genuine connection between man and the apes was Lamarck, whose theory of development naturally implied that man has evolved from a lower form. The orangutan was pointed out as the most likely ancestral form, but by the time Lamarck made this suggestion, the materialist outlook represented by his theory had gone out of fashion.

In fact, the closeness of the relationship between man and the orangutan already had been challenged. A study of the orang's vocal organs by Petrus Camper (1779) revealed that it was certainly incapable of speech. In later studies, Camper pointed out further differences, including the orang's inability to walk upright on two feet. This last point was seized on by Cuvier to make the distinction that Linnaeus had proclaimed impossible. Cuvier divided man from the apes by creating two separate orders: Bimana and Quadrumana. The apes, he argued, are "four-handed" in the sense that the structure of the feet hardly differs from that of the hand. Only man has a truly distinct foot adapted to his upright gait. This complete distinction between man and even his closest animal relatives still was being used as an argument against human evolution in Darwin's time.

Camper also made studies of the various human races. To measure the shape of the skull, he defined the "facial line" joining the jaw, nose, and forehead and the "facial angle" between this line and the horizontal. In the most perfect human features represented in classical Greek statues, the facial angle was almost ninety degrees. In the ape, this angle was much smaller. When Camper applied the technique to the different human races, he found that the European approximated most closely the classical ideal, while the facial angle decreased in other races, especially the Negro. In effect, he suggested that the Negro has a facial angle intermediate between that of the European and the ape. The anthropologist J. F. Blumenbach used his world-famous collection of skulls to arrive at the same result via a somewhat different technique (translation, 1865). Camper expressed the need for caution in interpreting these results: the ape was quite distinct from man, so there was no point in trying to see the Negro as a man-ape hybrid. Yet to many, it seemed inescapable that the Negro was the most apelike variety of man. Secure in a sense of their own racial superiority, Europeans began to use the naturalists' arguments to justify their exploitation of the "inferior" races (Snyder, 1962; Montagu, 1963, 1974; Barzun, 1965; Stanton, 1966; Mead et al., 1968; Haller, 1975; Stepan, 1982; Banton, 1987).

Far from accepting an evolutionary viewpoint, these early advocates of racial discrimination preserved the gulf between man and the animals by assuming that the inferior characteristics of some races were produced by degeneration from the most perfect form. Blumenbach followed Maupertuis and Buffon in arguing that the European was the original form of man from which the other races had degenerated through exposure to unsuitable conditions in certain parts of the world. The fact that the inferior characteristics thus produced were apelike was, apparently, a coincidence—or more probably an indication that thinking on this question was still influenced by the traditional notion of a unilinear chain of being. The resemblance of lower races to the apes certainly did not indicate an evolutionary link between man and the animal kingdom. Only in the post-Darwinian period would this paradox be resolved, when the inferior races came to be seen as living relics

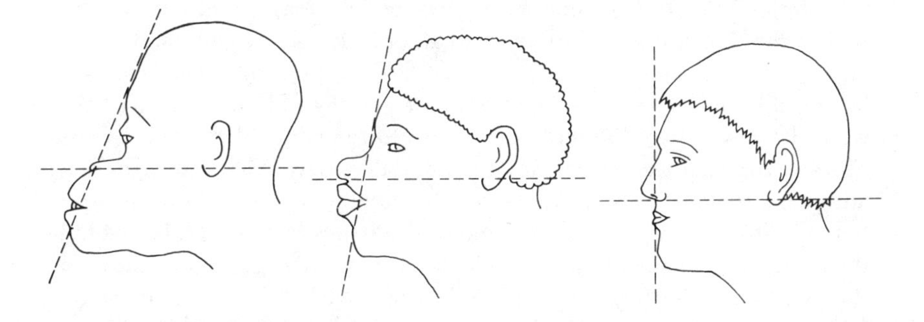

Fig. 12. Petrus Camper's "Facial Line" and "Facial Angle."

The facial line joins the jaw, nose, and forehead. The angle between this line and a horizontal joining the nose and ear is the facial angle. Note how the facial angle for the classical figure on the right is ninety degrees, whereas for the ape it is about sixty degrees. Camper and many of his later followers claimed that the Negro's facial angle was between that of the European and the ape.

of intermediate stages by which European man had progressed from his animal ancestors toward his supposedly more perfect form.

FRANCE: FROM THE ENLIGHTENMENT TO POSITIVISM

Enlightenment thinkers may not have believed that man had evolved from the apes, but they were determined to create a new foundation for morality and the social order (Willey, 1940; Cassirer, 1951; Hazard, 1963; Gay, 1966–69; Hampson, 1968). Their suspicion of Christianity ensured that moral laws could no longer be seen as absolute truths revealed by the Deity. Instead, the study of human nature would have to show what kind of behavior would be most successful in promoting a decent society. Because man can no longer expect reward or punishment in an afterlife, the principal goal must be to achieve happiness in this life. The "principle of utility" dictated that all actions should be aimed at maximizing pleasure and minimizing pain. The goal of new social policies was the creation of laws that, taking into account the basic nature of the individual, would ensure the greatest happiness of the greatest number of people. The ever-present danger was that this whole utilitarian approach would degenerate into personal hedonism. With the threat of divine judgment removed, why should the individual worry about the happiness of anyone else? How does the moralist persuade him that it is not in his own best interests to seek his own pleasure at the expense of everyone else, as advocated by the Marquis de Sade's moral nihilism (Crocker, 1963)?

The moralist had to propose a series of laws that would convince each person that he was better off living in an orderly society governed by these laws than in a state of anarchy. Successful framing of such laws would be possible only given a complete understanding of human nature. There were, however, two different approaches to the study of the human mind. The most radical was the complete materialism favored by Enlightenment atheists, a philosophy aimed at complete integration of man into nature. Potentially, this would anticipate some of the issues later raised by evolution theory. Although materialism would be driven underground at the end of the century, following the social upheavals sparked by the French revolution, politically radical thinkers continued to promote such ideas in the early nineteenth century. The less radical "sensationalist" philosophy derived from the writings of John Locke made no attempt to see mind as a mere by-product of matter. It accepted the existence of a purely mental plane of activity and sought to reduce this to a law-bound system by understanding how ideas and sensations are processed in the mind. This philosophical psychology was expanded by the French Enlightenment and proved far more resilient than outright materialism, surviving to form the basis of much nineteenth-century British thought about human nature. Its ramifications thus provided the intellectual framework within which Darwinism was eventually conceived.

Materialism stemmed from extension of Descartes's animal machine doctrine to man himself. Few could believe that animals were incapable of feelings, and once it was accepted that the physical organization of their bodies generated a certain level of awareness, it seemed only natural to assume that man's higher faculties were produced in the same way (Hastings, 1936; Rosenfield, 1968; Young, 1967). La Mettrie's *L'homme machine* of 1748 (reprinted 1960) first proclaimed elimination of the old idea of a "soul" providing the life-force of the body. Instead, life was now to be regarded as a product of the organized matter of the body itself, a view to which a number of scientific discoveries seemed to lend their support. In particular, La Mettrie referred to Abraham Trembley's discovery of the regenerative power of the "polyp" or freshwater hydra (translation, 1973; Vartanian, 1950; Baker, 1952) and Albrecht von Haller's work on the "irritability" of organic tissue (1755, 1786). Matter was no longer seen as the passive substance of Descartes's mechanism but as an active principle in its own right, as suggested in the thought of Spinoza and Leibniz (Vernière, 1954; Barber, 1955), by Newtonian physics (Hall, 1968; Heimann and McGuire, 1971), and by the animistic chemical philosophy of G. E. Stahl (Metzger, 1930; King, 1964; Callot, 1965; Naville, 1967). All of these developments encouraged the materialists to believe that the mind was merely a by-product of the body's physical activity. D'Holbach proposed a whole utilitarian social philosophy based on this view of man, an approach followed by Pierre Cabanis in 1802 (translation, 1981; Staum, 1974, 1980), and by the group known as the "ideologues" (Picavet, 1891). There was, however, a limit to how far this kind of materialism could go, given the limitations of biological techniques of the period. All too often it degenerated into a new kind of animism, because man only could be reduced to a machine by attributing the properties of life to matter itself (King, 1967; Hall, 1969; Schiller, 1974; Moravia, 1978).

John Locke founded the alternative approach to the science of man in his classic *Essay on Human Understanding* of 1690, by treating the mind as a system for integrating sensations generated by the nervous system interacting with the external world. The question of how the body generates mental sensations was sidestepped in order to concentrate on the laws governing the mind itself. This approach was extended by Condillac (translation, 1756; Knight, 1968) and by Helvétius (translation, 1810), who argued that by controlling the input of sensations via the educational process, the growing mind could be molded in any desired way. It was this belief in the malleability of the mind that would be seized upon by the utilitarian philosophers of early-nineteenth-century Britain under Jeremy Bentham. It promised the opportunity of shaping human beings by social legislation to ensure that they interacted in the most effective way. The emphasis was still on persuading individuals to behave responsibly, however, not on reducing them to mindless cogs in a state machine.

In the absence of an evolutionary view of man's origin, Enlightenment philosophers believed that human nature has remained constant throughout history. Because it was evident that mankind must have appeared at a certain point in the earth's past, it was possible for them to conceive of a time when men first came together to form societies. There was as yet no science of archaeology to give any real knowledge of prehistory, but it was thought that speculation based on an understanding of human nature might help to explain how societies were formed and thus throw light on the way any society must function. How had primitive man behaved in his original "state of nature," and why had he combined with others to form the first social groups? In his *Leviathan* of 1651, Thomas Hobbes had suggested that the state of nature was one of constant struggle and that men accepted the rule of a sovereign in order to escape the resulting miseries. Most Enlightenment thinkers preferred Locke's view of human nature, which did not require one to picture the state of nature as one of struggle (Locke, 1960). Locke held that our reason naturally leads us to recognize the rights of others. The first men entered into a "social contract" merely to ensure a more systematic enforcement of the natural laws of human behavior. Later on, Rousseau suggested that the state of nature was a happy one. Man was a self-perfecting creature, but his efforts to create civilization have imposed restraints on his behavior that compare poorly with the freedom of his original state. The cult of the "noble savage" emerged in the late eighteenth century to reinforce this view of human nature. The suggestion that society has emerged out of a state of nature this did not necessarily imply a theory of historical progress.

The social contract was a convenient fiction meant to reveal the underlying purpose of human interactions, but it was obvious that in many societies the contract was being broken by the ruling class. Reform was possible in light of the new knowledge of man, but it was necessary to know how societies had been corrupted to allow injustices to creep in. Here there was a role for the study of social history, which might reveal how earlier civilizations had collapsed. The growth of a sense of human history no doubt helped to create a climate of opinion within which evolutionary ideas could flourish (Toulmin and Goodfield, 1965; Rossi, 1984). Yet it is clear that only with the greatest difficulty did the Enlightenment historians break through to a concept of cumulative development or progress. The idea of progress had been unacceptable to traditional Christianity, which depicted man as a fallen creature. Renaissance humanists had added a new dimension to the past through their interest in ancient thought, but their exaggerated respect for Greece and Rome could produce only a sense of degeneration, not of progress. Renaissance man, at most, could hope to climb back toward heights already scaled in the past. A new sense of confidence in the future would be required before it became possible to treat social history as a record of earlier progressive steps.

Enlightenment social philosophers certainly believed that they had access to new levels of knowledge about man, but because they treated human nature as essentially constant, they tended to assume that all societies must work according to certain basic laws. Giambattista Vico hoped to found a "new science" of history as reliable as anything in the physical sciences (translation, 1948; Rossi, 1984). For those races outside the biblical story, he postulated invariable laws by which each civilization rose, flourished, and decayed. Such a cyclic view of history was the very opposite of a theory of progress, but the fate of Greece and Rome made it difficult to believe that any society could develop indefinitely. The collapse of the Roman empire was studied by the Enlightenment's most respected social thinker, the baron de Montesquieu (translation, 1965). In his more famous *Esprit des lois* of 1747 (translation, 1900), Montesquieu used historical material to understand the forces at work in every society. The breadth of his vision has led some modern commentators to regret that he did not arrive at a theory of progress, yet it was not Montesquieu's purpose to search for a cumulative trend in history. To rank societies in a temporal sequence of progress would imply that the forces governing each period of history are different. For Montesquieu, the historian's task was to discover universal forces controlling any society. By assuming the constancy of human nature, a steady-state view of history was defined in which the same forces are always at work creating and destroying civilizations.

In Montesquieu's system, societies differ not because of any historical progress but because each is a particular blend of human and physical factors. Climate and geographic circumstances, he felt, exerted considerable influence on society. In the tropics, for instance, any large-scale society would tend to become a despotism—a fact that he found morally repugnant but sociologically inevitable. Montesquieu's efforts to show that external conditions can influence human social behavior offer an interesting parallel to the ideas of Buffon and other naturalists who saw the production of human races or even of different animal species as the result of new conditions acting upon the original form.

Because they distrusted Christianity, Enlightenment historians were convinced that the medieval period represented a collapse back into barbarism. To believe that progress was possible, it was necessary for them to accept the idea that the modern world had at last climbed back to a level of civilization surpassing that of the ancient world. At the end of the seventeenth century, a famous literary battle was fought to establish the superiority of "modern" over "ancient" learning. It was not easy to establish a superiority in the arts; only in technology and natural philosophy had the progress become clear. Because human nature was assumed to be constant, it was only in areas of cumulative achievement that any progress was thought to be possible, each generation adding to the discoveries of the past. Enlightenment philosophers were most impressed by seventeenth-century thinkers such as Francis Bacon who appealed to inventions such as gunpowder and the printing press to encourage hope for further progress in the future (Bury, 1932; Zilsel, 1945; Jones, 1965; Pollard, 1968; Van Doren, 1967).

These hopes for the future gradually tempted historians to believe that the past reveals an irregular but ultimately cumulative progress toward the present. But the connection was not made easily. Voltaire presented the age of Louis XIV as a pinnacle of cultural achievement (translation, 1961; 1965), without sensing the possibility of a generally progressive trend. He thought that reason produced such triumphs whenever it overcame the forces of superstition ranged against it—but he was too well aware of the strength of those forces to believe that the developments would become cumulative. Others saw the possibility of future progress without connecting it with past trends, as when Sébastien Mercier (1770) for the first time placed a "utopia" in the *future* instead of some remote geographic location. Mercier saw the improvement of society not as an ongoing process that could be linked with the past but as the result of a single reorganization imposed deliberately on the existing state of affairs. All too often the Enlightenment pinned its hopes not on a process internal to society but on the rule of an "enlightened despot."

In France the first effort to propound a comprehensive theory of progress occurred in Turgot's lectures of 1750 (translation, 1973), but it was left for Turgot's friend and biographer, Condorcet, to provide the best-known expression of the new spirit after the revolution had finally broken the power of the old regime (Baker, 1975). Condorcet's sketch of human progress (translation, 1955) was written while he was in hiding from Robespierre and published in 1795. Despite personal difficulties with the revolution, Condorcet saw it as a sign that progress could no longer be resisted. This was not a matter of intellectual improvement but of inevitable social forces. Mankind advances in part through science and technology but also through increasing application of the moral sciences to ensure happiness for all. During the earliest phase of history, basic inventions such as agriculture were made: here the state of nature becomes a historical reality. Progress was slow at first, because few had time to think for themselves and the first priests created superstitions to consolidate their power. Similar forces undermined the achievements of Greece and Rome, although the middle ages did at least see elimination of slavery from western society. In modern times, the improvements of technology have given advanced societies a secure position, while inventions such as the printing press have made diffusion of knowledge inevitable. By following the cumulative, if somewhat irregular, advance of reason in the past, we can foresee that in the future man's increased understanding of the physical and the moral worlds will generate an indefinite series of social improvements.

Condorcet's optimistic view of progress was built into the foundations of the leading systems of nineteenth-century thought. During the early years of the new century, the marquis de Saint-Simon stressed the need for a scientific study of human affairs and called for social justice based on equality of opportunity (translation, 1952; Manuel, 1956). His disciple Auguste Comte adopted the law of progress as the driving force of his new philosophy of "positivism." His *Cours de philosophie positive* of 1830–42 (translation, 1975) argued that human nature is constituted so that a progressive improvement in the level of our understanding is inevitable. In its first stage, knowledge was constructed on a theological basis, assuming that all phenomena are the result of divine activity. In the next, or metaphysical, level, nature herself was endowed with mysterious powers. Finally, science emerged into its positive phase, in which the laws of nature simply are described without the need being felt to search for underlying causes. When such an approach is applied to human affairs, a true science of sociology will result and religious sanction for morality can be abandoned. Comte was aware that each generation can shape the attitudes of the next, but he could not escape the Enlightenment's belief that human nature is essentially fixed. Cuvier had disproved Lamarck's theory of biological evolution, so there could be no link between biological and social progress. Comte thus missed the opportunity of integrating the two phases of evolution, an opportunity that would soon be seized upon by a British philosopher with equally grandiose ideas for a new system of knowledge: Herbert Spencer (Green, 1959*b*).

BRITAIN: UTILITARIANISM AND LAISSEZ-FAIRE ECONOMICS

The atheism of the French materialists never had become popular in Britain. Political radicalism was even more suspect, especially in the aftermath of the French revolution and the Napoleonic wars. There *was* a radical tradition, however, that walked a political tightrope in an effort to retain influence on a society changing rapidly under the impact of the industrial revolution. The philosophy and social theory of this tradition were rooted in the less extreme form of Enlightenment thought and were adapted to the demands of the newly emerging class of capitalist entrepreneurs. Radicalism meant not revolution but reform of existing institutions to get rid of the straitjacket imposed by the legacy of aristocratic rule, thus freeing society to allow maximum exploitation of its economic potential. The belief that individuals left to themselves will interact to give the most effective social and economic order could be linked with the deist tradition in which the Creator's laws ensure the balance of nature. Individual effort is rewarded by nature herself, a philosophy that many assume to be the basic foundation of Darwinian natural selection. Yet it was only later in the nineteenth century that the evolutionists' exploration of the "struggle for existence" revealed the darker side of this individualistic policy, translating the harmonious balance of interests into the "survival of the fittest." (On the nineteenth-century thought, see Willey, 1949, 1956; Copleston, 1963, 1966; and Mandelbaum, 1971.)

The "philosophic radicals" derived their view of human nature from Locke via Helvétius (Halévy, 1955). To explain the malleability of the human mind, David Hartley (1749) emphasized the principle of "association of ideas" (Oberg, 1976). Once Joseph Priestley had freed it from the religious overtones retained by Hartley, this principle became the basis for a new science of man (Hartley, 1775). If two sensations were repeated together sufficiently often, the mind would automatically come to associate them, in something like a conditioned reflex. This could be used to mold a person's mind to ensure socially respectable behavior. In the early nineteenth century, Jeremy Bentham founded the utilitarian school of social philosophy by advocating this approach to the study of the mind as a means of ensuring the greatest happiness of the greatest number of people. The human mind was reduced to a law-bound entity open to rational study and social manipulation—but without the dangers of outright materialism.

Priestly argued that the new understanding of human nature opened the gates to a progressive future for mankind, a theme also taken up by William Godwin. The philosophic radicals certainly hoped for social progress, but in general they adopted a less enthusiastic tone. Their purpose was to remove obstructions that prevented free interaction of individuals; but once this state of freedom was achieved, they believed that economic progress would result from natural forces rather than deliberate human guidance. Bentham thus linked his utilitarian philosophy to the laissez-faire school of economics founded by Adam Smith. In his *Wealth of Nations* of 1776, Smith argued that prosperity depended on individual initiative and would be destroyed by any form of state intervention, however well-meant. Bentham wanted to use social manipulation based on the penal system to eliminate crime, but he too held that in the economic sphere free enterprise must reign supreme.

Utilitarian philosophy provided the "common context" of social and biological thought in the early nineteenth century (Young, 1969). The individualism of the laissez-faire school is almost certainly reflected in Darwin's decision to treat biological species as a population of diverse organisms rather than copies of an ideal type. Bentham's concern with the utility of all social habits (i.e., their usefulness for promoting happiness) is also paralleled in the natural theology that so influenced Darwin in his youth. William Paley (1802) interpreted the adaptation of each organic structure to its function as a sign of divine benevolence—God's attempt to ensure the greatest happiness of all living things (Le Mahieu, 1976). The controversial nature of Darwin's theory stemmed from the fact that he took so many of these cherished notions of his time and advocated a natural, rather than a supernatural, explanation of them (Cannon, 1961*a*). Yet this point emphasizes a major transition in nineteenth-century thought. Darwin reversed the logic of Paley's argument by making adaptation a natural process rather than a divinely ordained state. He did this by postulating a struggle for existence between individuals of the species, giving rise to a natural selection of the most fit. It frequently has been assumed that the concept of individual struggle lay at the heart of laissez-faire social theory and that Darwin merely translated the cutthroat ethos of capitalism into a natural principle. This assumption ignores the fact that for Adam Smith and his followers, the principle of laissez-faire itself was intended to allow the natural harmony of human interactions to flourish for the benefit of all.

Political economists of the period did not see individualism as a license for unlimited competition that would eliminate the weakest members of society. Although they believed that progress would result from the removal of all restrictions on economic interactions of individuals, they felt that this would work not by eliminating the least able but by encouraging everyone to contribute to the best of his ability. Accepting the idea that society is composed of self-seeking individuals, they nevertheless believed that free competition would work to everyone's advantage. Assuming a "natural identity of interests," when each seeks his own good he automatically benefits society as a whole. Some would be worse off than others—but they would be in even worse positions if anything interfered with the delicate natural balance of the economy. Such a view looks back to the old tradition of a divinely preordained balance of nature, not toward Darwinism. We may dismiss the whole exercise as a way of justifying a state of exploitation by those who benefited from it, but we must recognize that the logic of the justification was based on an attempt to reconcile individualism with the belief that man is a divine creation.

It can be argued that this optimistic interpretation of laissez-faire was already breaking down by 1800. Darwin derived the concept of the struggle for existence from Thomas Malthus's *Essay on the Principle of Population* (1797, reprinted 1959), which certainly had promoted a more ruthless image. Young (1969, 1985) sees Malthus's principle as a challenge to natural theology that paved the way for the Darwinian interpretation of struggle (see also Vorzimmer, 1969*a*). The *Essay* was a reply to the claims of writers such as Condorcet and Godwin that a better understanding of human nature would produce social progress. Malthus argued that the "passion between the sexes" was too basic a part of man to be eradicated, hence the production of children would always tend to a maximum. Potentially, this would give a geometric rate of increase in the size of the population, while the food supply at best could be increased at an arithmetic rate. Thus population would always tend to outstrip food supply, and starvation must be a permanent feature of the human situation. Poverty is *natural* and cannot be eradicated by social reform—a view that earned Malthus the hatred of all subsequent social reformers. He advocated a laissez-faire policy of no state support for the poor, arguing that while this might increase suffering in the short term, it would limit the number of poor people and hence work out best in the long run.

At first sight this seems a clear anticipation of the Darwinian concept of struggle: remove state support and the least able will be eliminated by starvation. Certainly, the optimistic image of the old balance of nature has been destroyed, and we now see that each species (including man) must be engaged at all times in a struggle against the limitation imposed on its numbers by the food supply. But it was by no means necessary to assume that the pressure of population must give rise automatically to the Darwinian level of struggle, that is, a struggle among individuals who make up the population (Bowler, 1976*b*). Malthus himself only used the crucial phrase "struggle for existence" in his discussion of primitive tribes, in which he realized that shortage of food would lead to conflict and elimination of weaker tribes. In discussing his own society, he reverted to the traditional approach by claiming that competition is best for all, including the poor. They are not poor because they have been forced to the bottom of society in a ruthless struggle, nor have the rich gained their wealth through superior ability. In general, wealth is inherited, and far from being the prize of achievement it imposes a responsibility because it is the only means for opening up new resources. The direction of Malthus's thought is obvious from the fact that in his first edition he saw the population principle as a divine institution, designed to prevent laziness by ensuring that a man must work or see his family starve. This was exactly the position adopted by Paley to reconcile the principle with natural theology. It ignores the real logic of the situation, which implies that eventually there will be so many people that some will starve even if they are willing to work. In later editions, Malthus even suggested that poverty could be eliminated by educating the poor to realize the dangers of having too many children. Clearly, he was not prepared to challenge the prevailing belief that nature is a divine contrivance (Santurri, 1982).

We now can see just how far Darwin went beyond Malthus to develop his theory of natural selection. He saw that among animals there must be a constant struggle for existence that will tend to eliminate the unfit. Social Darwinists of the later nineteenth century would use his theory to justify their claim that progress would occur only if the best individuals are allowed to fight their way to the top at the expense of the least able. This represents a massive shift in the attitude toward struggle in society, a move away from the old belief in harmonious interaction of individuals toward a more deliberately ruthless approach. Darwin paved the way for the new attitude by apparently showing that nature was not a harmoniously balanced system but a mechanism to ensure progress by rewarding ability and punishing failure. Not that he *created* social Darwinism, however; the more ruthless attitude already had begun to manifest itself unconsciously in Victorian society at large (Gale, 1972). Darwin may have absorbed this attitude and used it as a model for his theory. Yet in so doing he was going beyond the social thinkers of his youth, who still had been trying to justify a laissez-faire policy on the basis of the old view that it was best for all.

It is important to note that the idea of social evolution was conceived by a number of Victorian anthropologists without the stimulus of Darwinism (Burrow, 1966; Stocking, 1968, 1987). Utilitarian philosophy was based on the Enlightenment's faith in the universality of human nature. This meant that only one form of society was "perfect" (laissez-faire capitalism), and all others were distorted through the activity of small groups such as the aristocracy or priesthood seeking only their own good. The work of the social reformer consisted in merely destroying such archaic blocks to progress, after which any society should prosper automatically. By now, European culture was in contact with many different societies around the world, thanks to the activities of the empire-builders. The radicals were aware of the "backward" nature of these societies but at first tended to assume that their problems were equally amenable to rationally organized reform. As the century progressed, however, it became obvious that for all the reformers' efforts at education, colonized people stubbornly refused to throw off the shackles that bound them to their past. The only way of reconciling this with the constancy of human nature was to adopt an evolutionary view of society. It was not the case that any culture, however primitive, could be instantly translated into the highest form of society by simply educating people in the truths of political economy. To reach the highest level, a society must progress naturally through a series of intermediate stages, each with cohesive organization of its own. The obstructions to progress that could be removed so easily in Europe were still fundamental to the organization of the "lower" societies and could be removed only by a long period of development. Thus the essential stability of each social pattern was admitted and incorporated into a development framework as one stage in the overall pattern of social progress through which all must, in their turn, evolve. European society merely had progressed further than any of the others and had reached the high point of the scale first, perhaps because that part of the world enjoyed beneficial climatic conditions.

The attitude of the anthropologists who formulated this theory of social evolution may sound patronizing, but it was not necessarily based on a belief in the biological inferiority of other races. Use of biological evolution to explain the ranking of human races is essentially a product of the later nineteenth century, and one of its earliest pioneers was the philosopher Herbert Spencer (Greene, 1959*b*; Peel, 1971). Spencer was raised in the competitive atmosphere of the industrial revolution and remained one of the great champions of laissez-faire. It was he who coined the term "survival of the fittest" to describe natural selection, although Spencer himself was never a convinced Darwinist. He already had accepted a progressive view of human society *and* the idea of biological evolution during the 1850s, before Darwin's theory became public. For Spencer, Malthus's population principle was the dynamic agent of social development, constantly forcing societies to progress economically in order to escape the pressure of limited resources. At the same time, he was convinced by Lamarck's arguments for evolution and began to see the possibility of constructing a "synthetic philosophy" that would unite all aspects of natural and human evolution under the same laws.

By the time Darwin's theory appeared, Spencer already had begun to popularize his image of an essentially progressive universe, in which social development was an inevitable continuation of the biological processes that had actually raised man from the lower animals. In this system it was possible to argue that the technical and social achievements of the white race were products not just of a more advanced cultural evolution but also of a higher level of biological development characteristic of a "higher" race of mankind. Social progress actually improves human nature, so that those races that have not advanced as far as the Europeans already have been left behind as living fossils, doomed to extinction. Spencer believed that laissez-faire capitalism was the highest form of society, because it allowed each individual to exercise all his powers in the service of the community. The remaining problems of western society are caused by the fact that human nature has not quite caught up with social evolution, and the pressures of competition are needed to ensure that everyone will adapt to the new situation as rapidly as possible. This line of support for laissez-faire depended more on the hope of encouraging self-help than on the elimination of the unfit, but Spencer's notion of an evolution that generated a racial and social hierarchy would become an essential foundation of late-nineteenth-century "social Darwinism."

GERMANY: ROMANTICISM AND IDEALISM

If utilitarianism represented a continuation of eighteenth-century thought, there were other philosophies that reacted strongly against the spirit of the Enlightenment. John Stuart Mill (reprinted 1950) saw Jeremy Bentham and Samuel Taylor Coleridge as the chief exponents of the contrasting influences in Britain: Bentham was the utilitarian, while Coleridge stood for the Romantic reaction so clearly expressed in his poetry. The true home of Romanticism, however, was Germany, where J. W. von Goethe and his followers turned in disgust from the heartless materialism of the Enlightenment as expressed by d'Holbach. The Romantics wanted to see *spirit* as an active force imposing its will upon nature to create order and purpose. There long had been such an element in German thought, dating back to the time of alchemy and the mystical writings of Jacob Boehme in the seventeenth century (translation, 1912). Boehme maintained that God was the "soul" of the world, struggling to manifest Himself through spiritual activity within nature. This self-perfecting process worked through resolution of opposing trends or powers, already pointing the way to the dialectic. Now, in response to their dissatisfaction with the Enlightenment materialism, German thinkers around 1800 began to develop such themes into a whole new philosophy (Coppleston 1963; Jordan, 1967; Mandelbaum, 1971).

The novels of Goethe began to stress a more spiritual view of man struggling against limitations imposed upon him; but the Romantic viewpoint also gained access to formal philosophy through the movement known as idealism. David Hume had pointed out the paradox of sensationalist philosophy; if all our knowledge comes from the senses, we cannot have any absolute knowledge of the causes of sensation. We perceive the regularities in our sensations that we call the laws of nature, but we can see no guarantee that these regularities will always obtain. Immanuel Kant solved this problem by giving an active power to the mind. He argued that we do not just receive sensations; our minds must organize them to make them comprehensible. Thus the necessity of natural law is a necessity of the organizing process, a necessity of the mind rather than of nature itself. In morals, too, Kant followed Rousseau's reaction against Enlightenment utilitarianism and argued for a moral sense or conscience that makes our duty clear, independent of the prospects of pleasure or pain (Cassirer, 1945). Idealists such as J. G. Fichte and F. W. J. von Schelling took over this notion and maintained that the will is the primary reality, which in a sense creates the world it lives in. In Schelling's work, the individual will was translated into the universal Will, striving to manifest itself in the world of appearances. For G. W. F. Hegel, the universal Will, the Absolute, became the driving force of the universe, working itself out within human history toward achievement of its ultimate purpose.

Hegel's philosophy of history (translation, 1953) demonstrates the inherently developmental aspect of idealist thought. In the materialist scheme, progress seldom can be anything more than a by-product of nature's operations, because laws governing the behavior of matter are always the same. Social progress depended on simple accumulation of knowledge about man and the world. The idealists, by contrast, saw history developing through an inevitable sequence of distinct stages, driven by a universal spiritual force beyond the control of individual human beings. This approach had already been foreshadowed by J. G. Herder, in his survey of universal history published from 1784 to 1791 (translation, 1968). Herder unified nature and man via the idea of progress, adopting the temporalized chain of being as the unfolding of a divine plan of creation toward man (Lovejoy, 1959*b*). In addition, he saw human history as a progression but held that the kind of society existing at each stage was a unique structure quite capable of providing people with a satisfactory environment. Hegel took over this view of history as a series of distinct stages and explained it through the new logic of the dialectic (Jordan, 1967). Forces in nature or a society always tend to generate their opposites, and a transition into a whole new world occurs at certain points through resolution of this tension. The opposing forces were seen as necessary consequences of the Absolute's efforts to manifest itself in the world and thus could not be reduced to or explained in terms of activities of individual human beings.

For Hegel the ultimate human reality was the state, not the individual. The state expressed the form or spirit of the society, which may be symbolized by some great leader to whom all will naturally give allegiance. Here were the seeds of twentieth-century totalitarianism, in which the individual would be subordinated to the state and would be expected to find the meaning of his life not in activities for his own benefit but in supporting the whole of which he is part (Popper, 1962). For Hegel this subordination of the individual was justified because it allowed the development of mankind toward the spritual goal that was the final purpose of the Absolute.

Several aspects of idealist philosophy could be applied to biology. Just as Hegel's political theory subordinated the individual to the state, so the idealist approach to natural history subordinated the individual organism to the species. The species was the true reality, with a deeper level of meaning than the imperfect manifestations of its type in the physical world, and idealism thus stood opposed to any form of evolution that would tend to break down rigid distinctions between specific types. Conversely, idealists did accept the concept that all species were related, each forming an element or unit within a coherent pattern of natural order. Biological classification became the search for such a formal, harmonious pattern of relationships, to which end idealists invoked the notion of the "archetype" or basic form of which all members of a particular group were supposed to be but superficial modifications. Goethe sought the form of the archetypical plant, while Lorenz Oken postulated a vertebrate archetype meant to represent the underlying unity of all animals with backbones. The search for such relationships was not without some scientific value, because in some cases they could be translated into the kind of historical connection demanded by

Darwinism. But the idealist concept of a formal pattern of harmonious relationships that constituted a transcendental reality underlying all nature was fundamentally hostile to the spirit of Darwinism.

Perhaps the most important influence of idealist philosophy on biology was through its developmental picture of a world advancing toward a predetermined goal. This did not immediately give rise to a scheme of biological evolution—Hegel, for instance, did not believe that earlier stages in the pattern of universal development were manifested in a genuine temporal sequence. Goethe, nevertheless, explored the possibility of such a sequence unfolding in the course of the earth's history (Wells, 1967), and the idea became increasingly popular among German naturalists during the early nineteenth century. The original stimulus for this was successful application of idealist perspective to a more restricted and more obvious area of biological development: growth of the embryo. Under the influence of idealism, a generation of German embryologists from C. F. Wolff to K. E. von Baer destroyed the preformation theory and laid the modern foundations of their science (Adelmann, 1966; Oppenheimer, 1967; Temkin, 1950; Roe, 1981). Schelling and Oken ensured that the broader implications of these developments would be appreciated as they laid the foundations of a general *Naturphilosophie* (Oken, 1847; Gode von Aesch, 1941; Lenoir, 1978). Growth of the individual was now seen as a model for the evolution of life in the course of the earth's history via the so-called recapitulation theory (Russell, 1916; Meyer, 1935, 1939; Gould, 1977*b*). Progressive growth of the human embryo toward its final goal was merely a repeat performance in miniature of the grand upward progress of life following the universal plan of creation. Some German naturalists were even prepared to believe that the successively higher stages were produced by the actual transmutation of lower ones. Such a progressive, goal-directed concept of evolution was basically incompatible with Darwinism but would prove extremely difficult to displace even in the later nineteenth century.

The contrast between Darwin's theory of gradual, irregular evolution and that of a progressive, step-by-step ascent toward a predetermined goal is still appropriate when we turn to consider the influence of Karl Marx. Although he turned Hegel's idealism on its head by postulating that economic rather than spiritual forces are the true source of historical development, Marx's materialism retained the dialectical structure of Hegel's thought and hence its emphasis on progress through discontinuity (Jordan, 1967). Marx welcomed Darwin's theory because it undermined religion, but he and his followers could not accept the Darwinian mechanism of change based on struggle and individual variation. He was one of the first to recognize that natural selection could be seen as a transferral into the organic world of capitalism's ethos of individual competition. In contrast, Marxism stressed subordination of the individual to his class as the fundamental unit in the struggle through which history progresses. The final goal of social evolution, the communist revolution, was both desirable and predictable. For these reasons, the relationship between Marxism and evolution theory—particularly in its Darwinian form—has been one of suspicion arising inevitably out of their roots in the two opposing branches of nineteenth-century thought.

5

Geology and Natural History: 1800 – 1859

Darwin worked out his mechanism of natural selection in the late 1830s but did not publish the *Origin of Species* until 1859. Thus we can postpone our treatment of his work and concentrate first on the parallel developments made by those who were as yet unaware of this more radical alternative. Some of these advances contributed directly or indirectly to the foundations of modern evolutionism, but it is essential that we treat them within their own context, not as stepping-stones to Darwinism. Even those ideas that we know influenced Darwin directly had components incompatible with the complete Darwinian world view. It has been fashionable to trace the ancestry of Darwinism back to Charles Lyell's uniformitarian geology, in which all changes occurred gradually through the action of observable causes. Darwin was seen simply to have applied this element of continuity to the organic as well as the physical world. Lyell's opponents, the "catastrophists," frequently have been dismissed as reactionary thinkers committed to the subordination of science to religion, as in their efforts to prove the reality of the biblical deluge (Gillispie, 1951; for classic histories of geology, see Adams, 1938; Geikie, 1897; von Zittel, 1901).

We know now that this is far too simple a picture. Darwin owed a great deal to Lyell, yet the "principle of uniformity" was based on a steady-state world picture that neither Darwin nor any modern evolutionist could accept. Evolution contains an element of *development*, which Lyell denied but the catastrophists preserved and expanded. Thus it can be argued that the modern evolutionary viewpoint owes something to both catastrophism and uniformitarianism, although the latter connection is more immediately obvious (Hooykaas, 1957, 1959, 1966; Cannon, 1960*a* and *b*; Rudwick, 1970, 1971, 1972; Bowler, 1976*a*; Ruse, 1979*a*; Hallam, 1983; Gould,

1987; Laudan, 1987). Other recent studies have begun to deemphasize the significance of the uniformitarian-catastrophist debate in nineteenth-century geology (Greene, 1982; Rudwick, 1985; Secord, 1986). The sequence of events in the earth's history was worked out by geologists independently of their disagreement over the rate of change. We are thus led to the conclusion that the emphasis on Lyell's debate with the catastrophists represents an interpretation of nineteenth-century geology that has been shaped by our modern desire to see this episode as a prelude to the Darwinian revolution.

More militant writers of the post-Darwinian era tended to assume that there must be automatic hostility between science and religion (White, 1896). Thus the Lyell-Darwin axis was hailed as the key to scientific objectivity, while the catastrophists' interest in religious issues was thought to have hindered the development of science. Gillispie (1951) points out that what occurred was not so much a conflict between science and religion but a series of attempts to solve religious problems universally admitted to be relevant to science. More recent studies have pointed out the *scientific* achievements of the catastrophists. Their religious concerns did not prevent them from being good geologists, except in isolated incidents, such as the attempt to support the reality of the biblical deluge. Certain kinds of scientific development followed naturally from the catastrophist outlook, particularly the effort to divide the earth's history into a number of discrete episodes—which became the geological periods we still recognize today. Lyell's more "objective" approach also has links with his own rather unorthodox but deeply held religious views. On both sides of the debate there was an integral relationship between scientific and religious positions, each generating something of permanent value.

In many respects the early nineteenth century paved the way for the Darwinian revolution, yet cultural factors ensured that naturalists of the period had certain blind spots that would eventually create the need for a new initiative to clear their views. A positive aspect of this period was that it established the basic institutional framework of science, a framework within which Darwin's theory would be debated and which is ancestral to much of today's scientific organization. Scientific societies were founded or reformed along modern lines, and the first serious links between science and government were built up (Cannon, 1978). France and Germany led the way, and Britain tried desperately to catch up in mid-century, especially in scientific education.

Science certainly became an integral part of Victorian culture—we are dealing now with a period in which geology texts sometimes outsold popular novels. This gave science its strength and its prestige, yet also imposed restraints on how it could develop. Science was respectable as long as it did not appear to disturb religious and social conventions of the day, although such constraints were far more obvious in Britain than on the continent. This does not mean that early Victorian scientists were forced to pay lip service to the Bible, and they themselves, however sincere their respect for religion, realized the need for a sensible accommodation and sought more sophisticated ways of reconciling their work with their wider beliefs. Yet certain topics remained taboo, particularly anything that would threaten the status of man as a spiritual being. At a time when many German naturalists were already accepting transmutation as the means whereby the divine plan of creation unfolded, Robert Chambers's discussion of the same idea in his *Vestiges of Creation* (1844) had to appear anonymously to protect its author from the resulting scandal.

Although geologists and naturalists of the early nineteenth century were able to paint a revolutionary picture of how life had developed in the course of the earth's history, there were limits on how far they could go in explaining *why* the process occurred in this way. The basic problem was posed by fossil evidence for a progressive development of life from the simplest forms up to man in the course of successive geological periods. How had new populations been introduced, and why did they ascend the scale of organization in this way? The idea of a series of spontaneous generations was retained by a few German naturalists but now was abandoned generally. To most scientists, it seemed evident that some supernatural agency must be invoked to explain the appearance of totally new forms of life. Yet this did not necessarily mean the simple, biblical notion of a miracle, because it was now obvious that some kind of general law or trend governed the actions of this agency. Under the influence of William Paley, many British naturalists supposed that the Creator had produced successively higher forms of life as the earth's physical environment had improved in the course of geological time. Adaptation and divine benevolence explained the progress of life. German idealism encouraged another approach, in which progress was related to the unfolding of a rationally ordered pattern aimed at eventual production of nature's highest type: man. This view found its way into the English-speaking world through the writings of Louis Agassiz, who synthesized it with the more traditional Christian belief in divine creation.

Given the existence of such trends, there was an obvious possibility that the Creator had established some kind of teleological law by which His plan would unfold. In Germany, some naturalists boldly assumed that such a law would work through the discontinuous transmutation of existing forms of life into higher ones. This view was advocated in Chambers's *Vestiges,* and despite widespread opposition to that book, there is evidence that even in Britain such ideas were being taken with increasing seriousness in the 1850s. Lyell's views on the continuity of natural change made the question seem all the more acute, and for this reason Lyell himself rejected the whole idea of progression to leave man standing clearly above a totally steady-state material universe. The one thing that no one wanted to contemplate was evolution brought about solely by everyday laws of nature, a process all too easy to divorce from any form of teleology. Development, even transmutation, was acceptable as long as the basic explanatory function lay at the supernatural rather than the natural level, leaving the detailed question of the origin of species outside the scope of scientific investigation.

GEORGES CUVIER: FOSSILS AND THE HISTORY OF LIFE

If Lamarck's theory was the last of the Enlightenment's speculations on the origin of life, the work of his chief opponent served as a new starting point for many nineteenth-century views. Georges Cuvier detested Lamarck and attempted to use his position in the scientific and political hierarchy of post-Revolutionary France to ensure that evolutionism did not gain a fair hearing (Burkhardt, 1970). Yet the dislike was not just a matter of religious prejudice, because Cuvier quite reasonably saw his scientific findings as opposed to transmutation. Impressed by the complex relationships between the internal parts of any living organism, he felt that such a delicate balance made significant change impossible. Even when reconstructing extinct forms from their fossil remains, Cuvier insisted that they fell into distinct species that could not be linked by evolution.

Cuvier's first efforts were devoted to a study of the mollusks he found on the seashore in Normandy, where he isolated himself during the reign of terror (Coleman, 1964). On moving to Paris, he secured the position dealing with the vertebrates at the Museum d'histoire naturelle, which the revolutionary government had created from Buffon's old Jardin du Roi. Here Cuvier consolidated both his scientific and his political power, successfully adapting to the rise of Napolean and his eventual defeat. About 1800, he perfected the techniques of comparative anatomy, describing and comparing the structures of various animals available to him (Cuvier, 1805). From these studies emerged a new system of classification for animals (1812*b*) and a survey of the animal kingdom, its introduction a useful statement of Cuvier's scientific principles (1817; Outram, 1986).

The new comparative anatomy originated with Buffon's collaborator, Daubenton, and other late-eighteenth-century naturalists. But Cuvier placed a stronger emphasis on the need to study the internal structure of each animal and thus was able to appreciate the marvelous complexity of each organism, the delicate interactions that must exist between all the parts to maintain the functions of life. He emphasized the "correlation of parts"—the necessary relationships that must exist between the organs to create a viable whole. Similarly, the "conditions of existence" impose necessary links between the parts and the animal's environment. Once the anatomist had gained enough experience with different kinds of animals, he could begin to anticipate the kinds of relationships required to create a workable structure. If a certain form has sharp claws, it must be a carnivore and also must have teeth adapted for seizing and tearing at its prey. It was said that Cuvier had developed his understanding of these relationships to such a degree that he could in imagination reconstruct the whole animal from a single bone.

Better knowledge of the internal structures of animals yielded new insights into the relationships used to classify them. Cuvier declared that those characteristics essential to the animal's sensitivity and power of movement should be given more weight than mere superficial features when classifying, the principle of the "subordination of characters." This meant a strong emphasis on the structure of the nervous system, and Cuvier realized that the structure associated with the possession of a backbone represented a basic characteristic for unifying the Linnaean classes of mammals, birds, reptiles, and fish. Thus he created the vertebrate *embranchement* or "type" as a basic division of the animal kingdom. The invertebrates hitherto had received far less attention from naturalists because they were regarded as the inferior part of the kingdom, but Cuvier applied his method here and detected three additional types of organization. Each of the four types represented a fundamental group plan upon which an animal structure could be based. The invertebrate types were not necessarily inferior to the vertebrates just because they lacked an internal skeleton; rather, they were based on three quite different kinds of organization, which simply did not require a skeleton to function.

Such a division of the animal kingdom broke down the sense of a linear ranking inherited from the old chain of being. Naturalists might feel instinctively that some animals are more highly organized than others, but this was not a trustworthy guide in classification. Just because we ourselves are vertebrates does not permit us to assume that all invertebrates are to be treated as inferior types. A vertebrate is not necessarily superior to a mollusk, merely different—and the differences are so fundamental that it may be meaningless to rank one above the other. Cuvier was suspicious even of the attempt to rank classes within the vertebrate type, and he regarded fish and mammals as simply different kinds of vertebrate adapted to different habitats. It was many decades before the majority of naturalists could bring themselves to accept this complete breakdown of the old hierarchical viewpoint, but the implications for the rise of evolution theory were enormous. It no longer would be possible to think in terms of a linear progress through the animal kingdom; each group would have to be pictured as a separate branch in a treelike process of development.

Did the basic similarity of all species within a type indicate that all had descended from a common ancestral form? Cuvier resisted this interpreta-

Vertebrata (Vertebrates)	Creatures possessing a backbone: the four Linnaean classes of mammals, birds, reptiles, and fish. (The amphibians are now regarded as a separate class, but in the nineteenth century they were usually included with the reptiles.)
Mollusca (Mollusks)	Creatures with no backbone but sometimes an external shell: oysters, clams, etc.
Articulata (Articulates)	Creatures with articulated or segmented bodies: insects, spiders, worms, etc.
Radiata (Radiates)	Creatures with a radial or circular plan of organization: starfish, sea urchins, etc.

Fig. 13. Cuvier's Four Types of Animal Organization.

Each of the four types represents one of the basic patterns of animal organization and is subdivided into classes, orders, genera, and species as in the Linnean system. Cuvier's "type" is the origin of the modern "phylum"—still the most basic level of classification. But biologists have been forced to recognize far more than four phyla in the animal kingdom. The vertebrates have remained more or less the same, except that they are now called "Chordates" and include some animals with a spinal chord but no backbone. The mollusks also have remained with little change. But the articulates have been broken up into a number of separate phyla, with the insects, spiders, and so on, each given the status of a distinct phylum. The radiates had been used as a "dumping ground" for those creatures that Cuvier could not fit into any of the other three types and hence have been broken up into many different phyla (Winsor, 1976).

tion and argued against the possibility of one species transforming itself into another. He saw each species as a particular variant of the type, exploiting a unique set of harmonious relationships between parts of the body adapted to its own characteristic life-style. Bodily interactions were balanced so delicately that any significant change would upset the system and render the animal inviable. The ability of the environment to produce well-marked varieties within a species was limited by the necessity of preventing any significant disturbance of the basic pattern on which the species is modeled. Cuvier thus presented the fixity of species as a pragmatic consequence of his concern for the complexity of living things, and his writings only rarely expressed his faith in a supernatural Designer. In Britain, however, the greater public concern that science should not be seen to undermine religion ensured that Cuvier's followers there would seize on his views as an explicit confirmation of the traditional argument from design.

Cuvier's rejection of transmutation is all the more interesting because of his contributions to paleontology, which produced the first outline of the history of life based on solid evidence (Theunison, 1986). The techniques of comparative anatomy were ideally suited to reconstruction of fossils, where often only an incomplete skeleton was found. Using his experience gained with living animals, the anatomist could study fossil bones and visualize how they must have fitted together in the whole animal. From this he could attempt to reconstruct the outward appearance of the original form. Cuvier soon applied himself to the study of fossil bones in this way and became the acknowledged authority in the field. His collected papers (1812*a*) became the foundation of modern vertebrate paleontology.

The discovery of fossil bones had aroused wide interest at the end of the eighteenth century (Greene, 1959*a*; Rudwick, 1972; Buffetaut, 1986). From Siberia came the remains of the woolly mammoth, an elephantlike creature so recently extinct that its bones were not truly fossilized, and in some cases, the flesh still was preserved in the ice. From America came the remains of an even stranger kind of elephant, the mastodon, whose teeth resembled those of a hippopotamus. At first it had been argued that these forms still might be alive somewhere in the world, but as exploration proceeded the chances seemed ever more remote. Once Cuvier had confirmed that these were indeed the remains of creatures quite unlike any that are known today, the fact of extinction became inescapable. Cuvier already had shown that the African and Indian elephants were so different that they must be treated as distinct species, and in the same way he established the mammoth as an equally distinct species within the same genus. The mastodon had to be placed in a separate genus, because it differed more widely from living elephants. Successful identification of these spectacular creatures confirmed the general interest in fossils and led to widespread searches for more remains.

As Cuvier reconstructed the ever-widening range of extinct forms, he at first assumed that they all had formed a single ancient population. The Wernerians, however, already had shown that the earth was covered with a succession of rock formations laid down one on top of the other over a vast period of time. The mammoth came from superficial gravel deposits that were very recent by geological standards, only a few thousand years old; but other fossils came from lower (and hence older) deposits, building up a picture of a *sequence* of extinct populations corresponding to each period of rock formation. To confirm this kind of relationship, Cuvier collaborated with Alexandre Brongniart in a study of strata making up the Paris basin (1811; new edition, 1825). Brongniart used marine invertebrate fossils to establish distinctions between successive formations, one of the earliest systematic uses of fossils in stratigraphy, while Cuvier reconstructed the corresponding vertebrates. They established a sequence stretching down through the Tertiary to underlying chalk deposits, known to form the upper limit of the more ancient Secondary series. The older the period of formation, the more bizarre the vertebrate fossils, and the less similar to any form still alive on earth today.

Cuvier's belief that the fossil animals now are extinct was based on the assumption that they could not (as Lamarck supposed) have evolved into

Recent Deposits:	Woolly mammoth (*Elephas primigenius*) from Siberia. Mastodon (*Mastodon americanus*) from America, and also some European species of the same genus.
Tertiary Formations:	*Palaeotherium*, several species of the same genus, a mammal unlike any known today but with vague affinities to the tapir, the rhinoceros, and the pig. (Actually from the oldest Tertiary deposits, later named "Eocene" by Lyell.)
Secondary Formations:	*Mososaurus*, a giant marine lizard from Maestricht in Germany. (From the chalk or Cretaceous deposits of the upper Secondary.)

Fig. 14. Examples of Fossils Described by Cuvier and Their Geologic Relations.

Examples are shown with the most recent at the top. Note that the more recently extinct forms, such as the mammoth, are more similar to living animals than the older ones, such as *Palaeotherium.* Vertebrate fossils of the Secondary formations turned out to be almost exclusively those of fish and (more spectacularly) reptiles, such as *Mososaurus*, hence this era of the earth's history was later called the "Age of Reptiles."

their modern counterparts. Ancient species proved to be just as complex and well balanced in their structure as living forms, confirming Cuvier in his belief that each species is a functional whole that cannot be disturbed by significant variation. In any case, the French expedition to Egypt had brought back mummified animals thousands of years old which were identical to those now living. The stability of species thus was established, and the extinction of ancient forms became a verified fact. It should be added that a modern evolutionist still would accept the latter part of this proposition. Most fossil species indeed have died out, while a small number of their more successful contemporaries have diversified into many later forms.

Why, then, did each of the ancient populations die out, and how were new forms introduced to replace them? In his *Discours sur les révolutions de la surface du globe*, Cuvier discussed the geological events that may have affected the earth's inhabitants (1812*a*, introduction; translation, 1813; new edition, 1825). Exploration of the Paris basin had revealed an alternation of deposits from fresh and salt water, indicating a series of major changes in the relative position of land and sea. Cuvier tended to assume that the change from one set of conditions to the next was a comparatively sudden affair, because there were abrupt breaks between fossil populations. There was also a good deal of evidence for a sudden change in the conditions of Europe in the not-too-distant past (which later geologists would interpret as the end of the last ice age). In any case, massive alterations of the position of land and sea must have been brought about by causes far more powerful than any we now observe. Without actually suggesting that these events were instantaneous, Cuvier thus prepared the way for the school of geological thought that would later be named catastrophism.

Cuvier saw the geological revolutions as the cause of extinction. An invasion of Europe by the sea would dispose of all the animals living there, if they could not migrate elsewhere. How then was a new population introduced at a later date? It would be easy to jump to the conclusion that Cuvier must have accepted miraculous creation. Yet he insisted that a new creation was not necessary to explain repopulation of a particular area, because "new" animals could have been living in a part of the world unaffected by the revolution, migrating from it to take the place of those now extinct. For this reason, Cuvier refused to support those who identified the last revolution with the universal deluge of the Bible; he wanted strictly localized catastrophes. His theory implied that at some earlier time all extinct and living animals had coexisted around the world and that elements of this vast ancient population were wiped out by successive revolutions. Wider exploration, however, gradually made it clear that there was no part of the world in which fossilized remains of *modern* animals could be found. Cuvier's migration theory thus was rendered implausible, leaving geologists with the obvious conclusion that new forms of life had been introduced on the earth at many different points in its history.

The only case in which Cuvier left room for supernatural creation was at the beginning of the Tertiary, when he supposed that all mammals were introduced simultaneously. He remained, nevertheless, an implacable opponent of transmutation. At the end of his career he engaged in one last debate on the subject, with Etienne Geoffroy Saint Hilaire (Isidore Geoffroy Saint Hilaire, 1847; Bourdier, 1969; Appel, 1987). For Geoffroy, the unity of structure linking the species within a type had an idealized significance. It was a relationship with a transcendental reality of its own, not just a pragmatic consequence of the limited options available to nature in the construction of viable organic forms. This belief in turn encouraged him to explore the possibility that one form actually might be derived by transmutation from another within its type. After studying some of the extinct reptiles, he went on to apply his research in embryology to provide an explanation of how these ancient forms could have been transformed into their modern counterparts, or even into birds and mammals (Geoffroy, 1833). A change in atmosphere might affect growth of an embryo so that new organs were developed. The offspring thus would appear with a totally new character, a kind of monstrosity, but would be able to survive and reproduce if its organs were adapted to the new conditions. This was evolution by sudden mutation (to use a modern term), not by gradual adaptation. Inevitably, Cuvier rejected the idea, as he had rejected Lamarck's earlier speculations. He argued that a monstrosity could not be expected to have the harmoniously balanced structure necessary for survival and thus could never form the basis of a new species.

It has often been assumed that Cuvier's influence prevented the idea of transmutation from gaining any real influence in the early nineteenth century. But we now know that his control over French science was not absolute (Outram, 1984). Lamarck's ideas were in fact taken up by a few more adventurous thinkers (Corsi, 1978). Even in Britain, a handful of radicals maintained enough support for Lamarckism to require the conservative social establishment to take active steps against them (Desmond, 1984, 1987). Materialism was an ever-present ideological threat that, although stifled in these early decades, preserved attitudes and values that were to emerge more openly in later evolutionary debates. On a less radical plane, many continental naturalists seem to have followed Geoffroy in his belief that sudden transformations of individual growth could produce new variants on nature's basic plan (Temkin, 1959; Lovejoy, 1959*d*).

Geoffroy scored another victory in terms of how the unity of type within the animal kingdom was to be explained. For Cuvier, each type defined a group of successful solutions to the problems encountered by nature in the construction of viable organic forms. There was no mystical unity, only the pragmatic fact that certain basic structures more easily satisfied the "conditions of existence" imposed by the requirements of internal balance and external adaptation. For Geoffroy, by contrast, unity of type had a higher meaning. The fact that many diverse forms could be linked by a central theme in their structure was a sign that nature expresses a fundamentally orderly pattern. The archetypical form can be varied in many ways to meet the demands of the environment, but its imprint remains unmistakable. For all its links with idealism, this was a powerful idea that excited many naturalists in the mid-nineteenth century (Ospovat, 1981). It implied that nature was not a haphazard collection of individually designed forms and encouraged scientists to search for groups unified by genuine morphological resemblances. It thus could be argued that Geoffroy's interpretation of the type concept, rather than Cuvier's own, stimulated the revolution in morphology and taxonomy that helped pave the way for Darwinism. Groups that at first were thought to be unified by a common underlying form eventually were taken over by the Darwinists, and their unity was explained not in terms of a mystical archetype but as the result of common descent from a single ancestor.

CATASTROPHISM AND NATURAL THEOLOGY IN BRITAIN

Neptunism enjoyed wide support in Britain at the turn of the century. Playfair's account (1802) of the work of Hutton kept alive the spirit of Vulcanism but gained little wide support, because the theory all too frequently was identified with infidelity (Gillispie, 1951). Werner's mineralogical techniques were exploited with great success, particularly in the work of Robert Jameson (e.g., 1804–08). As the leading spirit of the Wernerian Society of Edinburgh, Jameson excluded Hutton's ideas from wide influence even in their native city. By the 1820s, however, geologists were beginning to realize how much evidence there was to suggest that movements of the earth's crust had played a role in shaping the surface we see today. No one was prepared to accept the Huttonian steady-state interpretation, so the gradual infiltration of Vulcanist principles occurred without open controversy. Even Jameson added notes to the English translation of Cuvier's *Discours*, which made it clear that he too now was prepared to accept movements of the earth as a factor in geological change. In London the Geological Society was founded in 1807, ostensibly to encourage a spirit of empirical enquiry that would be free from the acrimony of the Neptunist-Vulcanist debate. But under its auspices, a new breed of geologist emerged (Laudan, 1977), committed to a philosophy of the earth's history that eventually was to be called "catastrophism."

At this stage, most geologists simply were unable to believe that normal earthquakes could raise a whole mountain range, even if they had been prepared to allow the vast amounts of time required. Nor were they able to accept the idea that a river could gradually erode a large valley. Thus the catastrophists postulated massive earth movements in the past, far beyond the scale of anything known today. These upheavals would raise mountain ranges in a short period of time and would cause gigantic tidal waves capable of large-scale erosion. There was hard evidence to suggest that a great deluge indeed had swept across the whole of Britain. Great boulders occurred in certain areas, formed of rock quite unlike that on which they rested. It seemed obvious that a tidal wave had torn them from their original position and dashed them across the face of the earth. Only in the 1840s did an alternative explanation based on the greater extent of glaciers in a recently ended ice age begin to emerge. In the meantime, "diluvialism"—the postulation of an extensive flood in the recent past—seemed a reasonable explanation of evidence such as the erratic boulders. Far from being a purely religious speculation, it could be defended and attacked on scientific grounds (Page, 1969). The danger was that such an idea could all too easily be associated with the biblical deluge by those determined to subordinate science to religion. "Scriptural geology" was certainly popular in Britain (Millhauser, 1954), although its most enthusiastic exponents were not professional geologists. Many clergymen felt that diluvialism did not go far enough; they wanted to turn the clock back to a situation in which the whole Genesis account of creation could be accepted as literally true.

Like their colleagues on the Continent, British geologists wanted to set up an autonomous science, but they had to balance their professional interests against the need to minimize the apparent challenge their science offered to religion. The fact that some geologists made limited concessions to the Genesis account has misled some historians into assuming that catastrophism was little more than an exercise in scriptural geology. The shortsightedness of this assumption can be seen as soon as we turn to the foremost scientific advocate of the deluge's reality, the colorful and eccentric Reader in Geology at Oxford, William Buckland (Rupke, 1983). When inaugurated to his position, Buckland delivered an address (1820) defending geology against the charge that it tended to undermine religion. Yet only in limited areas did he allow his theology to distort his scientific work. Buckland did not want to prove the literal truth of everything in the Genesis story of creation but merely to show that geology confirmed the occurrence of a deluge as reported in the story of Noah. He discovered what he took to be the best evidence of this position at Kirkdale in Yorkshire where workmen uncovered a cave filled with dried mud; embedded in the mud were the bones of animals quite unlike those that inhabit England today. Buckland showed that the cave had been a hyenas' den, into which various forms of prey had been dragged to be eaten. In his *Reliquiae Diluvianae* (1823), he argued that the period in which the hyenas had lived in England had been terminated by a great deluge, which had filled the cave with mud and had been associated with the change in conditions that made the country suitable for its present inhabitants. As similar caves had been found in other parts of the world, it could be argued that the whole earth must have been covered with water, just as the Bible claimed.

If the diluvialist explanation of the caves was accepted, was it valid to claim that the catastrophe had really been worldwide? Cuvier himself had argued against this, and it soon became apparent to others that on this question Buckland had allowed his religious concerns to get the better of his scientific caution. By the end of the decade, it was generally acknowledged that evidence derived from various parts of the world could not be connected with a single event. The extreme diluvialist version of catastrophism was abandoned, and most supporters of the theory began to scale down the extent of the upheavals they envisaged. The collapse of diluvialism clearly illustrated the dangers of linking science too closely to the details of the biblical story. If such a link was exposed as fallacious by further research, the resulting controversy would be more damaging to religion than if the reconciliation never had been attempted.

Even in the case of the deluge, serious geologists had not intended to suggest that miraculous forces had been at work. All were convinced that the earth's history has been shaped by divine Providence, yet the Almighty could surely achieve His ends by applying the laws He had built into the universe (Hooykaas, 1959). By 1830, catastrophism was beginning to acquire a coherent directionalist framework in which the cooling earth theory was exploited to give a natural explanation of the gradual decline in the level of geological activity (Rudwick, 1971). If the center of the earth was once hotter than it is today and the solid crust thinner, then we should expect past earth movements to have occurred on a scale much greater than that of modern earthquakes. In France, Elie de Beaumont suggested a catastrophic theory of mountain building in which the earth's crust "wrinkled" as it cooled. Such ideas gave catastrophism a sensible conceptual foundation and were accepted widely. Even Buckland eventually accepted the cooling earth interpretation, while admitting that he had originally overestimated the extent of the last deluge (1836).

Within the framework of this theory, the catastrophists went on to complete some of the most significant geological work of the century. The sequence of geological formations as we recognize it today began to take shape in the course of the 1830s, and some of the most difficult parts were worked out by catastrophists. The Tertiary formations already had been established by Cuvier and Brongniart, but now the more ancient Secondary and Transition series were subdivided into distinct formations and given the names we still use today. Each formation represented a complex of rock strata with underlying similarities, in general quite clearly demarcated from the formations above and below. Following the principle of stratification established by the Wernerians, it was assumed that each formation was laid down during a particular period of the earth's history, the lowest being the oldest. By establishing the sequence of rock formations, the geologists created a picture of the earth's historical development.

Fossils now became the key to establishing the sequence of formations. The Wernerians had established the principles of stratigraphy but had assumed that the succession of formations could be established on the basis of their mineral character. Now it was realized that rocks of the same type could be formed at different periods in the earth's history. Only the fossils gave an unambiguous clue to the sequence, because each period had been inhabited by its own unique population of living things. The most powerful advocates of this new technique were members of the "English" school of geology (Rupke, 1983). Ostensibly, this school had been founded by William Smith, the "father of English geology," whose geological map of the country (1815) had established the principles of the new technique (Eyles, 1969). Smith was a canal builder who had begun to study fossils for purely practical reasons, but his followers were gentleman-scientists who had little interest in economic geology (Porter, 1973, 1977). Indeed, it is probable that Smith's name was invoked merely to head off the claim that the real pioneers of the new stratigraphy were French. Recent studies have stressed the extent to which the effort to map the sequence of formations was undertaken with little reference to the debate over geological dynamics, uniformitarians and catastrophists both making contributions (Rudwick, 1985; Secord, 1986).

The most difficult rocks were the oldest, because these had been subjected to the greatest amount of distortion through earth movements. In Britain, examples of such rocks could be found in Wales, and the story of their unraveling is one of the great episodes in the history of geology. Both men involved were catastrophists who regarded the divisions between the formations as evidence of sudden changes in the conditions under which the rocks were laid down. They were an unlikely pair: Adam Sedgwick had been elected Professor of Geology at Cambridge although he knew nothing about the subject, while Roderick Impey Murchison was an ex-soldier persuaded by his wife to take up geology as a substitute for fox hunting (Clark and Hughes, 1890; Geikie, 1875, 1897). Both mastered the techniques of their chosen field and descended on Wales in the 1830s, Murchison to the south and Sedgwick to the even older rocks of the north, to work out the Silurian and Cambrian systems (Murchison, 1839). Although they later quarreled over the intermediate strata, the basic sequence of the oldest fossil-bearing rocks now was established beyond dispute (Secord, 1986).

At last the scattered parts of the earth's crust—and hence of its history—were beginning to fall into place. Many agreed with Sedgwick and Murchison that the new picture substantiated the catastrophist-directionalist view of the earth's development. Sudden breaks between the formations seemed to indicate equally sudden changes in the conditions under which the rocks were laid down. Even more significant were the changes in fossil populations in successive formations. The fossils used in Smith's technique were generally those of invertebrates, and even here there were major changes in the course of time. Far more spectacular was the development of the vertebrates. As Cuvier's methods were applied to the reconstruction of an ever-increasing number of fossils, it seemed that a steady-state view of the earth's history could be shown to be demonstrably false. There was clear evidence of a systematic development in the history of life through successive periods of geological time.

Cuvier had known that there are few mammalian remains below the Tertiary, but now it became obvious that a wide range of bizarre reptilian forms had flourished during the Secondary era. In 1824, Buckland described the first-known dinosaur, a giant carnivorous form that he called *Megalosaurus*. The actual name "Dinosauria" was coined in 1841 by the great anatomist Richard Owen, and ever since, these creatures have remained symbols of the exotic nature of prehistoric life in the public mind (Swinton, 1970; Colbert, 1971; Delair and Sarjeant, 1975; Desmond, 1976, 1979, 1982). In lower formations, however, even reptiles were absent, and the only representatives of the vertebrate type were strange, armored fish (Miller, 1841). At the very bottom of the whole sequence, Sedgwick's Cambrian, not even the fish remained. In this ancient period the only living things were invertebrates such as the Trilobites. A basic outline of the history of life on earth thus had emerged. It began with an age of invertebrates, followed by the successive introduction of fish, reptiles, and finally mammals.

Despite Cuvier's reluctance to accept a hierarchical classification of living things, most naturalists believed that the order of creation was progressive, an ascent toward ever-higher forms of life. It began with the lowly invertebrates and then rose to ever-higher levels of organization through the hierarchy of vertebrate classes leading up to the mammals. Man, of course, was the highest form of all, combining an advanced physical structure with a new kind of spiritual character. It seemed significant that absence of human fossils suggested a very recent origin for this highest product of creation. What purpose, then, did the Almighty serve by creating this whole sequence of populations leading up to the appearance of man? Eiseley (1958) has written of the "transcendental, man-centered progressionism" of the catastrophists, but it can be argued that there were two quite different interpretations of the fossil sequence (Bowler, 1976*a*). Idealists could indeed see the advance as the unfolding of a transcendental plan symbolizing man's position at the head of creation. But British catastrophists were not in general idealists; they were more at home in the utilitarian tradition and were able to adapt this to their own interpretation of progression.

William Paley's classic *Natural Theology* (1802) restated the utilitarian version of the argument from design used earlier by Ray (Edmund Paley, 1825; Le Mahieu, 1976). Each part of an animal's body is useful to it in its mode of life, and this universal adaptation of structure to function illustrates the wisdom and benevolence of a God who cares for His creatures. Just as the intricate structure of a watch implies a watchmaker, so the incredible complexity of living things proclaims the power of their Designer. In this tradition, each case of adaptation is considered individually, and the argument gains strength from the sheer number of examples cited. The movement reached its climax in the *Bridgewater Treatises* of the 1830s, a series of eight works commissioned in the will of the Earl of Bridgewater to illustrate the "Power, Wisdom and Goodness of God in the Works of Creation." A number of eminent scientists contributed to the series but with mixed success. The problem was that this endless catalog of adaptations produced not a sense of divine benevolence but sheer boredom in the reader.

It was easy for paleontologists to adapt this argument from design to their fossil discoveries. Cuvier's techniques inevitably stressed adaptation of the whole organism to its way of life. When applied to the reconstruction of fossil species, each was revealed automatically as a form adapted to the conditions of the time in which it lived. The results could be taken as evidence of design just as easily as Paley's examples drawn from living organisms. Even catastrophic extinction could be interpreted as a merciful wiping of the slate, leaving the way clear for the creation of a new set of species designed for the quite different conditions of the next epoch. In re-

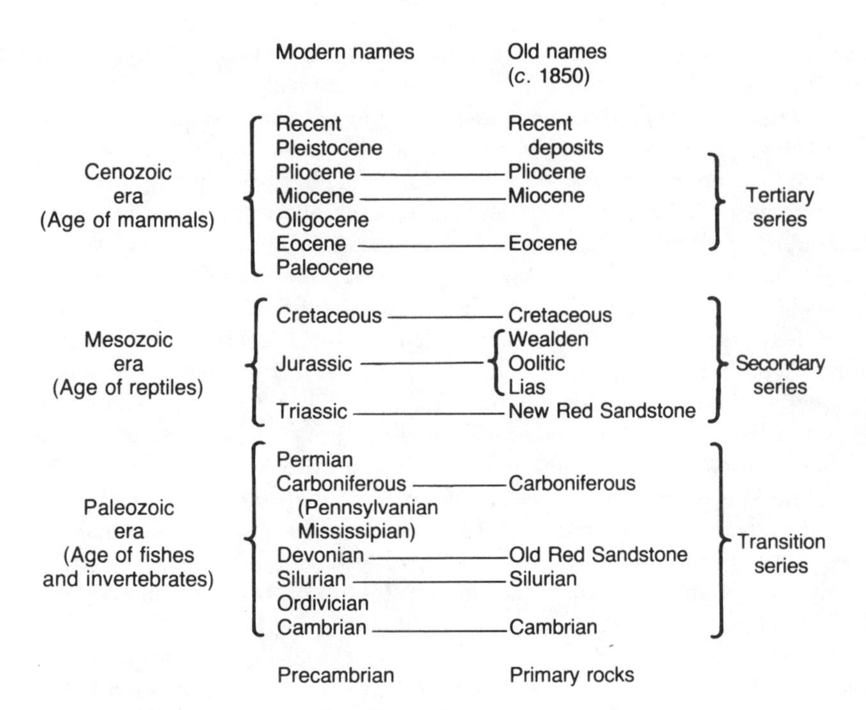

Fig. 15. Sequence of Geological Formations.

This diagram shows the sequence of formations established by the mid-nineteenth century and the one known today. The sequence of formations corresponds to the succession of geological periods in time, the oldest at the bottom. The complete sequence is never observed in any one location but was built by fitting together an order of superposition based on parts of the whole observed in many different areas. The three great epochs in the history of life—Paleozoic, Mesozoic, and Cenozoic—were established by John Phillips (1841) on the basis of invertebrate fossils. Dates were not usually estimated in the mid-nineteenth century—most catastrophists would have preferred a time scale much shorter than that we accept today, although not just a few thousand years. Today the beginning of the Tertiary is dated at seventy million years before present, the beginning of the Paleozoic at over half a billion years ago (from Bowler, 1976*a*).

jecting Cuvier's migration theory, later catastrophists were free to envisage a series of fresh creations, each adapted to successive periods of the earth's history.

Why did the character of the successive creations change in a progressive sequence leading from the invertebrates to the mammals? This could be answered once the directionalist interpretation of catastrophism was established, because the requirements of adaptation would ensure that a directional change in the physical conditions must produce a corresponding development in the earth's inhabitants. If the earlier periods were suited only for lower forms of life, these would be all that God would create. As conditions improved, He would be able to create successively higher forms until at length the world became suitable for man and the rest of the present population. When he wrote the *Bridgewater Treatise* devoted to geology (1836), Buckland suggested that the armor of the earliest fish protected them from the high temperature of the ancient ocean, as predicted by the cooling earth theory. In France, Adolphe Brongniart (1828*a*) used fossil plants of the Carboniferous to argue that the whole earth then had enjoyed a tropical climate. In a paper of the same year (1828*b*), he also suggested that the level of carbon dioxide in the earth's atmosphere had declined gradually as it was fixed in the rocks to form coal and peat. This would explain the later production of higher animals, because only in later periods had the air become pure enough to support them. Such an idea illustrates the sophistication of the directionalist approach, and to British naturalists such a trend was exactly what they wanted to show the superintendence of divine providence throughout the earth's history.

Progress, then, was a sign of divine benevolence, ensuring adaptation to steadily improving conditions. As Christians, British naturalists believed that the ultimate purpose of the universe was to provide a theater for the drama of human sin and redemption. Providence must have ensured that the end product of the earth's physical development was an environment suitable for man. Earlier populations were created to allow us a glimpse of the Creator's benevolence, not to symbolize our supreme position in the plan of creation. Man's uniqueness was determined by his spiritual character, and this could not be predicted on the basis of a merely biological progression.

Within the context of Paley's natural theology, catastrophists formulated a rational explanation of the historical development of life. Their account left one question open: exactly how did the Creator accomplish the introduction of new forms at appropriate times in the earth's history? The most obvious answer was by miracles, although their creationism was not necessarily of the most simpleminded variety (Cannon, 1960*b*, 1961*a*; Ruse, 1975*d*; Gillespie, 1979). New species certainly did not appear through operation of everyday laws of nature. The acts of creation, furthermore, were discontinuous, occurring only at the beginning of each geological period. Yet "creation" did not necessarily imply a miracle in the strict biblical sense, and it is most unlikely that the catastrophists pictured the Creator as a white-haired old gentleman rushing around the world forming species out of dust. Creation was a systematic process, revealed by the gradual progress of life, and it might be legitimate even to talk of "laws of creation"; but because these laws must involve the Designer's intelligence, they were not a fitting subject for scientific investigation. Details of the process were left vague deliberately, except for the negative requirement that it could not occur through gradual transmutation of previously existing forms.

IDEALISM IN BIOLOGY

The English catastrophists had established a developmental view of the earth's history and had linked this to the progress of life through the assumption that higher forms appeared as the conditions improved. But there was an alternative philosophy that could generate a significantly different—though no less teleological—version of the developmental world view. The idealist philosophy of early-nineteenth-century Germany was based on the assumption that the physical universe is a manifestation of the divine Mind. This encouraged a belief that the apparent diversity of natural forms concealed an underlying unity representing a rational plan of creation. In its most extreme form, this philosophy gave rise to the frankly mystical speculations of German *Naturphilosophie* in which nature was seen as striving to produce the perfection of the human form (Gode von Aesch, 1941; Lenoir, 1978). Lorenz Oken (translation, 1847) was the most influential spokesman of this philosophy. His excesses were repudiated by more sober idealists who adopted a pragmatic approach to the search for an underlying unity of organic nature. Geoffroy Saint Hilaire's transcendental anatomy was one expression of this approach, while in Germany a synthesis of teleology and mechanism underpinned an equally ambitious research program in biology (Lenoir, 1982). In Britain, too, a generation of "philosophical naturalists" now emerged, committed to the search for a rational order among the diversity of living things (Rehbock, 1983). Here idealism combined with the native tradition of natural theology to give a new version of the argument from design (Bowler, 1977*a*; Ospovat, 1978, 1981).

One of the most important applications of the idealist philosophy was the creation of an overtly developmental view of the history of life on earth. Here the concept of a rational plan of creation was given a dimension of development through time: the plan unfolded in a purposeful way toward a predetermined goal. This model exploited the early-nineteenth-century biologists' fascination with embryology; the growth of the individual organism was thought to provide a perfect example of nature's goal-directed constructive powers. Because the embryo acquires an ever-increasing complexity of structure, it seemed natural to link the process with the hierarchical view of classification to give the "law of parallelism." In the course of its growth, the human embryo was thought to pass through the ranks of the natural hierarchy, starting as an invertebrate and then ascending to the level of fish, reptile, and finally mammal (Russell, 1916; Meyer, 1935, 1939, 1956; Temkin, 1950; Oppenheimer, 1967; Gould, 1977*b*). The development of the human form thus served to link together the whole of the animal kingdom.

The law of parallelism soon was connected with the history of life on the earth to give the foundations of the "recapitulation theory." In it, the ascent of the human embryo through the classes was seen as a repetition of the historical development of life through geological time. This symbolized the universality of nature's plan and confirmed again that man was its central goal. Such a possibility was hinted at by C. F. Kielmeyer (Coleman, 1973) and received its clearest development in the work of Johann Friedrich Meckel in 1821. At first only a speculation, it was inevitable that the idea would achieve greater prominence once the fossil record had begun to confirm the progressive introduction of the classes. Thus the idealist viewpoint helped to popularize the concept of organic development, although within a philosophical framework quite alien to that of Darwinism.

The leading proponent of the idealist interpretation of fossil progression was the Swiss naturalist Louis Agassiz (E. C. Agassiz, 1885; Marcou, 1896; Lurie, 1960). Educated at first in Germany, Agassiz absorbed the idealist philosophy of nature from Oken. A short spell in Paris under Cuvier taught him the necessity of tempering broad speculation with careful observation and led to the study of fossil fish (1833–43) that brought him fame. In 1846, Agassiz traveled to America and was persuaded to stay on as Professor of Zoology at Harvard. He went on to become one of America's leading naturalists, helping to spread the idealist approach into the English-speaking world.

Agassiz believed that the true significance of the fossil progression from fish to mammals could be seen only when it was compared with the process of embryological development. He had studied embryology under Ignatius Döllinger and remained fascinated by the process of growth throughout his career. The development of the human embryo was a goal-directed process by which the organism passed through the hierarchy of classes on the way to final perfection. A lower animal merely passed a shorter distance along the same preordained path. The goal-directed nature of the process gives the naturalist a clue to understanding the same sequence of forms in the fossil record.

> The history of the earth proclaims its Creator. It tells us that the object and term of creation is man. He is announced in nature from the first appearance of organized beings; and each important modification in the whole series of these beings is a step toward the definitive term of the development of organic life. (Agassiz, 1842, p. 399)

Progress is not a response to changing conditions, nor is adaptation the best indication of the Creator's intelligence. Progress is a transcendental symbol of man's unique position in the world, showing him that he is the final goal of a carefully preordained and harmoniously structured plan of creation.

In later writings such as his *Essay on Classification* of 1857 (reprinted

1962), Agassiz qualified his acceptance of the law of parallelism, but he never wavered from his belief that the history of vertebrate life represents the unfolding of a plan of development aimed at man. Of course one could not arrange all the species into a linear sequence, yet there was a central thread linking the hierarchy of classes to man himself. The side branches in the history of life could be seen as variations on the main themes of the vertebrate type, embellishing the basic plan just as a composer will elaborate the theme in a piece of music. For Agassiz, the Creator was a rational, almost artistic, being whose main concern was the overall structure of the plan of nature, not the adaptation of individual species to their environment.

By interpreting the pattern of development as a divine plan of creation, Agassiz made the idealist interpretation of progression acceptable to the English-speaking world. Hugh Miller's *Old Red Sandstone* (1841) expounded the same theme at a popular level. Yet both Agassiz and Miller were convinced that the advance of life worked by discontinuous steps of supernatural origin. They believed that successive populations had been wiped out in turn by geological catastrophes. Agassiz even added a new kind of catastrophe to the geologists' armory—the ice age (translation, 1967). For Agassiz, species were fixed absolutely and only could have a miraculous origin. Yet this commitment to the fixity of species hardly was derived from religious fundamentalism. Mayr (1959*a*) argues that the true source of his typological view of species was his link with idealist philosophy. In such an interpretation, individual organisms are merely physical manifestations of an ideal form representing the essence of the species. Natural variation can never modify this ideal form, so that the only source of new specific forms will be the Creator's will. Other idealists, nevertheless, were able to accept a discontinuous or saltative version of transmutation, and Winsor (1979) has argued that Agassiz's opposition to evolution was stimulated by the observed fixity of modern forms, coupled with his belief that each level of classification formed a distinct category of intellectual analysis forced on the naturalist's mind by the study of nature. Such categories paralleled those in the mind of the Creator and thus were fixed and eternal. At an even deeper level than the typological view of species, idealism shaped Agassiz's view of natural development.

A much deeper level of discontinuity in the advance of life was introduced by Agassiz's insistence that the appearance of each new class represented a step forward into a whole new world. There was no continuous link between the last members of the previously dominant class and the first members of the new one—it was only *within* each class that patterns could be traced out in detail. This confirmed the need for supernatural creation to explain the origin of each new class. Yet the fact that trends could be seen linking the species within each class easily could be taken as evidence for a continuity of development resembling evolution. This was the great paradox of the idealist view of development. Although every effort was made by Agassiz to stress a discontinuous and hence antievolutionary interpretation of how the plan of life unfolded, the very fact of a coherent plan encouraged the belief that it might unfold gradually. If such a continuous development were imagined to work through the preordained transmutation of each form into the next, a kind of evolution theory would result. Although a far cry from the Darwinian view, such an interpretation in fact was to play a significant role in popularizing the basic idea of continuous development.

Robert Chambers's *Vestiges of Creation* (1844) proposed just such an extension of the idealist philosophy of nature, and although this highly controversial book certainly helped to popularize the idea of transmutation, it adopted a very simple interpretation of the divine plan that was supposed to govern the process. Exploiting the law of parallelism, Chambers presented the advance of life as an essentially linear development toward man. Even Agassiz had realized that this was not appropriate, because the branching developments within each class clearly did not lead necessarily toward the next highest type. In fact, the whole basis of the linear view of development now was beginning to break down, allowing greater emphasis to be placed on the image of branching growth. A more sophisticated version of the idealist view of nature played a major part in this revolution, contributing to the development of evolutionism in a subtler, yet equally important, manner. Superficially, the new image of branching development resembled that being developed by Darwin, and the evidence brought forward in its favor was later incorporated into Darwinism. Yet, to the idealists, the more complex process of development they envisaged was still seen as part of a slowly unfolding divine plan.

The first steps in the new direction were taken in embryology, when Karl Ernst von Baer attacked the law of parallelism in the fifth scholion of his *Über Entwickelungsgeschichte der Thiere* (1828; translation in Henfrey and Huxley, 1853). Von Baer maintained that in their efforts to unify nature by analogy with human development, Oken, Meckel, and others had exaggerated the resemblances between early stages of the human embryo and adult structures of the lower classes. There was no real resemblance between a mammalian embryo and the adult form of any lower organism, and the human embryo does not pass through stages in which it is a fish and then a reptile. It *is* true that at an early stage of development the embryos of a man and (say) a reptile will be difficult for the naturalist to distinguish—but this is a resemblance to the *embryonic* not the *adult* reptile and is a far cry from the original claim that the human embryo passes through a reptilian phase. Von Baer's real point was that, at this early stage, the embryos cannot be described as either reptilian or mammalian in form. They simply are not differentiated enough for the distinguishing marks of the adult classes to be visible. At a later point in their growth, these characteristics will appear, and it will be seen that the embryos have developed along different paths. Only at an even later stage will the human embryo take on the characteristics that distinguish it from other mammalian species.

For von Baer, development was a process of specialization, not the ascent of a linear hierarchy toward man. The embryo of any species begins as a simple, very generalized structure and then progressively acquires a higher degree of complexity by adding the specialized organs that will eventually define its adult state. This is why embryos of widely differing forms are similar at first—they do not as yet possess the degree of specialization necessary to identify them. Von Baer had shown that it was impossible to unify all living things by treating them as different stages in the unfolding of a single progressive plan. Nature is far more complex than the law of parallelism allows, although this does not mean that the whole is not governed by a far more complex pattern. There is a directing force driving each embryo toward its final goal, but the goal is different for each species, and the plan of nature must take this into account.

Because lower animals are not immature forms of man, the human embryo cannot recapitulate the history of life on earth. Indeed, von Baer realized that his system threatened the whole hierarchical view of nature on which the very idea of progress rested. Progressionists had assumed that there is an unequivocal hierarchy stretching up through the types and classes to man. This continued evasion of the real implications of Cuvier's taxonomy seemed justified by the law of parallelism, because no one could deny that the final form of the human embryo was "higher" than its earlier stages. Von Baer now had shown that this effort to retain the idea of a man-centered hierarchy was misguided. It was impossible to construct a scale of organization by measuring the degree of similarity to man of different organisms. The human form is just one among many products of embryological specialization, different from but not necessarily superior to any other. Von Baer continued to believe that there *was* a purpose underlying the complex pattern of natural forms, and for this reason he could never accept Darwin's theory of natural selection. It could be argued that von Baer's system, nevertheless, played an important role in preparing the view of nature that made modern evolutionism possible. By undermining the concept of a linear chain of being, he complemented the work of Cuvier and laid the foundations on which later theories of divergent evolution could be built.

Was it possible that von Baer's concept of branching development might provide a better model for understanding the history of life in the fossil record? An early suggestion along these lines was made by W. B. Carpenter (1851) and was exploited further by the great anatomist Richard Owen (Ospovat, 1976, 1981). Of all British naturalists, Owen was most strongly influenced by idealist thought. But his attempt to unify the animal kingdom broke away from the old linear model toward a more flexible system that would prove to be compatible with von Baer's concept of development. For Owen, the rationality of nature's plan could be demonstrated not by comparing everything to man but by seeking the underlying unity beneath the diversity of living forms. He sought the "archetype" or ground plan on which all forms of life, or at least all the vertebrates, are modeled. The archetype was an idealized vision of the simplest form of living structure, from which the anatomist's mind had stripped all the specialized organs required by real living things. Goethe had sought the archetypical plant form in the heyday of German Romanticism and had speculated about the possibility of natural development (Wells, 1967). In France, Geoffroy Saint Hilaire had opposed Cuvier's pragmatic taxonomy by insisting that the similarities of living things had a transcendental significance (1818–22; Isidore Geoffroy Saint Hilaire, 1847; Cahn, 1962; Bourdier, 1969; Appel, 1987). Now Owen set out to justify the same approach on the basis of his own anatomical skills.

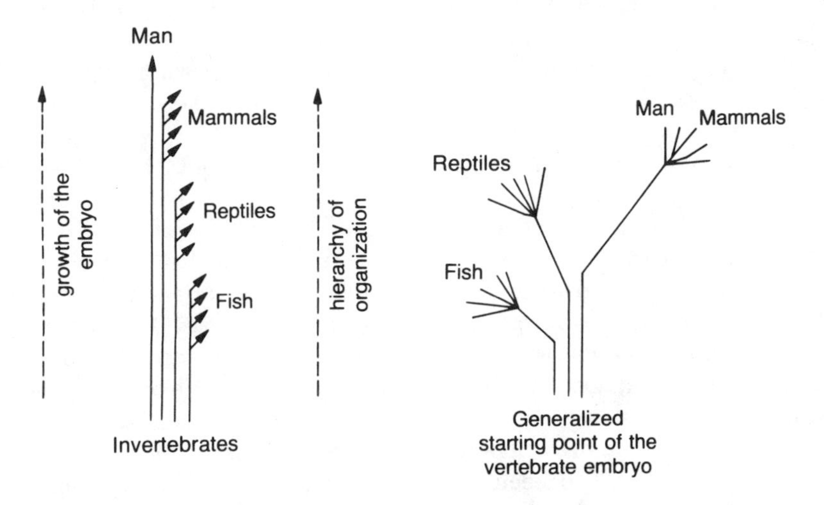

Fig. 16. The Law of Parallelism and von Baer's Law.

In the law of parallelism (left), the stages of embryological growth correspond to a supposedly linear scale of organization leading through the classes up to man. The lines representing the growth of different embryos run in parallel along the same hierarchy, but those corresponding to the lower classes terminate at an earlier stage in the process. In von Baer's law of development (right), there is no linear scale. All vertebrate embryos begin at the same point, but as they differentiate into classes they branch off in separate directions of growth. Further branchings correspond to orders, genera, and species, as the specialization proceeds. Man is not the goal of development, nor is a fish or reptile merely a permanently immature form of man. One class could only be defined as "higher" than another by claiming that as a whole its embryos develop farther beyond the simplest embryonic stage, as indicated here by the longer line drawn for the mammals.

In his *Archetype and Homologies of the Vertebrate Skeleton* (1848), Owen described his idealized vision of the simplest vertebrate form, an imaginary creature that had the essence of the type without any of the specialized modifications required by every real animal. In this way, he tried to emphasize the existence of a principle of transcendental unity within the type, a unity existing at a deeper level of reality than that of the physical world. Yet the result was not just idealist moonshine, because this approach helped Owen to formulate the important concept of homology. He realized that it was necessary to distinguish carefully between what he called "analogies" and "homologies." Analogies are cases in which structurally unrelated organs happen to serve a similar purpose in different animals. Such superficial adaptive similarities have no significance for classification because they are not based on any structural resemblance. Homologies, by contrast, are cases in which the same basic structure is used by various species for quite different purposes. Thus almost every bone in the human hand can be identified with an equivalent in the wing of a bat or even the paddle of a whale. There is a genuine relationship, despite the superficial differences of function, indicating that all three examples belong to the same class, the mammals.

For Owen, the fact that many different forms could be linked by homologies, and the whole type via the archetype, was indicative of the unity underlying the plan of creation. In *On the Nature of Limbs* (1849), he adapted his idealism to the British tradition of natural theology to give a new and more sophisticated version of the argument from design. Followers of Paley studied each form separately to find the adaptation of structure to function that they took as evidence of divine benevolence. Owen now pointed out that the great variety of adaptive structures, nevertheless, was based on a single structural plan for the vertebrates. Indeed, the unity of the type seemed a more important principle, because in some cases efficiency of adaptation had been sacrificed to maintain it. This fundamental unity could not occur by chance, he maintained: it must indicate that the Creator is a rational being who wishes to demonstrate to us the logical pattern of His handiwork (Bowler, 1977*a*).

Owen's understanding of homologies prepared him to accept a concept of branching, as opposed to linear development. Homologous structures are simply different modifications of the same basic pattern, which do not necessarily fall into a linear hierarchy aimed at man. Why should a whale or bat be treated as higher or lower than one another, just because one is adapted to water and to air? Owen thus was able to appreciate the significance of von Baer's view that embryological development is a process of specialization in many different directions. As a paleontologist, he was drawn naturally to the possibility that the same principle might be used to explore trends emerging in the fossil record. Instead of a linear ascent through the classes, perhaps the most important process operating in the history of life was the branching out that took place within each class as it began to explore the many different adaptive possibilities open to it (Bowler, 1976*a*; Ospovat, 1976, 1981; Desmond, 1982).

Owen now could appreciate that there were degrees of specialization, something ignored by Paley's followers in their effort to show that each species was an equally perfect illustration of God's handiwork. Some members of a class are highly specialized for a narrowly defined life-style, while others retain a more general structure allowing a more flexible way of life. Owen's wide experience with fossils enabled him to show that the earliest members of any class to appear in the record are usually of a generalized structure. The subsequent history of the class consists not of a linear ascent toward its "highest" member but of a radiation outward of many lines of development, each leading toward increased specialization (Owen, 1851, 1860). The history of the classs resembles the picture we get by comparing the embryological development of its modern members as described by von Baer's system. There no longer can be a single goal of development, and if one were to equate specialization with "progress," this now would have a very different sense to that implied by the old man-centered transcendentalism.

In 1860, Owen conceded that his work on specialization lent support to Darwin's newly published theory of natural selection, in which the driving force of adaptation would be expected to produce exactly this kind of pattern in the fossil record. Yet Owen did not accept natural selection and wrote such a bitterly critical review of the *Origin of Species* that many historians have dismissed him as an outright opponent of evolution. There is still some debate over the true state of Owen's opinions on this score. Clearly, his earliest efforts were devoted to attacking transmutation because of its apparently materialistic implications (Desmond, 1985). By the 1850s, though, he had begun to appreciate the possibility that the unfolding of the divine plan might occur through a natural process involving an element of transmutation (MacLeod, 1965; E. Richards, 1987; for a contrary view, see Brooke, 1977). As early as 1849, in the conclusion of *On the Nature of Limbs,* he had written of "intermediate causes" or "laws" responsible for production of new forms of life. This suggests that he certainly had rejected the idea of miraculous creation and saw God as delegating this responsibility to some force or cause embodied in nature. Trends in the history of life represented the unfolding of a divine plan of nature, so the "law of creation" would have to embody the Creator's wisdom and foresight. No amount of random variation and selection could explain the existence of such a constructive directing force in nature, and hence Darwin's theory missed the point entirely. Despite the superficial similarity of their views on the fossil record, Owen's idealist conception of a "cause" as an active power in nature was poles apart from Darwin's naturalistic, empirically grounded mechanism of selection.

THE PRINCIPLE OF UNIFORMITY

All the theories we have discussed were built around the general theme of *development* through time. Whether this development was caused by an inherently progressive plan of creation or merely followed from changing conditions in the earth's physical history, there was still a direction in the process defined most clearly by sequential introduction of the vertebrate classes. It was a challenge to this prevailing concept of directionalism which provoked the hottest geological debate of the 1830s. In order to uphold his "uniformitarian" methodology based on gradual change, Charles Lyell found it necessary to revive the steady-state world view proposed earlier by Hutton. Lyell's emphasis on gradual change brought about by observable causes eliminated the need for catastrophes and has been hailed as the foundation stone of modern geology. The most eloquent modern advocate of this view (Wilson, 1967, 1969, 1972, 1980) has suggested that efforts by some recent historians to "rehabilitate" the catastrophists have gone too far. Clearly, Lyell's method of using observable causes wherever possible was a great step forward, because the catastrophists all too often had been tempted to invoke mysterious past causes instead of trying to work out a natural explanation. Yet the steady-state world view gained a few converts and remains unacceptable to modern geologists. Similarly, Lyell's method had great influence on Darwin—yet the modern theory of evolution still contains the concept of development exploited more by Lyell's opponents. Undoubtedly, the principle of uniformity was a major development in nineteenth-century science, but the modern literature on Lyell shows how careful we must be in evaluating its influence (Hooykaas, 1957, 1959, 1966; Cannon, 1960*a*, 1961*b*; Rudwick, 1970, 1971; Fox, ed., 1976).

Lyell came from a wealthy Scottish family and originally was trained for the law. During the 1820s, he became interested in geology and took exception to Buckland's extreme diluvialism. By the end of the decade, his experiences had led him to develop the complete uniformitarian position. An important influence at this time was the work of George Poulett Scrope on volcanoes (1827; Rudwick, 1974a). Scrope was not a uniformitarian because he accepted the cooling earth theory, but he did apply the method of "actualism" in an effort to refute the diluvialists. Actualism claims that causes such as those we can observe in its present state are sufficient to explain the past history of the earth. In the extreme form of claiming that only observed causes *at present-day intensities* are necessary, it is the basic method of uniformitarianism. Scrope found that the (now extinct) volcanoes of central France seemed to have erupted occasionally over a long period of time, just like modern volcanoes. More important, their lava flows had preserved evidence that the valleys below had been eroded gradually—not all at once by a gigantic tidal wave, as the diluvialists asserted. This, of course, would have required vast amounts of time, and this was the lesson that Lyell learned from Scrope: normal causes can produce great effects if only they have enough time to operate.

To test this hypothesis, Lyell traveled to Sicily to investigate the volcano Etna. There he found evidence that the massive cone of the volcano had been built up gradually through a long series of comparatively small eruptions, only the last few of which have been recorded in historic times. The whole structure thus must be immensely old; yet the volcano rests on sedimentary rocks that are very recent by geological standards, because they contain the fossils of mollusks almost identical to those now living in the Mediterranean Sea. Elsewhere on the island the same strata have been elevated to form hills. Given the age of the volcano, Lyell concluded that enough time had elapsed for this elevation to be the result of ordinary earthquakes, not a single catastrophic episode of mountain building. Furthermore, if these were only the most recent sedimentary rocks, the total geological record must span a period of time almost unimaginable in its immensity. If the imagination could be persuaded to take this leap, it would become possible to explain away all of the supposedly catastrophic events that had hitherto been invoked as causes of every large-scale geological change. Convinced of the validity of this view, Lyell returned to England and began writing his classic *Principles of Geology* (1830–33).

Lyell quite deliberately set out to reform geology's scientific methodology (Laudan, 1982). The *Principles* was full of detailed arguments to show that hypotheses based on observable causes offered a better explanation of the phenomena than those based on catastrophes. Lyell's real purpose, though, was demonstrated in his survey of the history of geology contained in the book's introduction. Here he identified his own approach as the inevitable goal of scientific geology, in effect, as the only truly scientific way of studying the earth. The advance of geology had been retarded constantly by temptation to speculate about causes beyond those observable today, and Lyell tried to label catastrophism as an excuse to retain a role for supernatural events in the earth's past. This was most unfair, because in fact, the catastrophists were developing a perfectly natural mechanism based on the cooling earth theory. From their point of view, the catastrophists could admire Lyell's skill in showing how earthquakes and erosion could account for features now present on the earth's surface, but they could see no point in arbitrarily limiting nature to those intensities observed in the comparatively short span of human history. The principle of the uniformity of natural laws was admitted by all as a means of eliminating the supernatural, but Lyell was mistaken in claiming that the same principle could be extended to the complex effects shaping the earth's surface. Geological agencies do not have the same status as laws of physics, because in addition they are controlled by the particular structure of the earth at the time. The laws of nature do not change, but the forces governed by those laws may produce changes in the earth's structure that will affect the level of geological activity—as in the cooling earth hypothesis.

The real crux of the debate comes to light when we perceive the full implications of Lyell's methodology. What was presented as the only truly scientific way of doing geology in fact required postulation of a steady-state world view totally opposed to the directionalism of the catastrophists. If the extreme actualist method is to be applicable, we must be able to assume that no part of the earth's crust will reveal rocks laid down under conditions different from those of today. The only way Lyell could guarantee this was by setting up a steady-state picture to cover all of the time accessible to the geologist. Even the oldest rocks available for study must have been formed under conditions uniform with those of today. For all practical purposes, this implies that the earth can never have changed significantly. It must be treated as a self-regulating system that has been able to maintain itself indefinitely throughout all of the time period into which it is meaningful for us to enquire.

Thus Lyell revived Hutton's steady-state system by suggesting that the earth is maintained in a dynamic equilibrium by a perfect balance of constructive and destructive forces. Erosion by weathering and constant flow of water wears down the land surface to give sediment that will form new strata when it is consolidated on the floor of a lake or ocean. Earthquakes provide a compensating force capable of elevating new mountains and even new areas of dry land to replace those destroyed by erosion, thus preserving the perfect balance through time. For Lyell, who was a Unitarian rather than an orthodox Christian, this ahistorical world view gave a better illustration of the Creator's power and benevolence than the alternative directionalist approach. By eliminating the notion of a beginning identifiable with the biblical creation, one could imagine the Creator as the perfect workman who could design a totally self-sustaining system capable of maintaining itself forever as a habitat for living things.

The steady-state theory did not imply that the earth had to be exactly the same in every period of its history, only that any changes would be cyclical in character—merely fluctuations about a mean. Lyell was prepared to admit that Adolphe Brongniart's evidence from paleobotany showed that the Carboniferous had been a period in which tropical conditions had prevailed over much of the earth; but this was just a temporary effect, not a sign that the whole earth was steadily cooling down from an originally molten state. The kinds of changes envisaged by the uniformitarian could easily generate such fluctuations in climatic conditions, because geological forces could alter the relative *positions*, if not the actual *proportions*, of land and sea (Ospovat, 1977; Lawrence, 1978). If during one period all major landmasses happened to be concentrated around the equator, the world's overall climate would be much warmer than if they were equally distributed between polar and equatorial regions. Constant fluctuations in the positions of landmasses thus would produce changes in climate, but these could never add up to a directional trend.

There has been some dispute among historians over which part of Lyell's viewpoint was more fundamental: actualist methodology or steady-state cosmology. Certainly, the method played an important part in his writings, and there can be no doubt that Lyell felt a genuine need to revolutionize geology by committing it more firmly to use of observable causes. If this is taken to be the most important part of his intentions, it could be argued that he was not really committed to the steady-state world view—he merely adopted it as the only way of providing a foundation for his methodology. It was not that the earth *cannot* have had an origin under conditions different from those of today, only that speculations about such an earlier state were likely to be unfruitful and hence should be avoided by the careful scientist. It must be emphasized, however, that there are some aspects of Lyell's system that seem to imply a more significant commitment to steady-state cosmology, a genuine preference for an ahistorical view of nature.

Lyell assumed that the earth's central heat powered forces responsible for elevating new mountains and dry land. Yet he seemed totally oblivious to the logic of the cooling earth theory: a hot earth must radiate energy out into space and thus cool down. If the earth is hot now, it must have been even hotter in the past, unless one could postulate a source of energy capable of replenishing the internal heat and thus balancing the cooling trend. Such a source of energy would have to operate on a permanent basis, but of course Lyell had no satisfactory way of turning the earth into a perpetual motion machine. His system thus was open to an attack based on the law of cooling and the principles of thermodynamics. Opposition was slow to develop on this front because the science of thermodynamics was still in a primitive state during the 1830s. By the 1860s, however, physicists had made great advances in this area, and in 1868, Lord Kelvin launched a belated but extremely effective attack on the steady-state system (Burchfield, 1975). Kelvin pointed out that because no known mechanism could provide the earth with a constant supply of new energy, the planet *must* cool down. He tried to estimate how long it would take to reach its present condition from an originally molten state and allowed at most a hundred million years. Lyell had made little effort to provide solid dates for geological periods within his system, but Kelvin's estimate was far too low for gradual changes to have produced the results we observe and thus threatened the uniformitarian methodology itself.

Leading catastrophists had always held out against Lyell's uniformitarianism, and now their position seemed vindicated. In the late nineteenth century, the majority of geologists acknowledged the logic of Kelvin's argument and once again began to compress the earth's history into a much shorter time span. Indeed, the logic of that argument was sound and illustrates the artificiality of Lyell's steady-state hypothesis. No modern scientist could accept the idea of an earth that maintained itself *indefinitely* in the same state as today. Only the factual basis of Kelvin's calculations was wrong: he grossly underestimated the age of the earth because he was unaware of the one factor that could provide a really long-lasting source of new energy. In the early twentieth century, it was discovered that radioactivity of certain elements is capable of maintaining the earth's central heat and thereby of providing stable conditions for billions of years. Ever since, geologists again have felt free to postulate very slow processes acting over vast periods of time.

In some respects, Lyell's uniformitarian approach has thus been revived as the modern paradigm in geology. The twentieth century has, however, seen a major revolution against the image of the earth built up by nineteenth-century geologists (Hallam, 1973, 1983; Wood, 1985). The old theories allowed only for vertical movements of the earth's crust, but the modern science of geophysics has confirmed the existence of major horizontal movements. The theory of plate tectonics has confirmed Alfred Wegener's speculations about continental drift—dismissed as ludicrous by geologists as late as the 1930s. Such large-scale movements of the crust are, of course, supposed to take place very slowly. In this sense, uniformitarianism is preserved, although in a context very different from anything that Lyell could have anticipated. Modern theories do, however, limit the extent to which a steady-state model of change can be extended into the past. We now believe that the continents and indeed the earth itself had their origins under conditions very different from those prevailing in recent geological periods. We can even estimate a date for the earth's origin (4.5 billion years ago)—something Lyell would have insisted was quite beyond the bounds of scientific geology.

Of more direct interest to this study is Lyell's extension of his steady-state viewpoint to the history of life. For catastrophists, progressive development of life up to man was the clearest evidence that the earth has not remained in a steady state. This advance, they believed, must signify an irreversible development in the earth's physical conditions. It thus was essential for Lyell to provide an alternative explanation of fossil evidence consistent with his claim that there has been no such directional trend. He had two lines of attack. The first was to point out that his theory of fluctuating climates also could account for basic changes in the kinds of animals and plants living at different times. The Carboniferous indeed could have been a time of tropical, swampy conditions over most of the earth, and under these circumstances the Creator might have decided that a population composed mainly of reptiles was more suitable than one dominated by mammals. The "age of reptiles" thus would be the consequence of a mere temporary fluctuation in conditions, not part of an overall progressive sequence. Perhaps the age of reptiles will return in the future, when the earth's continents once again are positioned to allow the recurrence of similar conditions (Rudwick, 1975)!

The second line of attack questioned the value of fossil evidence for progression. Lyell was able to point out that at least a few mammalian fossils had been found in the very heart of the so-called age of reptiles. The remains of primitive, opposumlike mammals had been discovered in the same bed of oolitic slate that had yielded *Megalosaurus,* the first-known dinosaur. After considerable debate, this evidence was accepted at its face value, making it clear that mammals had existed at least in small numbers during the age of reptiles. Lyell thus could argue that there had been only a change in the *proportions* of the two classes, not an absolute progression from reptiles to mammals, just as his own theory of climatic change would predict. If this point is accepted, how can we be sure that even the age of fish was not inhabited by a few reptiles and mammals? Lack of any fossils to confirm this is only negative evidence and proves nothing, because Lyell could hold out the hope that such finds still might be made in unexplored locations. He thus pioneered the view that the fossil record is only an imperfect representation of the history of life and that significant parts of that history might be concealed from us by lack of evidence. Although he exaggerated the extent of this, Lyell had hit upon a valid point that was crucial for explaining lack of evidence for transitional points in a continuous series. His argument would be exploited in a very different context by Darwin.

In Lyell's steady-state view of the history of life, all classes, high and low, have been present in at least small numbers in every geological period. At first, new evidence seemed to support this radical view. The number of Mesozoic mammals increased, and a reptile was found in what appeared to be the lower Paleozoic. In 1851, Lyell still was able to defend his position in an address to the Geological Society of London, but except for the Mesozoic mammals, his best evidence eventually was shown to result from mistaken identifications. In 1863, Lyell at last conceded defeat, accepting progression as the one factor that would give meaning to the new evolutionism Darwin was popularizing.

The one progressive step Lyell was anxious to admit was the first appearance of man. His religious convictions gave him a profound aversion to the belief that our spiritual character might have been derived from the merely physical world of the brutes (Bartholomew, 1973). Thus it was necessary to treat the appearance of man as a recent event of a unique character. The last thing Lyell wanted was that the introduction of man be downgraded to merely the last step in a progression through the animal kingdom. Physically, he argued, we are not necessarily the highest form of life—our superiority lies in our mental and moral powers, not in our bodily structure. By rejecting the normal idea of progression, he was able to reinforce this gulf between man and the animals, because if there was no real advance in earlier periods, it would not be possible to connect the eventual appearance of man with earlier stages of the history of life.

Lyell was sensitive on this issue because his rejection of catastrophes removed the most obvious guarantee that the appearance of all new species involves a direct act of creation. The catastrophists postulated mass extinction of whole populations followed by creation of new species on a scale that could only be the result of supernatural power; but there were no catastrophes in Lyell's system, so that extinction had to become part of the normal processes of nature. As conditions gradually changed, species would have to migrate or, if geographic barriers prevented this, face slow extinction as they gradually became less able to cope with new conditions. In a steady-state system, such gradual extinctions must be occurring all the time.

Where do the new species come from, to replace those that die out? If extinction is a gradual process, then only occasional creation of a replacement species would be required. It has been said that uniformitarianism in geology cries out for evolution in biology: if the physical conditions change slowly, why not assume that at least some of the species will be able to modify themselves accordingly? This was the conclusion to which Darwin was driven, leading him to search out a mechanism based on natural causes to explain the process. But Lyell applied his actualistic method in a different way. After careful examination, he rejected the one theory of natural transmutation available to him, that of Lamarck. All evidence derived from study of domesticated varieties of plants and animals suggests that natural causes can produce only a very limited amount of change, even when artificially enhanced. Thus evolution cannot be observed in the present, and it should not be assumed to have occurred on a larger scale in the past. In effect, Lyell accepted the traditional view that each species has a carefully balanced structure that cannot be disturbed significantly (notebooks, Lyell, 1970; Coleman, 1962). To break this barrier, Darwin would have to assume (as Lyell had done in geology) that a natural process might operate so slowly that we cannot observe any significant change in a human lifetime.

Lyell's own opinions on the origin of new species made use of the deliberate vagueness of the idea of creation typical of the pre-Darwinian period. He certainly introduced an element of continuity into the process by suggesting that new forms appear from time to time in the regular course of nature. In an exchange of letters with the astronomer Sir J. F. W. Herschel in 1836, he declared his support for "intermediate causes" in the origin of new species (K. M. Lyell, 1881, I, p. 467). This would imply a lawlike process as opposed to direct intervention of the "first cause," that is, the Creator. Yet Lyell had decided that species could not be modified by natural causes and in addition was convinced that adaptation of every species to its environment reflects the Creator's wisdom and benevolence. In modern terms, it sounds as though he was trying to have his cake and eat it: the creation of new species was not a miracle, yet it occurred in distinct acts embodying design. Lyell seems to have been implying that the Creator could build powers into nature that would fulfill His wishes without His direct intervention, although these powers would manifest themselves in a series of discrete events not linked by ordinary laws of nature. The vague notion of such a "law of creation" may seem meaningless today, but it was to become an important vehicle by which many mid-nineteenth-century naturalists were brought closer to the nonmiraculous origin of species.

In Lyell's view, the appearance of man would lie outside the scope even of those special powers delegated for the creation of new animal species. But because the latter process now was to be seen operating from time to time in the regular course of nature, there was a potential danger if it were ever combined with the idea of a progressive plan of creation. Someone who did not share the general reluctance to associate man with the animals might combine the concepts of continuity and progression, giving a system in which man was merely that last stage of the ongoing progression, rather than a totally distinct creation. Lyell was in a secure position since he had denied progression—and now we can see how important it was for him to do this, because it made sure that the process of continuous creation was not aimed at man. Catastrophists, by contrast, accepted progression but denied any continuity in the series of creations. Because neither side could be accused of implying that man was just the highest animal, issues could be discussed in a gentlemanly manner, and there was no question of Lyell's ostracism for unorthodox beliefs. In the 1840s, though, Agassiz's idealist version of progression began to popularize the view that man *was* the final goal of progression. Agassiz himself insisted on the discontinuity of the ascent and the need for distinct creations at all levels. It was only a short time before the elements of continuity and progression were combined in a new way that would destroy the uniqueness of man exactly as Lyell had feared.

THE VESTIGES OF CREATION

The author of this highly controversial synthesis of continuity and progression was Robert Chambers, cofounder of the famous Edinburgh publishing house. Chambers had a strong amateur interest in science and felt that professionals concerned themselves too deeply with details and were incapable of working out a truly comprehensive theory. He sought a grand synthesis that would pull together the whole of geology, natural history, and the moral sciences. Eventually, he found the key to such a synthesis in the belief that everything in nature is progressing toward a higher state, which in biology he took to imply transmutation. Chambers's system was published anonymously as *Vestiges of the Natural History of Creation* in 1844. There was an immediate outcry from the scientific establishment, but the book sold well and there was much speculation about its authorship. To defend his position against the attacks, Chambers drastically rewrote later editions and published a sequel, *Explanations*. The whole affair brought the basic idea of evolution up to man to public attention fifteen years before Darwin's *Origin of Species* appeared (Lovejoy, 1959*c*; Millhauser, 1959; Yeo, 1984).

Although Chambers's book promoted the basic idea of evolution, his conception of *how* the process occurred was in no way an anticipation of Darwinism (Hodge, 1972). Darwin provided a mechanism based on natural causes to explain how a species can adapt to a changing environment. Chambers tried to link the whole of natural philosophy under the all-embracing concept of progression and in biology fell back on the vague notion of a law of creation. He pictured the universe developing inexorably toward a preordained goal in accordance with laws established by its Creator. His book began by explaining the earth's origin in terms of the nebular hypothesis (Ogilvie, 1975), thereby establishing the theme that development could occur as a consequence of natural law. The nebular hypothesis was, in fact, a powerful tool in the hands of nineteenth-century thinkers promoting a developmental model of evolution (Numbers, 1977). Chambers then supposed that the earth's physical conditions had steadily improved through time. Clearly, he was not a follower of Lyell's steady-state cosmology. Yet unlike the catastrophists, he did not translate his directionalism into a claim that the earth has been racked by great upheavals. In order to leave room for gradual evolution, he supposed that the conditions had changed gradually, adopting the idea of continuous change (if nothing else) from Lyell.

Even the first appearance of life was a purely natural process. Chambers revived the old idea of spontaneous generation, citing as evidence certain soon-to-be-discredited experiments in which small insects apparently had been produced by electricity. The law of progress required that there must be certain conditions under which simple living things could be formed from nonliving matter. This particular claim branded him as a materialist in the eyes of most contemporaries and helped to fuel the charge that he was an atheist dedicated to eliminating the Creator's role in nature. Such charges were not really justified, although Chambers's approach to the argument from design was certainly unorthodox. From the simplest forms of life there would have to be a progression toward the higher animals, and a study of Chambers's explanation reveals that he saw the process as the gradual unfolding of a divine plan.

The gradual progression of life was revealed by the fossil record. Chambers devoted a third of his book to a survey of paleontology, slanting all his descriptions to make the advance of life seem as continuous as possible. He argued that each class began with its lowest forms (i.e., those related to the previous class) and gradually rose to the highest (i.e., those anticipating the next class). The first fish were only primitive vertebrates, with armored skin and a vertebral column of cartilage rather than true bone. This showed their relationship to the invertebrates from which they had evolved. The geological history of the reptiles was unclear and did not fit the pattern very clearly, but the mammals began with the primitive marsupials of the Mesozoic (which had so interested Lyell) and rose at last to the highest form: man. Assuming a certain amount of imperfection in the record, Chambers argued that the history of life showed a gradual progression in which each new form could have evolved by a small change from the one immediately below.

This picture of the history of life was based on a sequence of the highest forms known from each geological period and created a linear pattern leading toward man. This resembles Agassiz's idealist system, except for the obvious point that Chambers saw the whole process as a gradual one. The advance of life was aimed in a certain direction, toward a predetermined goal—although Chambers was prepared to speculate that this might lead beyond man toward even higher forms. The idealist element in his thought is also indicated by the whole chapter devoted to W. S. MacLeay's "circular" system of classification, which attempted to show that nature was constructed on a plan of almost mathematical regularity. Ignoring the question of adaptation, Chambers expounded a principle of unity linking all forms of life into a pattern revealing the rational plan of creation through time.

Chambers's primary concern with the ascent toward higher forms left him with a number of problems. Why do we still have such a diversity of forms, high and low, still alive in the present? Darwin solved this problem by picturing evolution as a branching process with no central progressive theme; but Chambers was so fascinated by the idea of linear progress that in his first edition he did not even tackle the question of diversity. In later editions, he admitted that there must be many separate lines of development, but instead of picturing these branching away from one another, he imagined them running in parallel, at different rates, through the same basic hierarchy. This led him to suggest some strange evolutionary connections, particularly as he now insisted that each line of development within a class must begin with a "lower" aquatic form. Dogs, for example, were supposed to be derived from seals, something quite unthinkable in modern evolutionism.

Like Agassiz, Chambers built his view of the history of life on an analogy with embryological growth described by the law of parallelism. He believed that the human embryo passed successively through phases corresponding to the fish and reptiles before turning into a mammal and finally a man. This goal-directed process he saw mirrored in the progression shown by the fossil record. Unlike Agassiz, however, Chambers saw a material connection between the two areas of development, not just a symbolic one. The history of life was a continuous progression like the growth of the embryo and actually had been unfolded via a series of extensions to the growth process. A species with a particular grade of organization has a natural period of gestation during which its embryos advance to the appropriate point on the scale of development. If something could *extend* the growth process for a short period, the embryo would develop a little further up the scale and at birth would appear as a member of the next highest species. The process of evolution thus consists of a long series of small extensions in the period of gestation, each allowing life to advance one step further along the hierarchy of complexity. This explains why that hierarchy serves as the pattern for both embryological growth and the overall history of life.

For Chambers, instances of transmutation were distinct events leading to the sudden appearance of new species. A species does not change by gradual accumulation of minute variations. In fact, it breeds true despite such variations until something sparks off the extension of growth that will allow members of the new generation to appear as the next highest form. Chambers implied that external conditions might catalyze these sudden mutations, yet it is clear that the modifications were not supposed to be adaptive responses to new conditions. The crucial question is: what directs growth during the increased period of gestation? No natural directing force was suggested, leaving us to assume that the change is somehow predetermined. There is only one hierarchy of organization possible within this universe, so that each step in evolution must be the unfolding of one more phase of the plan. Because there is no natural explanation of why such a linear pattern exists, we can see how Chambers's theory emerges as a contribution to the argument from design. The orderly plan of nature's advance is mapped out from the beginning by the Creator. Chambers's main break with Agassiz was to insist that it was not necessary to consider the unfolding of the plan as a series of miracles. Surely a truly rational God would not involve Himself personally in every trivial act of creation—an insect here, a worm there, and so on. Does it not give a more exalted picture of Him if we accept His establishment of the means within nature to fulfill His will without His constant supervision?

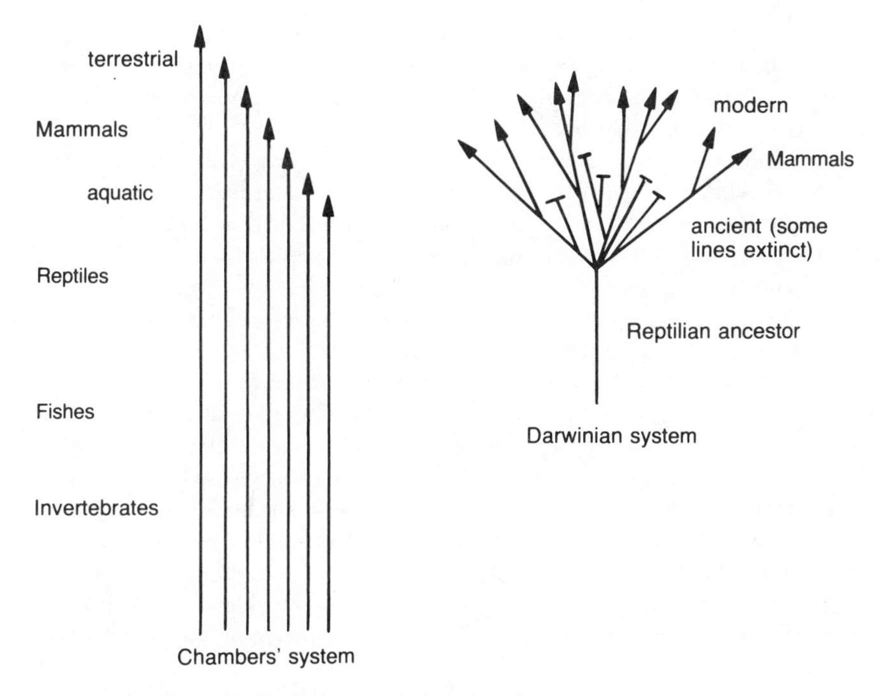

Fig. 17. Chambers's System of Linear Development

In the Darwinian system, all mammals, living and extinct, are derived from one (or at the most, a very few) reptilian ancestor. In Chambers's system, by contrast, the mammals are not directly related to one another—they consist of a series of separate lines that have reached approximately the same level of organization on the scale of development. This is rather similar to Lamarck's concept of progress, except that Chambers was not clear as to why some lines have lagged behind others. Nor did Chambers admit a separate adaptive trend that would account for the diversity of life among many different habitats. Instead, to preserve his basically hierarchical viewpoint, he simply assumed that aquatic forms are the more primitive and come first in the scale of development. How aquatic mammals were derived from reptiles never was explained.

If evolution is really a *designed* process, what is the nature of the guiding powers that God has built into the universe? Sometimes Chambers implied that these powers can be seen operating in the normal course of nature, as when he cited a popular superstition that oats, if ploughed into a field and left over winter, would grow next year as rye. But his more detailed discussions imply that the constructive forces are not observable. During the acts of transmutation, a higher force steps in and produces a change that could not have been predicted on the basis of regular observation. Because this change follows the will of the Creator, one is tempted to dismiss the "mechanism" as only a series of disguised miracles. Yet Chambers was able to argue that such events could be treated as manifestations of a "higher" law built into nature and capable of interfering from time to time with the laws we observe normally.

His authority for this, quoted at some length, was Charles Babbage's unofficial *Ninth Bridgewater Treatise* (1838). Babbage had argued that an apparent miracle could in fact be a product of some higher law, rather than a purely supernatural act. He had invented a "calculating engine"—a forerunner of the modern computer—and noted that he could program his machine to change its operations according to a predetermined plan. Surely God could build such a preordained pattern into the universe which could change the normal laws of nature from time to time in a way that would *appear* miraculous to the casual observer. Chambers's series of transmutations was ideally suited to such an interpretation; the individual acts fitted together to give a rational pattern. By adopting Babbage's position, Chambers turned God into the Great Programmer, who has built the law of progression into the universe where it can unfold through a series of changes to the normal law of like reproduction. The scientist is helpless to investigate the *cause* of such changes; he can see the overall pattern and call it the "law of progression," but he cannot understand the programming mechanism itself.

Had *Vestiges* merely postulated this view of the Creator acting through a mysterious law of development instead of by miracles, it might have been accepted by the more liberal thinkers of the time. But one further implication that Chambers refused to conceal damned him in the eyes of most of his contemporaries. Because the law of creation was progressive, man had to be included as the last step in the process, the highest product of the animal kingdom. Chambers argued quite openly that man's character did not stem from a spiritual quality marking him off from the animals but was a direct extension of faculties that had been developing throughout the evolutionary process. Man does not stand above natural law, since he is both product and member of a totally law-bound universe. His great intelligence is a function of increased brain size, an inevitable product of the general progression in animal organization. To argue that the brain is the organ of the mind, Chambers seized on the ideas of "phrenology," a popular movement in the early part of the century (Young, 1970a; Cooter, 1985). Phrenologists such as George Combe had argued that if the brain is responsible for producing the qualities of the mind, it should be possible to work out which part of the brain is responsible for each mental character. By the 1840s, their system had degenerated into a parlor game in which a person's character was described from the superficial bumps on his skull, on the mistaken assumption that these revealed the well-developed parts of the brain underneath. Chambers, nevertheless, used the basic concept of phrenology to illustrate the materialistic implications of his connection between man and the animal kingdom.

Vestiges sold well, communicating its heresies to a wide audience, but the reaction from the scientific and religious establishments was almost universally negative. Many journals published reviews condemning the crude materialism of the book as a force that should be stopped before it undermined the foundations of religion and the social order. Sedgwick wrote a rambling eighty-five page critique for the *Edinburgh Review* (1845) proclaiming the need to protect "our glorious maidens and matrons" from such poisonous ideas. Hugh Miller, who had already challenged the "development hypothesis" in his *Old Red Sandstone,* wrote *Footprints of the Creator* (new edition, 1861) to refute this latest version of the heresy. It was the status of man that proved the real stumbling block. Even Miller claimed that he would be prepared to accept the notion of God working through law rather than miracle, if it did not inevitably lead to a connection between man and the animals. Man must be a distinct creation, and any theory tending to subvert his unique position was automatically the enemy of true religion.

The first edition of *Vestiges* was easy to attack because of its scientific blunders and oversimplifications, but the chief scientific argument against the book was based on a serious point with important implications for the history of evolutionism. To create his picture of continuous development, Chambers greatly distorted the fossil record as it was known in the 1840s. His opponents pointed out that the progression of life appeared to be *discontinuous*, with new classes appearing suddenly on the scene at certain points in time. There were no signs of intermediate stages by which the first members of a new class could have evolved from the one below. Such instantaneous leaps in the level of organization proved that here, at least, some extraordinary creative power had been exerted. Historians have sometimes dismissed this argument as a product of religious hysteria, as though once the basic outline of life's history had emerged it should have been obvious to everyone that gaps in the overall progression were due to imperfections in the record. But this oversimplifies the situation (Bowler, 1976*a*). While the fossil record has many gaps today, the delight of modern creationists, these discontinuities were far more pronounced in the 1840s, and opponents of *Vestiges* only needed to describe the facts as they seemed then in order to make their case. Darwin took careful note of Sedgwick's attack, because it pinpointed these weak spots in the argument for transmutation (Egerton, 1970*b*). T. H. Huxley, later one of Darwin's chief supporters, wrote a critical review of a later edition of *Vestiges* (Huxley, 1854), objecting to the vague notion of a "law of creation" but also pointing out the fossil difficulties. The discontinuity of the fossil record was real, not just a product of the imaginations of natural theologians.

The situation improved considerably with new fossil discoveries in the 1850s. Partly as a result of this evidence, the possibility that creation might follow some continuous trend became more popular, although few were prepared to argue openly for transmutation. We have noted already that by 1849, Richard Owen was writing of a continuously operating law of development. Baden Powell, professor of geometry at Oxford and a noted commentator on science and religion, ridiculed the idea of miraculous creation and advocated development by law as a better illustration of the Creator's real power (1855). The possibility of continuous development was gaining ground in Germany also, as reaction against the idealist philosophy led to demands for a more realistic view of nature (Lovejoy, 1959*d*; Temkin, 1959). Human evolution still was unacceptable, but the general idea of lawlike development was widely known to the public and was becoming more acceptable in scientific circles.

There was, however, an important modification in the kind of development that now was being recognized in the history of life. Chambers pioneered the idea of transmutation using the outdated linear model of development based on the law of parallelism, but paleontologists themselves began to recognize that the new evidence could not be fitted into a linear model. Some other kind of law would have to be found to describe the complex trends now emerging in the fossil record. Owen's adoption of von Baer's concept of specialization as a model for the history of life already has been described. A similar interpretation was also proposed in Germany by the leading paleontologist H. G. Bronn (1858; partial translation 1859). According to this new approach, the advance of life was an ever-branching process in which each line specialized in some particular way of exploiting the environment. Thus no single modern form could be regarded as the goal of the whole process, nor could the plan of creation be seen as a harmonious pattern.

It would be easy to present these developments as paving the way for Darwinism. Up to a point, that is exactly what they did, but we must not be carried away into assuming that Darwinism was "in the air" and inevitably would have emerged out of the general scientific activity of the 1850s. Certainly, there was now a greater willingness to admit the possibility of gradual development, perhaps even through some form of transmutation—a significant advance beyond the old idea of a series of quite distinct creations. More sophisticated naturalists had even realized that the pattern of development was a complex and branching one, with no fixed goal. Yet all of those we have dealt with so far shared a common belief that the law of creation represented the unfolding of the Creator's purpose. The vague notion of a law of creation was suited exactly to this continued acceptance of the argument from design, and no one was worried that it did not specify the details of *how* life evolved. This was the attitude that Darwin challenged. He had started off in a totally new direction, attempting to explain the process of adaptation through blind operation of everyday laws of nature. The mechanism of natural selection made nonsense of the claim that the end product of universal development was of personal concern to any superintending Power. This is why the eventual publication of his theory came as such a shock to the scientific world and the general public. It raised up a whole new level of materialism to threaten the edifice of natural theology, which up to that point had been modified but never seriously challenged.

At this point, we break off to study the emergence of Darwin's radical new approach to the question. But in so doing, we should ask ourselves if this procedure does not place an artificial emphasis on the role played by Darwin's discovery of natural selection (Bowler, 1988). In many respects, the *Origin of Species* was interpreted in terms of the developmental world view already popularized by Chambers. Whether the model was linear or branching, many naturalists still thought that the purposeful development of the individual organism provided the best analogy for understanding the history of life on earth. As we shall see in our survey of the post-Darwinian debates (chaps. 7–9), such ideas remained popular long after Darwin converted the scientific world to an acceptance of transformism. It is thus possible to argue that the publication of Darwin's theory stimulated but did not really deflect the growth of non-Darwinian, developmental ideas. His more radical insights anticipate the concepts used by modern evolutionists, but in their own time, they were overwhelmed by the general preference for a teleological model of development.

The *Origin of Species* precipitated the conversion of many scientists to evolutionism because it shattered the stalemate that had blocked further elaboration of the developmental view. Pre-Darwinian efforts to construct a theory of gradual or lawlike development had evaded the crucial question of how one species might suddenly be transformed into another. Conservative thinkers preferred the vagueness of "creation by law," while radicals such as Huxley saw no future in the only available theory of natural transmutation, Lamarck's inheritance of acquired characteristics. The philosopher Herbert Spencer had begun to promote Lamarckism in the 1850s to back up his social evolutionism, but significantly, he was unable to interest working biologists in the idea. Some new initiative was needed if scientists were to unite behind the philosophy of natural development. Darwin's theory provided this initiative, allowing radicals such as Huxley to argue that naturalistic evolutionism was possible after all—even if they could not really swallow the whole of Darwin's materialistic vision.

The "Darwinian revolution" may thus have to be reinterpreted as a conversion to evolutionism within a still largely non-Darwinian conceptual tradition. The continuity between pre- and post-Darwinian developmentalism can all too easily be overlooked in our desire to investigate the origins of the selection theory. From the perspective of the many late-nineteenth-century naturalists who rejected natural selection, the "modern" aspects of Darwin's thought represented only an unwelcome stimulus that forced them to accept a developmental model of evolution in order to preserve the traditional view that nature was designed to achieve a meaningful goal. Far from converting the world to materialism, Darwin merely precipitated the general acceptance of the developmental world view pioneered by Chambers, Spencer, and other pre-Darwinian thinkers. The more radical aspects of Darwin's theory would only become widely accepted after the introduction of Mendelian genetics had destroyed the developmental model that Darwin had challenged but failed to overcome.

6

The Origins of Darwinism

Few figures in the history of science have received the same degree of critical attention as Charles Darwin (for surveys, see Loewenberg, 1965; Ruse, 1974; Greene, 1975; Kohn, ed., 1985). In recent decades, Darwin's published works have been reprinted, while historians have edited and published many of his journals, notebooks, and incomplete manuscripts (see bibliography). This, in turn, has stimulated the flow of historical interpretation, as has the availability of the Darwin papers at Cambridge University Library. A project is now under way to reprint the whole of Darwin's correspondence (Darwin, 1984, 1986; Burkhardt et al., 1985). The *Origin of Species* is available both as a variorum text (Darwin, 1959*b*) and with a concordance (Darwin, 1981). The amount of material produced by this "Darwin industry" (to borrow Ruse's phrase) makes the task of providing a coherent, brief account of the origins of Darwinism a difficult one. The difficulty is compounded by substantial disagreements among contributing historians, including a major debate over the influence of external (i.e., nonscientific) factors on Darwin's thinking.

The greatest obstacle to interpretation of Darwin's work is its association with revolutions both in science and in Western culture as a whole. Symbolic of this are the two different kinds of historians who get involved. Some are concerned mainly with scientific developments, often because their own original interests lay in the sciences. Others are historians of ideas who may have varying degrees of patience with scientific details but whose real concern is to assess the role of Darwinism in the overall development of modern thought. From these two backgrounds come two different kinds of history and—because the backgrounds represent alternative values and prejudices—conflicting interpretations of the revolution's significance.

The historian of ideas sees the revolution in biology as symptomatic of a deeper change in the values of Western society, as the Christian view of man and nature was replaced by a materialist one. Willingness of biologists to consider an evolutionary theory is a scientific expression of this wider movement. Darwin frequently is presented as a product of the laissez-faire individualism typical of Victorian capitalism and his selection theory as a biological application of this value system. Such insights are important, but they create problems because these issues still influence our emotions today. To those who approve of the general decline of dogmatic religion, Darwinism is to be greeted with enthusiasm; but those who disapprove of materialism and the rat-race society may judge the scientific developments in a negative way. Two notable modern accounts have presented such a negative image of Darwinism as a movement and Darwin as a personality (Barzun, 1958; Himmelfarb, 1959). Historians more concerned with the scientific issues have argued that these accounts present a distorted picture of Darwin's achievement by failing to take proper notice of his primary concerns.

Not that scientist-historians always treat Darwin as a hero figure. There are complaints about his reluctance to acknowledge the sources of his ideas, which in some cases amount almost to a charge of plagiarism (Darlington, 1961; Eiseley, 1959). In general, though, historians from a scientific background have adopted an oversympathetic attitude. For them, Darwin becomes the ideal researcher, a practitioner of sound scientific method, and the founder of important concepts still proving their worth in modern biology (De Beer, 1963; Ghiselin, 1969). The danger of this approach is twofold. It is all too easy to read unconsciously into the past those concepts that are now familiar and to produce a modernized version of Darwin ignoring the confusion of nineteenth-century thought. In his zeal to show the objectivity of Darwinian method, furthermore, such a historian may find it difficult to admit the influence of any source external to science in the creation of the theory.

If the "internalist" loses sight of the wider implications of Darwinism, the "externalist" may become so involved with the cultural impact that he ignores the scientific issues. Fortunately, there always have been attempts to steer between these two extremes (e.g., Eiseley, 1958; Greene, 1959*a*). The time for a general synthesis may be approaching, heralded by the more recent work that has used hitherto unavailable Darwin material to show just how varied were the influences on his thinking. Darwin was a naturalist, and any effort to understand him that does not take this into account will fail. But a scientist may be open to any kind of influence while he is constructing the hypothesis he measures against the facts. He may thus consciously or unconsciously import philosophical or social concepts into his thinking. Whether or not the result is fruitful will depend on the reaction of both the scientific community and society at large.

A far more serious problem is that of historical causation. The externalist approach has a strong tendency to imply that scientific details are not crucial in determining how events unfold. It becomes easy to imagine that as culture became more materialistic, scientists automatically translated the new values into an interpretation of nature. The scientist reflects the ideology of his social group and creates a world view that will justify it (e.g., Young, 1969, 1971*a*, 1973, 1985). He thus becomes a puppet in the hands of historical forces, free to be creative only within the smaller details of his work. Evolution by natural selection must have been "in the air" in the mid-nineteenth century, waiting for someone to work out the details so the whole scientific community could recognize it. From this viewpoint, it is no surprise to find that a second naturalist, Alfred Russel Wallace, hit upon the theory independent of Darwin. This coincidence reveals the extent to which cultural pressures were steering scientists in this direction.

The internalist all too easily can go to the opposite extreme, treating Darwin as the great hero whose insight finally revealed the true structure of nature. While admitting that lessening of religious pressures created a climate in which the search could begin, this approach stresses the objectivity of the scientific method as a means of gaining pure knowledge. It was Darwin who did the kind of research that led him to put all the facts together, and to claim that he was influenced by ideological factors is pure heresy. If another naturalist hit on the same idea, this shows only that the way to truth is open to all. Once published, the great idea can then play its role in convincing the world that the scientific way of looking at things is better than any outdated religious notion.

Again, it ought to be possible to create something between the extremes. Even the most convinced exponent of the sociology of knowledge can admit that the scientist is just as likely as the social theorist to have the key insight that allows an ideology to be translated into a coherent world view. Perhaps Darwin went beyond his contemporaries in his efforts to create a new materialism, in which case it should be important to know exactly why he did it. Conversely, the internalist should be able to allow for cultural factors in the formulation of a scientific hypothesis—whatever its success as science. By allowing Darwin a degree of creativity within the cultural development of his time, we can set up a balanced view of the revolution he precipitated. An insight out of nowhere could not have had such a great effect. Yet Darwin was working along lines that we can show to be vastly different from those pursued by the majority of his contemporaries. It certainly could be argued that if Darwin had drowned on the voyage of the *Beagle*, the subsequent history of biology—and Western thought—

would have been different. This implies that the scientific details he studied cannot be dismissed as unimportant in the overall march of events, because they are in fact participants in that march along with all the other factors involved.

The attempt to build up a balanced picture of Darwin's work must confront the following issues (for a more detailed analysis, see Oldroyd, 1984):

1. Two of the most important characteristics of Darwin's theory are its utilitarianism and its use of "population thinking" to replace the old belief that each species is modeled on an ideal type. It is utilitarian in the sense that it sees adaptation as the sole driving force of evolution—there is no ultimate goal, just a day-to-day scrutiny ensuring that only those creatures with useful characteristics survive. To what extent does this reflect the utilitarianism of laissez-faire economic theory? Similarly, does Darwin's treatment of each species as a population of diverse individuals reflect the individualism of this economic theory? Does his use of competition as the mechanism of selection show that the theory is modeled on the cutthroat ethics of Victorian capitalism? Questions such as these are characteristic of the old internalist-externalist debate outlined above and at one time had to be discussed almost solely on the basis of indirect evidence. Direct input from external sources is more acceptable now that study of Darwin's unpublished papers has shown the extent of his reading in philosophy and political economy. The same papers also reveal, though, the extent to which his views were shaped by his biological concerns, and there is some evidence that a more synthetic interpretation of the origins of selectionism is beginning to emerge. Thus Schweber (1977) allows a considerable role for Adam Smith's economics in showing Darwin that a mechanism based on individual activity could give a purposeful trend, yet concedes that Darwin's biological thinking had already convinced him that a successful theory would have to take into account individual variation.

2. A major disagreement still remains concerning Darwin's religious views and the rapidity with which he threw off the traditional belief that nature is designed by a benevolent God (Brooke, 1985). Many historians who are studying Darwin's early papers are convinced that by 1838 he already had recognized the materialistic implications of his theory (Schweber, 1979). From this point on, Darwin was doubtful about the prospect of reconciling natural selection with even a watered-down version of natural theology. Any comments in his writings that still refer to the "purpose" of nature's laws are to be interpreted either as wishful thinking or as attempts to conceal the full implications of the theory from his wife or his colleagues. This interpretation of his position must explain how he came to make so radical a break with his cultural environment on this issue or must point to hitherto

unrecognized forces in that environment capable of encouraging a more radical viewpoint. Here again, the sheer extent of Darwin's reading is important (Manier, 1978). In disagreement are a smaller number of historians who are convinced that Darwin did *not* immediately become a materialist (Gillespie, 1979; Ospovat, 1981) and that he continued to try to reconcile natural selection with design by showing that the purpose of the apparent harshness was to ensure the long-term well-being of each species and the overall progression of life toward higher forms. Ospovat argues that the early form of Darwin's theory differed radically from the mature version published in the *Origin* because of this influence. Even these historians accept the fact that as Darwin gradually explored the implications of his theory, he became increasingly pessimistic over the prospect of a reconciliation with natural theology. The advantage of their interpretation is a lessening of the sharpness with which Darwin must be assumed to have broken with conventional views of his time; it allows him to be portrayed as someone who only gradually recognized the full implications of the scientific theory he had devised.

3. What exactly was Darwin's scientific method? At one time it was fashionable to picture him as a simple observer of nature. Opponents of selection regarded him as incapable of deep thinking, while supporters maintained that he was drawn inexorably toward the true explanation by the facts themselves. In later life, Darwin himself encouraged this "patient observer" image, because it would make his discovery seem more legitimate by the scientific standards of his time. We now know that he was anything but a simple fact gatherer. Not only was he well versed in the scientific literature of the period but he also read deeply in philosophy and social theory as a means of grappling with the consequences of his new ideas. The claim that Darwin was essentially a follower of the modern hypothetico-deductive method has been used by Ghiselin (1969) to show that many of the apparently irrelevant branches of research Darwin addressed himself to were in fact test cases for the theory of evolution. During the early years of discovery, we find that Darwin's path toward natural selection consisted of a complex and highly creative process in which ideas were synthesized and tested (Gruber, 1974). There is little doubt that he was deeply impressed by current discussions of the scientific method and did his best to match accepted requirements of a good theory (Ruse, 1975*b*, 1979*a*).

4. Finally, there are debates centered on the detailed scientific factors that shaped Darwin's thought. Traditionally, the key influences were supposed to be (a) the voyage of the *Beagle*, especially the discovery of the Galápagos finches, (b) his acceptance of Lyell's uniformitarian geology, and (c) the analogy provided by the selective process used by artificial breeders. The latest research shows that these influences need to be reexamined and

throws doubts on Darwin's own account of the discovery in his autobiography. Even Lyell has been submerged in the wealth of additional influences recorded in Darwin's early papers, while Sulloway (1982*a*) has exploded the myth of the Galápagos finches. The role of artificial selection also has been disputed by historians who claim that Darwin could have been led to his theory without this analogy (Limoges, 1970; Herbert, 1971). Few scholars doubt that Darwin's unique combination of biogeography and the study of animal breeding helped to generate his radical vision of evolution as an adaptive, open-ended process. Recent work has shown, however, that he explored these apparently "modern" concerns within a context defined by contemporary interests in sexual reproduction as the source of variability (Hodge, 1985). His thought thus had deep roots in an older (and by modern standards) non-Darwinian view of nature. The theory of natural selection did not represent a complete break with the past, however radical some of its components may have seemed. Darwin's "failure" to anticipate Mendelian genetics can now be seen as a direct consequence of his involvement in a pre-Mendelian view of sexual reproduction.

DARWIN: EDUCATION AND VOYAGE OF THE *BEAGLE*

Charles Robert Darwin was born in 1809. His father was a successful physician, while his mother came from the Wedgwood family of pottery fame; and his grandfather was Erasmus Darwin, author of the *Zoonomia*. In his autobiography (first published with some passages omitted in F. Darwin, ed., 1887; completely, in Darwin, 1958), he made no claim to have been a good scholar, although he admitted to an early interest in natural history. Darwin was sent to Edinburgh University to follow the family tradition of medicine, but he was nauseated by the operating theater and soon gave up this plan. It was decided that he should become ordained into the church, and as a step in this direction he entered Christ's College, Cambridge, late in 1827.

It is usual to trace Darwin's real intellectual origins to his Cambridge years: his reading of Paley's *Natural Theology* and his contact with the botanist, John Henslow, and the geologist, Adam Sedgwick. Yet his notebooks reveal a number of earlier influences (Hodge, 1985). It would be ridiculous to claim that Darwin simply borrowed his theory from his grandfather's *Zoonomia* or from the Lamarckism expounded by Robert Grant at Edinburgh. These early contacts with the last remnant of the Enlightenment's more speculative approach to nature, nevertheless, may have alerted him to the problem of generation (sexual reproduction). This would influence his early speculations on species and contribute directly to his own theory of heredity, pangenesis. His environment both at home and at Edinburgh linked him with a more radical intellectual tradition and may account for the comparatively superficial effect of the more orthodox education he received at Cambridge.

Darwin's decision to train for the ministry was neither based on wild enthusiasm nor was it hypocritical. At this time he still accepted the literal truth of the Bible and was quite prepared to follow Paley's reasoning. He read with delight the examples of adaptation used by Paley to demonstrate the Creator's wisdom and benevolence. His attention was fixed on the significance of adaptation in nature, leading him naturally toward the utilitarian school of thought (Cannon, 1961*a*). Within a few years, Darwin would turn the logic of Paley's argument on its head. For Paley, adaptation was a fixed state in which the connection between structure and function was explained by supernatural design. For Darwin, adaptation became a *process* by which species adjust to changes in their environment by purely natural means. Adaptation in this more flexible context was to remain a central feature of the selection theory, although Darwin's sense of the complexity of evolution prevented him from insisting that every structure of every species must have developed for a utilitarian purpose.

Darwin's scientific training at Cambridge was thorough, even if it was gained outside the curriculum. He built up a close relationship with Henslow, while Sedgwick took him on a geological excursion to Wales in 1831 (Barrett, 1974). Darwin seems to have absorbed Sedgwick's catastrophism, although this would soon change dramatically. He had read Alexander von Humboldt's accounts of his travels to various parts of the world (1814–29) and was excited by the prospect of studying natural history in the tropics. Late in 1831 an opportunity arose. The British navy was sending a small ship, H.M.S. *Beagle*, to chart the waters of South America. Her captain, Robert Fitzroy, wanted a gentleman-companion to relieve the monotony of the voyage—hence the excuse for inviting a naturalist who could describe the areas visited (Gruber, 1968; Burstyn, 1975). Darwin was proposed for this position, and after misgivings by both his family and Fitzroy were ironed out, he set out with the *Beagle* on her five-year voyage of discovery (Darwin, 1839, 1845, 1933; Barlow, 1946; Moorehead, 1969; Keynes, 1979).

While the ship's company surveyed the South American coast, Darwin made numerous trips to the interior. He thus built up a body of information that was to change his whole outlook on both geology and natural history. He had been given the first volume of Lyell's *Principles* at the beginning of the voyage, and the second volume reached him in South America. Darwin's observations of South American geology meshed with Lyell's opinions in a way that soon made him a convinced uniformitarian. The one thing that Darwin could not accept was Lyell's complete steady-state world view. Although never a simple progressionist, he did not doubt that there has been an overall direction in the history of life on earth. The vertebrate classes had been introduced one after another in the course of time, and this fact would have to be built into any theory of evolution. Yet Darwin found much to support Lyell's claim that the earth's surface features were formed by natural causes operating at present-day intensities. He thus became a uniformitarian in the modern, rather than the strict Lyellian, sense.

What most impressed Darwin was evidence that earthquakes can produce permanent effects on the land surface. He saw at first hand the ruinous effects of the great 1835 earthquake on Concepción, Chile, and noted that at the same time the neighboring coastline had risen ten feet from its original level. Even more striking was a series of raised beaches, still lined with shells, to be found well above sea level along the coast. Elsewhere, Darwin found fossils of Tertiary shells similar to living species but covered with a considerable thickness of overlying rock, which showed that they once had been carried to great depths. Obviously Lyell was right: earthquakes not only shook the land, they also raised or lowered it to a significant degree. Over a period of time a series of such movements would produce large-scale effects, such as the building of a mountain range. Darwin went on to use Lyell's idea of gradual elevation and subsidence of the earth's surface in a particularly constructive way. He proposed a successful theory of coral reefs (1842) based on gradual subsidence of the Pacific Ocean floor. He was less fortunate in his attempt to explain a curious formation in Scotland known as the parallel roads of Glen Roy (Rudwick, 1974*b*).

In the course of his explorations, Darwin also discovered numerous vertebrate fossils, including giant relatives of the modern armadillo, sloth, and llama. The resemblances of extinct and modern forms showed that there had been a continuity in the development of South American life. To express this fact, Darwin later introduced the "law of succession of types," which stated that there is a close resemblance between the animals inhabiting a particular area in the course of its geologic history. Such a law would be inconsistent with any theory of transmutation in which life was forced to mount a predetermined ladder of creation, as in Chambers's *Vestiges*. When Darwin came to formulate his own theory, he was forced to regard each group as a separate "branch" evolving in its own particular way. Even within each group, there would have to be further branching, because the smaller modern creatures could not have evolved directly from their giant forebears. The larger forms would have to be seen as a now-extinct side branch of the group's evolution, while other branches have continued to evolve into smaller, modern forms. Darwin's theory thus would be conceived within the context of branching rather than linear development. What other naturalists would recognize in the 1850s, Darwin already had accepted in the late 1830s.

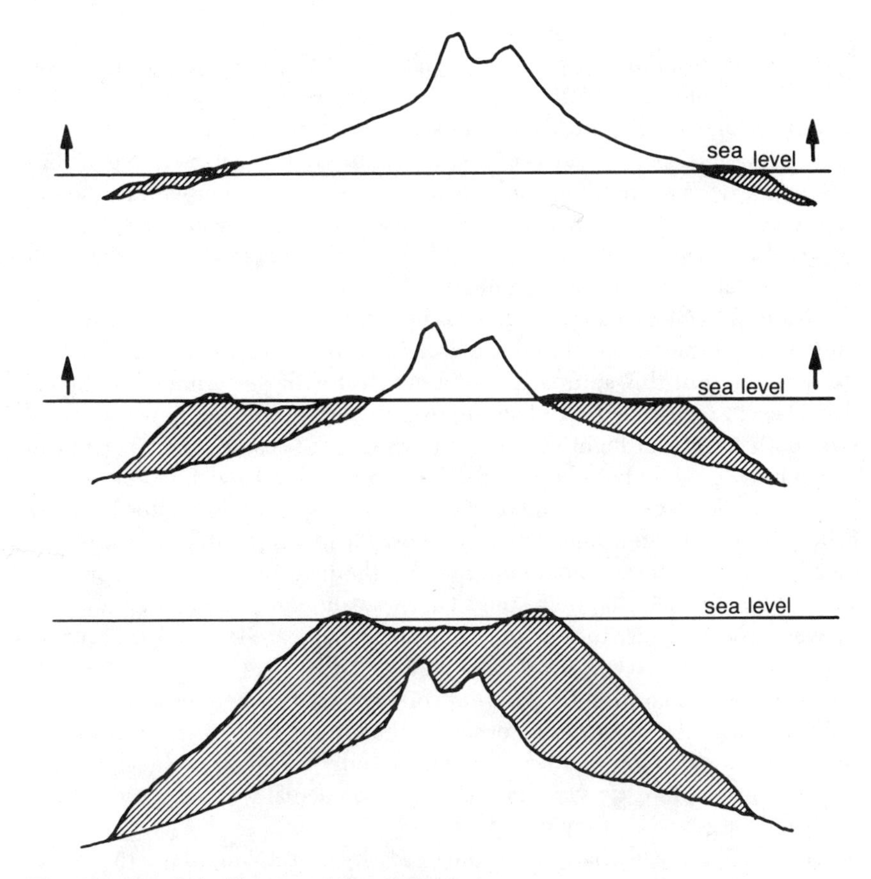

Fig. 18. Darwin's Theory of Coral Reefs.

Coral is formed by minute sea creatures that can live only in fairly shallow water. Darwin's theory assumes that reefs are formed in an area where the land is subsiding, because only in these circumstances could the corals build up large masses whatever the size of the original land foundation. Subsidence must be gradual if the corals are not to be killed off by submersion to a great depth. In the uppermost diagram, an island stands well above sea level and corals begin to build around its shores (dark shading). As the land subsides, the sea level rises against the island, but the corals continue to build up to the surface and eventually form a reef enclosing a lagoon and what is left of the original island. Finally, the whole island is submerged, but the corals have continued to build up to sea level and have now formed a simple reef surrounding an empty lagoon—an atoll. (Adapted from Darwin, 1842)

Fossils were an important clue, but the evidence that actually convinced Darwin of transmutation came from his study of geographic distribution of species (Sulloway, 1982*b*). Darwin was alerted to this topic in reading Humboldt and Lyell, but his own observations on the *Beagle* voyage were to be the crucial factor. Darwin was the first to see that the facts of distribution might be explained by combining a theory of evolution with a study of how species are able to migrate around the world. A new approach to biogeography and ecology thus emerged to form the unique framework within which his theory would be formulated.

Natural theology had provided a simple explanation of geographic distribution. Each species inhabited the area for which it had been designed by the Creator. But this approach would not deal with the complex problems that Darwin recognized. He had discovered, for instance, a new species of the South American flightless bird, the rhea. He became familiar with the common rhea while traveling on the open pampas of Patagonia but did not realize that he had entered the territory of a new species until after his party had cooked and eaten one. Once the new form was identified, a problem emerged: why at some undefined point on the plains did the common rhea begin to give way to the new kind? There was no sharp line of demarcation between the two territories, and the species appeared to coexist over an intermediate area. Even if one assumed that they had been created with life-styles adapted to slightly different conditions, it seemed inescapable that they were competing with each other for the intermediate territory. Darwin was forced to abandon the old idea of a carefully balanced ecology and recognize a less optimistic view in which species actually struggle with their rivals for the territory they occupied.

Humboldt already had shown the necessity of studying the factors limiting a species' population in a given area (Vorzimmer, 1965; Egerton, 1970*a*). Lyell had made an even greater contribution toward establishment of a modern view of ecology (Egerton, 1968). The old idea of a static, harmoniously designed balance of nature was being replaced by the concept of the ecological "niche" that each species exploits within its environment. Lyell saw that the environment could not be uniform over any large area because of geographic factors and postulated that conditions must also change slowly through time due to geological activity. Thus a species could not be adapted perfectly to a single environment, nor could a group of species inhabiting a particular area form a perfectly harmonious system. As Darwin himself saw, the conditions at one point might favor a particular species, while a short distance away a rival species with a slightly different life-style might have the advantage. In between, the two forms would be struggling to occupy territory that was open to either of them. Lyell quoted the botanist Alphonse de Candolle to the effect that all species in a given area are at war with one another. Long-term climatic modifications might alter permanently the balance of power in the area and could lead even to the extinction of an unfavored species (Kinch, 1980).

It was not always easy to see why one species began to be replaced by another at a certain point, as Darwin discovered in the case of the rhea. Presumably, some gradual change in conditions began to favor one over the other. But the belief that species are in competition to occupy as much territory as possible was fundamental in shaping Darwin's later attitudes. Perhaps such a view was encouraged by his observations of how the land-hungry Europeans were exterminating the native South American Indians. Such an input from the social conditions of the time would certainly pinpoint the most important aspect of this new concept of struggle. The most bitter competition in nature is not between a species and its food supply or its predators. It is between the species and its closest ecological rivals—those other forms in the neighborhood (or migrating into it) which have similar life-styles and thus will threaten to exclude it from its livelihood if some slight change in conditions gives them an advantage.

It is a common assumption that nineteenth-century Europe was fascinated by the image of competition (Gale, 1972). Yet there are many different concepts of struggle that must be distinguished if we are to understand their role in the Darwinian theory. When Alfred Lord Tennyson wrote in his *In Memoriam* of "nature, red in tooth and claw," he was certainly challenging the traditional viewpoint of natural theology. His words imply acceptance of struggle as an integral part of nature, not something to be rationalized to preserve the image of a benevolent God. This represents an important change in the intellectual climate, which may have helped naturalists to speculate along new lines, but it serves as only an indirect clue to the origins of Darwinism because it concentrates only on the predator-prey relationship. A lion killing a zebra is "red in tooth and claw"—but the zebra eating grass is also feeding, although no one worries about cruelty to the grass. More important still, Tennyson's words do not reflect the sense of ecological competition that Darwin absorbed from Lyell and his South American experiences. Competition between rivals is truly relentless, with extinction the penalty for failure, yet it may take place without actual bloodshed. Awareness that nature and society might be based on struggle rather than harmony may have served as an important foundation for the changes in natural history. But naturalists such as Lyell and Darwin had to exploit this atmosphere in a constructive way to solve problems, such as those of geographic distribution. They could not simply borrow a poetic metaphor and apply it directly to nature.

The new ecology based on struggle would become an important part of the Darwinian attitude, but it did not provide evidence of transmutation, nor did it supply the exact concept of struggle that Darwin would use as the driving force of natural selection. Despite the breakdown of the balance of nature, Lyell still believed that species were fixed and thus were invariably driven to extinction by changes in environment. Nor did Lyell realize that de Candolle's "war of nature" might be applied *within* each species, allowing struggle between individuals to select out those best capable of dealing with the changes. Something else was needed, first to convince Darwin that transmutation had occurred and then to suggest an extension of the struggle concept that could be applied within the population of a single species.

Darwin eventually was convinced of evolution by his study of how physical barriers such as oceans affected the geographic distribution of species. He was puzzled by the fact that the rhea, for instance, differed significantly from the African ostrich, although both were adapted to a similar life on the open plains. Why do continents such as South America and Africa have such entirely characteristic populations? If adaptation were the Creator's sole concern, one would have expected Him to employ the same species in all areas subject to the same conditions. Lyell suggested that oceans served as barriers limiting the possible migration of species, hence each continent was populated by forms characteristic of its own "center of creation"; but Darwin still could not understand why the Creator acted differently in various parts of the world. Certainly, the oceans acted as isolating barriers, but Darwin wanted a natural explanation of why the populations were different in the first place. Eventually, he was to realize that evolution would provide a real explanation of the phenomenon. Once a continental area was isolated from the rest of the world, there was nothing to force its inhabitants to evolve in the same way as in any other place. They would adapt to changing conditions in their own unique way and gradually would acquire characteristics distinguishing them from the inhabitants of other regions.

It was the study of isolated oceanic islands that really brought the point home to Darwin. Had the Creator performed separate miracles to stock such islands with animals and plants, or had the inhabitants migrated across the ocean from the nearest landmass? The fact that the population normally resembled that of the nearest continent suggested the latter possibility. To confirm it, Darwin later was to begin a long study of dispersal mechanisms, to show that species sometimes could be transported across vast reaches of ocean. Birds could be carried by storm winds and other animals by rafts of vegetation swept out to sea, and by such accidents, remote islands eventually would acquire a population. But the real question arose when it was recognized that the isolated species were not identical with those of the nearest continent, merely related. This would require some form of transmutation by which the isolated population changed, once separated from the parent form.

Darwin's experiences in the Galápagos archipelago were to prove crucial. This group of volcanic islands lies several hundred miles off the Pacific coast of South America, straddling the equator. The *Beagle* spent some time there, allowing Darwin to collect specimens from many of the islands, but it was not until he was on the point of leaving that Darwin became aware of the real puzzle: he was told that local people could tell the island to which a giant tortoise belonged by looking at its shell. On the long voyage home he puzzled over this curious fact, wondering if it could have any more general significance. Ever since David Lack's (1947) study of "Darwin's finches," it has been assumed that these birds, in particular, supplied the crucial stimulus to convince him of divergent evolution under conditions of geographic isolation. According to this view, Darwin realized that a parent South American finch must have diverged into a number of separate species when populations derived from it became isolated on the various Galápagos islands. Each population adapted to its new environment by taking up a different life-style, and as a result it evolved a characteristic shape of beak. Eventually, the differences between the separate populations became so great that each formed a distinct species.

Sulloway (1982*a*) now has shown that the Galápagos finches could not have played such a crucial role in Darwin's thinking. Although they indeed constitute an excellent example of speciation, the situation was sufficiently complex to prevent Darwin from becoming immediately aware of what had happened. In fact, the various beak structures were so distinctive that Darwin did not even recognize the close affinity of the finches when he collected them. It was John Gould, the ornithologist of the Zoological Society, who correctly identified them as a group of closely related species after the *Beagle*'s return. Far from revealing the significance of geographic isolation, Darwin's own specimens were not labeled to show the island from which they came. He was forced to try to reconstruct the group's history from other collections *after* he had become convinced of the truth of evolution. Even then, he tended to oversimplify the situation, because the species in fact are no longer confined to single islands. A further complication obscuring the significance of the case is the absence of any candidate for a hypothetically ancestral form of finch in modern South America.

It was the mockingbirds of the Galápagos that provided the clue to evolution, because Darwin himself could identify several species with obvious resemblances to American forms. Once he was convinced that similar forms on the Galápagos comprised not varieties of a single species but a group of distinct (but closely related) species, Darwin was placed in a quandary. It seemed unreasonable to suppose that each island should have received its own visit from the Creator, who had, for no apparent reason, formed a different species on each. It was far more plausible to believe that a parent South American form had been transported to the islands and had evolved in different directions on each. In the absence of normal competitors, a wide variety of ecological niches had been open, and each population had adapted by specializing for a different way of life. In accepting this, however, it was necessary for Darwin to admit that the process of adaptation to new conditions could go beyond the level of mere varieties to produce totally new species, incapable of breeding with each other or with the parent form.

THE CRUCIAL YEARS: 1836–39

The *Beagle* reached England in October 1836. In March of the following year, Darwin took lodgings in London and began what he later described as the two most active years of his life. He read papers to the Geological Society and engaged in the scientific life of the capital (Rudwick, 1982), and at the same time he pondered the implications of his discoveries on the voyage, especially those of the Galápagos islands. It is suggested occasionally that Darwin had come to accept the transmutation of species during the final stages of the voyage itself. The majority of modern historians, however, believe that the conversion took place after his return, probably between March and July of 1837. Mayr (1977) calls this the first Darwinian revolution: at this point Darwin had accepted the idea that new species are formed by natural processes but had yet to discover the actual mechanism of change. Over the next few years, he recorded his changing views on the possibility of transmutation and its implications in a series of notebooks (now published: Darwin, 1960–61, 1967, 1974, 1979; collected ed., 1987). In September 1838 his reading of Malthus on population finally enabled him to put the basic idea of natural selection together, and from this point on he was committed to a comprehensive reinterpretation of natural history in the light of his new theory.

Darwin later described the events leading to the discovery in his autobiography, although recent scholarship has shown that this account must be revised in light of the notebooks. In the autobiography, Darwin implied that he began collecting facts in the traditional manner of Baconian induction, gathering information at random in the hope that eventually a pattern would emerge. We now know that he always had some hypothesis in mind, trying out several possibilities before hitting upon a successful one. In the autobiography Darwin also claimed that he came across the idea of selection in his study of animal breeding. He then searched for a natural equivalent of this selective action and found it in the struggle for existence caused by population pressure. Some modern studies have suggested that the autobiography greatly exaggerates the role played by artificial selection, while even those admitting its importance agree that the process of discovery was by no means simple. The part played by Malthus's principle also has been debated, some treating it as the crucial insight, others as merely the last step in an almost complete deduction.

As Mayr (1977) points out, there are two ways of trying to reconstruct Darwin's thought processes. One is through a logical analysis of the argument for natural selection, deciding which steps must have been crucial in allowing Darwin to put such a radical new concept together. The other works through a study of the notebooks themselves, attempting to pinpoint what most impressed Darwin at the time and how he used his sources of information. The first approach is useful in that it allows us to see the structure of the theory that eventually emerged and the fundamental assumptions it makes about nature. The orthodox story of the discovery can lay a conceptual foundation on which to build, although such an abstract account misses the enormous complexity of Darwin's actual thinking as recorded in the notebooks and prevents us from being fully aware of the wide range of influences affecting him. Thus the second phase of our study must be a survey of the latest modern scholarship based on actual details of the process by which Darwin was led to the idea of natural selection.

One thing is now clear: Darwin did not borrow the idea of natural selection from an earlier writer. Several naturalists have been credited with anticipating the discovery of natural selection, principally, William Charles Wells, Patrick Matthew, and Edward Blyth (Eiseley, 1959; McKinney, ed., 1971; Wells, 1973*a*, 1973*b*; Beddall, 1972, 1973; Schwartz, 1974). The notebooks confirm the fact that there was no crucial input from these sources, and it is doubtful if any of these so-called precursors of selectionism anticipated the true spirit of Darwin's theory.

We begin with an outline of the basic argument on which natural selection is based. Despite the apparently rambling character of Darwin's later writings, the theory indeed has a logical structure resting on certain verifiable assumptions about nature (Ruse, 1971*b*). Perhaps its most important foundation is a new interpretation about exactly what a biological species is. Darwin pioneered what Mayr calls "population thinking" to replace the old typological view of species. This is a willingness to treat the species as a population of unique individuals united only by the fact that they are potentially capable of breeding together. The species *is* the population, whatever the amount of variation in physical structure between the individuals concerned. It is not defined by some ideal type on which the individuals are supposed to be modeled. Variation is not a trivial disturbance of the ideal form but an essential character of the population and hence of the species. If an external factor favors certain individuals in the population at the expense of others, then the average nature of the population will change and, by definition, so will the species.

The second major factor is a suitable concept of heredity. The individual must be able to pass on his unique characteristics to his offspring. Darwin's early interest in Lamarckism showed a willingness to accept what is now known colloquially as "soft" heredity: the belief that what the parent transmits to its offspring is subject to modification by external causes. But if the environment can control heredity, there should be little individual variation, because all members of a population will have absorbed similar influences. To make a selection theory plausible, Darwin had to abandon "soft" in favor of "hard" heredity: the belief that what the parent transmits to its offspring cannot be affected by outside influences and depends only on what *he* inherited from his parents. This gives a deeper significance to individual variation, allowing Darwin to appreciate how selection could change a population by picking out a certain kind of variation that then will be preserved by heredity.

In his efforts to understand the variability of animals and plants, Darwin began to collect information from breeders and horticulturists. His interest was natural. As a follower of Lyell, it was inevitable that to study the past development of life he would turn to the one area where organic change could be observed today. Animal and plant breeding offered an experimental way of studying the effects of variation. But it is no accident that the significance of both variability and hard heredity were recognized first by breeders, who knew that they owed their success to manipulation of these factors by selection. Darwin always presented natural selection through an analogy with the artificial form. The breeder picks out those individuals in his group which possess something of the characteristic he seeks and breeds his next generation solely from these. He thus isolates the desired characteristic and by selecting further variations in the same direction can improve it in later generations.

The only problem with this analogy is that some force in nature must be found which can substitute for the element of choice exercised by the breeder. Darwin eventually was satisfied that the struggle for existence would play this role, by selecting out those individuals best adapted to the environment. This would provide a natural explanation of adaptive evolution, with no implication of divine supervision. Darwin inferred the selective power of struggle from Malthus's principle of population, coupled with the observed fact that the population of any species must remain more or less constant. Malthus's principle showed that numbers of any species potentially can increase at an exponential rate. Yet both observation and common sense tell us that the population of a wild species cannot increase significantly from year to year, because of the limited nature of the food supply. Many individuals born in each generation must die before their time, because available resources simply cannot support the potential increase in population. From this, Darwin deduced that there must be a constant struggle for existence in nature, as individuals compete to see who will get enough of the limited food supply to stay alive and breed.

By putting all these points together, the argument for natural selection emerges. Given a degree of variation among individuals, it seems obvious that some will do better than others in the struggle for existence and will pass on their advantageous characteristics through breeding. We can develop the argument by reworking Lamarck's famous example of the giraffe evolving its long neck to feed off trees. In the original population of grass feeders, some individuals would by chance have longer than average necks, others shorter. When the grass began to disappear, those with longer necks would be able to reach leaves on trees more easily; because they could exploit the alternative source of food more effectively, they would be healthier and able to breed more readily; their offspring would be more numerous and inherit the extra length of neck. Conversely, those animals with shorter necks would get less food and not breed so easily; in the extreme case they would die of starvation, although a difference in rate of reproduction is all the mechanism requires. It then follows that in the next generation more individuals will come from long-necked parents than from short; the average length of neck in the population will have increased because of the preservation of this characteristic by heredity. If there is continued variation in future generations, the selection process will continue, resulting eventually in the major change in the species, such as the modern giraffe's long neck.

The argument for natural selection seems clear enough in outline, but as presented above it misses one of Darwin's chief concerns: the problem of how a single species divides itself into a number of "daughter" species. In addition, the notebooks themselves reveal that Darwin only came to the idea after a major intellectual odyssey. Once he had become convinced that new species must originate from old ones by natural means, he began to try out a number of hypotheses on how the process might work, measuring each against his own experiences and information derived from his wide reading of scientific literature. At the same time, he was aware that any theory of natural evolution ultimately would have to be applied to man, and he began an extensive study of psychology and social theory in an effort to assess the wider implications of his views. His reading in political economy, including Malthus, formed part of this deliberate program of research. In describing Darwin's path toward natural selection, a principal task must be to determine the relative influence of scientific and nonscientific factors on his thinking. First, however, we must try to identify Darwin's state of mind on approaching these questions, particularly his attitude toward the philosophical and religious implications of his quest for a natural mechanism of evolution.

With his original faith in Paley's argument from design, Darwin must have believed that the Creator was directly responsible for the appearance of each new species. His *Beagle* experiences had turned him away from miracles; the Galápagos discoveries provided a kind of *reductio ad absurdum* of simple creationism. He now was convinced that natural laws were responsible for creating and designing new species to fit new environments. This does not necessarily imply that he became an atheist, however, or that he necessarily must have rejected any concept of design. Many naturalists continued to believe that the laws of nature were themselves designed to bring the Creator's plans to fruition within the material universe. Darwin probably began with such a belief; one of his earliest hypotheses for a mechanism of evolution supposed that variations were produced in direct response to the demands of the environment. Living things automatically changed in an adaptive, progressive manner, a view that would certainly be consistent with the claim that the laws of variation were designed by a benevolent Creator. But what happened to Darwin's faith when he began to realize that there was no such direct response to the environment and that a far harsher relationship must exist between living things and the world they inhabit?

Most historians studying the notebooks have concluded that Darwin rapidly came to realize that the kind of system he now was developing would be difficult, if not impossible, to reconcile with the belief that the laws of transmutation were designed by a benevolent God (Limoges, 1970; Gruber, 1974; Schweber, 1977; Manier, 1978; Brooke, 1985). An early step in this direction may have come when he acknowledged the full implications of the ecological insights gained on the *Beagle* voyage. If there was no balance of nature, only unrelenting competition between rival species that tried to occupy the same territory, extinction would be inevitable for the less successful. When Darwin also began to realize that transmutation must come about through the differential survival of essentially random variants struggling to cope with the environment, he was again forced to question whether a benevolent God would rely on so harsh and apparently uncoordinated a process. Individual competition did not benefit the whole population, because the fruits of progress could be enjoyed only by those lucky enough to survive. This ruthless selection gave a new kind of determinism that did not seem reconcilable with design, and from this point on it can be argued that Darwin became an agnostic, if not an outright atheist. At best, he continued to have occasional lapses of wishful thinking, when he expressed the vague hope that some ultimate good might, after all, come out of all this struggle.

Atheist or not, Darwin certainly realized from the beginning that this new materialism had drastic implications for the status of man (Gruber, 1974; Herbert, 1974, 1977). He began to collect information on psychology and on social questions with the aim of understanding how man's supposedly unique characteristics could have been produced by natural evolution. He moved straight to the position that human nature is not fixed but has been produced by a natural extension of the powers already possessed by animals. His readings in political economy were thus no accident—as he claimed in the autobiography—but part of a deliberate campaign by which he hoped to gain insight into the evolution of human society.

Why should Darwin have been able to develop such a thoroughgoing materialism at a time when the vast majority of his contemporaries still were committed to design and the unique spiritual status of man? The radical intellectual tradition of his family may have helped—ever since Erasmus, the Darwins had been skeptical of formal religion. Recent scholarship also has identified elements within his wide reading that may have encouraged the break with design. Manier (1978) notes the resonance between Darwin's view of nature and the pessimism expressed in a more romantic vein by Wordsworth's poetry. At a crucial moment, his resolution may have been strengthened by an encounter with the Positive philosophy of Auguste Comte (Manier, 1978; Schweber, 1977). Comte argued that science would reach its highest state when it abandoned theology and tried to understand the universe solely in terms of observable causes. Although Darwin was introduced to this philosophy through a disapproving review by David Brewster, it well may have encouraged his own break with religion.

An alternative interpretation of Darwin's religious views is expressed in Gillespie (1979) and Greene (1981). According to this reading, Darwin did not abandon design but realized that the Creator had decided to work in a less directly obvious way. His references to the laws of nature achieving a higher purpose through evolution are genuine expressions of his faith that, despite its superficial air of harshness, natural selection does work for the good of all living things. Adaptation, even progress, *are* brought about in the long run, and the Creator's intentions fulfilled. Ospovat (1979, 1981) even has suggested that in Darwin's early formulations, the selection mechanism was intended explicitly to bring about a state of "perfect" adaptation, in which there was no need for further struggle until the next change in the environment occurred. Several writers have drawn attention to Darwin's use of the analogy of a "Being" who superintends the operations of natural selection (Young, 1971*a*; Manier, 1978). This Being takes the place of the human breeder in artificial selection, and by using such language Darwin invites us to imagine that the Creator is really in charge of natural evolution. It was all too easy to think of nature as a conscious selecting agent and thus as an arm of divine providence. In his more hard-headed moments, Darwin certainly appreciated that this was only an anthropomorphic way of describing the effects of struggle, yet his decision to use this analogy may reflect an inability to throw off the influence of the design argument.

Whatever position we take on this issue, it is important to note that by the late 1830s, Darwin had broken with the interpretation of natural theology still accepted by the majority of his contemporaries. Even if he did believe that selection could be reconciled with design, his mechanistic interpretation meant that one no longer needed to use the Creator's intentions as an explanatory tool. Natural selection would work whether or not one believed that the Deity continued to keep an eye on things, because it depended solely on the operations of the everyday, deterministic laws of nature. To find a theory with this characteristic, in fact, had been the driving force of Darwin's investigation. At an early stage in his research, he had decided that the origin of species must be treated as a purely scientific problem that could be answered without appealing directly to the Creator's guiding hand. The search that ended with the discovery of natural selection was inspired by his desire to find a theory that would provide a scientifically acceptable solution to this problem. If this sounds today like a natural extension of Lyell's uniformitarian method into the organic world, it was an extension that Lyell himself deliberately refused to allow. Indeed, the majority of naturalists at the time believed that some form of higher power *did* interfere with the normal operations of nature to generate new species, putting the phenomenon outside the range of scientific investigation.

Looking in more detail at Darwin's discovery, one must be aware that he was acutely conscious of the need to follow a method of investigation that would be seen as fully scientific. Precisely because he was extending science into an area that his contemporaries thought unsuitable, he was determined to minimize the risk of being criticized on grounds of inadequate methodology. His later emphasis on the fact-gathering side of his work was intended to show that he was not a mere speculator, someone who rushed into wild theorizing with an insufficient basis in hard facts. The methodological debates of the time clearly acknowledged a role for theorizing in science, and Ruse (1975*b*) has argued that the writings of Sir J. F. W. Herschel and William Whewell were particularly important to Darwin. Herschel (1830) emphasized the need to balance theory and experimental work. Whewell pointed out that a powerful scientific theory achieved its status through its ability to link diverse areas of study, as in the case of Newton's theory of gravitation. This made Darwin aware of the need to expand his own theory's explanatory power, especially once he realized that he had no adequate understanding of the causes of individual variation.

The creative nature of Darwin's search for a mechanism of transmutation has been stressed by Gruber (1974). The most original aspects of his thought clearly derived from his efforts to grapple with the problems of biogeography (Richardson, 1981; Hodge, 1982). Darwin began with the conviction that evolution is a branching process, that one species gives rise to a number of descendants under the influence of geographic isolation. Grinnell (1974) suggests that Darwin's first hypothesis was simply to assume that isolation by itself produced speciation. But this ignored the need for a mechanism that would make the changes adaptive for new environments to which isolated populations were exposed. By July 1837, Darwin was determined to explore the possibility that ordinary individual variations are the key to evolution, by a process of accumulation over many generations. He suspected that such variations indeed are inherited, and to confirm this he decided to investigate their cause.

One of the most important new insights into Darwin's thinking has emerged from the work of historians investigating his early writings on "generation," or reproduction (Kohn, 1980; Hodge, 1985; Sloan, 1985). Without denying that the most original aspects of his thought derived from his confrontation with biogeography, it is now possible to see that throughout his career he treated evolution as a process mediating between the environment and variations produced by the reproductive systems of the organisms making up the population. Darwin's own theory of heredity, "pangenesis," was conceived during these early speculations, although it was not published until much later. Modern biologists draw a fairly sharp distinction between growth and heredity: evolution is a function of how new genetic characters enter the population, and the question of how those characters are produced in the growing organism is of secondary importance. Darwin, by contrast, saw variation as a disturbance of the organism's growth process. Like most of his contemporaries, he saw both variation and inheritance as functions of an integrated process of reproduction and growth. Evolution was an interaction between the environment and the generation of new organisms.

As Hodge (1985) points out, this gives us a very different picture of the kind of science Darwin was doing—a picture that emphasizes the pre-Mendelian character of his thought. An older generation of historians expressed dismay at Darwin's failure to discover the laws of genetics, but we must now accept that his work was conditioned throughout by a conceptual system that did not make distinctions we take for granted today. It is no longer possible to treat Darwin's theory as a complete break with the older tradition in which naturalists such as Buffon had investigated generation as a clue to the origin of life. Early-nineteenth-century speculations on generation may also have helped pave the way for Darwin's materialism (Sloan, 1986). His lifelong commitment to "blending" inheritance and his refusal completely to break with Lamarckism must be seen as inevitable by-products of this pre-Mendelian approach to reproduction. Looking forward to the reception of the theory, it may also be possible to argue that Darwin's failure to dislodge the view that evolution is a purposeful process modeled on individual growth may have been a consequence of his own loyalty to the belief that growth is the ultimate source of variation.

Among the ideas that Darwin explored before he came to natural selection, one of the strangest was the possibility that new species are "born" with a built-in life span, after which they become extinct. He soon moved on to the assumption that external conditions affect the reproductive system of individuals so that their offspring are adapted to any changes in the environment. He also speculated that new habits might affect the species (Grinnell, 1985). We can no longer rule out the possibility that these ideas may have been inspired by his knowledge of Lamarck's theory. The most crucial step in the development of his own theory came when he abandoned such notions in favor of a less direct link between the environment and the organism. He never gave up the belief that variations are caused by external conditions disturbing the process of individual growth, but he soon began to suspect that the majority of such changes would be random rather than purposeful.

Darwin thus abandoned the view that individual growth can direct evolution along purposeful channels and began to search for a mechanism that would pick out those variations that were by chance of some use to the organism. Forced to admit that his study of the cause of variation was blocked, he simply accepted the existence of random individual differences as an observable fact that legitimately could be built into a theory. By speaking of variation as "random" he signified not only that it occurred in all directions (useless as well as useful) but also that its causes were beyond the scope of immediate investigation. Darwin thus pioneered a new approach to scientific explanation that would become characteristic of the later nineteenth century (Merz, 1896–1903). Use of the populational approach required acceptance of an explanation based on factors that only could be described statistically, bypassing the need for reducing everything to absolutely fixed laws. As with the kinetic theory of gases, the Darwinian theory required that the laws used by the scientist must be seen as merely the average effect of a vast number of individual events, each of which is subject to causation, but at a level impossible to describe.

Darwin's conversion to a hypothesis based on random individual variations was associated closely with his move to investigate the work of animal breeders (Secord, 1981). Here he found confirmation of his belief that individual differences are inheritable. He also must have become aware that the breeders' success lay in their ability to pick out or select just those variations that happened to suit their purposes. In the autobiography, Darwin claimed that this was the key that opened up his search for a natural equivalent to artificial selection. It has been argued recently that the analogy cannot have been crucial, because the notebooks themselves make only a few references to the selection process (Limoges, 1970; Herbert, 1971). According to this view, Darwin later found artificial selection so useful an analogy when describing this theory that he eventually came to believe that it had helped in the original discovery. There *are* a few references to "picking" in the notebooks, however, and it is difficult to believe that Darwin's mind was not prepared to accept a mechanism based on the same model (Ruse, 1975*a*; Mayr, 1977; Cornell, 1985; Evans, 1985; Hodge and Kohn, 1985).

By the summer of 1838, Darwin's biological investigations had brought him a long way toward the theory of natural selection. He knew that transmutation must occur through changing the proportion of varying individuals in an isolated population, presumably through the environment picking out those with a useful character. It was at this point, according to the autobiography, that he read Malthus and suddenly realized that the pressure of population must create a struggle for existence in which only the fittest individuals would survive and breed. This leads us to the hottest debate over the development of Darwin's ideas. Did Malthus merely serve to drive home the effectiveness of a selection mechanism that Darwin had already put together on the basis of purely scientific influences (De Beer, 1963)? Or does the population principle symbolize the ideological content of the theory, showing how he translated the competitive ethos of Victorian capitalism into a principle of nature (Young, 1969; Gale, 1972)?

The situation is now complicated by the fact that the notebooks have revealed other nonscientific influences in Darwin's reading at the time (Manier, 1977; Schweber, 1977). Brewster's review of Comte showed him the need for a theory based on mathematical principles, thus creating a role that would be played by the arithmetical logic of the population principle. Darwin also was reading other material on political economy, and the literature on Adam Smith and laissez-faire seems to have helped convince him that it was possible for a mechanism based on individual interactions to give a purposeful overall trend. Looking for a way of measuring variation, he turned to the work of the Belgian anthropologist Lambert Quetelet (translation, 1842), who pioneered the application of statistics to the human population. For any characteristic, Quetelet showed that there was a range of variation between two extremes, with most individuals clustered around the center of the range—the phenomenon we now represent with a frequency distribution curve. Here was an excellent illustration of variability and population thinking, treating a species (man himself) as a group of diverse individuals rather than a single unified type. In addition Quetelet discussed Malthus and thus triggered Darwin's crucial reading of the population principle.

Those historians who portray Malthus as merely a catalyst, pulling together various strands of thought already in Darwin's mind, argue that it was the mathematical force of the population principle that converted his vague ideas into a coherent theory of natural selection. Darwin already realized that adaptive variants must do better than others, and his ecological insights had made him aware of the competitive element in nature. Malthus enabled him to see that a perpetual struggle for existence must go on *within* each species, eliminating the less fit individuals (Herbert, 1971). It was not just a case of the fit having a minor breeding advantage: the struggle for existence exerted a major selective pressure at all times by preventing the unfit from ever becoming mature enough to breed. The strong emotional effect of this insight on Darwin was confirmation of his long-standing suspicion that death indeed played a creative role in the world (Kohn, 1980).

The need for a balanced account of the factors influencing Darwin is stressed even by those modern accounts that lay most emphasis on "external" forces (e.g., Schweber, 1977). The notebooks confirm just how far he had come toward natural selection on the strength of his biological investigations. His readings in laissez-faire economics merely reinforce the population thinking already derived from animal breeders. Malthus did not come as a sudden insight that revealed the extent to which struggle was inevitable within society and hence within nature. Yet those who have suggested a more significant role for Malthus have a case that does not rest on the notebooks alone, because Malthus symbolizes what is perceived as a more pervasive ideological influence on Darwin's thought. Even if only a catalyst, the concept of individual struggle seems to have gained great strength in Darwin's mind as a result of reading Malthus. Whatever its origins, we still may ask why the struggle metaphor appealed so strongly to him and whether its appeal owed anything to a sense that his own society itself was governed by individual competition.

The claim that Malthus did not create the image of nature based on struggle in Darwin's mind is reinforced by a more careful study of Malthus's own account of his principle. Many historians have assumed that Malthus himself deduced that population pressure must lead to a struggle for existence (Vorzimmer, 1969*a*; Young, 1969). Yet the link is not a necessary one, because there are other conceivable ways in which excess population could be eliminated. Malthus used the phrase "struggle for existence" only in his discussion of savage tribes—in his own society he thought the pressure of population could be avoided by educating the poor to have fewer children (chap. 4; Bowler, 1976*b*). Thus Darwin seized upon a comparatively small aspect of Malthus's system and ignored the chief lesson of the *Essay on Population,* which was that the principle could be reconciled with the traditional laissez-faire assumption of a balanced society.

If Malthus did not intend to stress the competitive nature of capitalist society, we must ask why Darwin assumed that struggle must be the inevitable consequence of population pressure. One answer to this question is a subtler version of the ideological connection that has long been urged by those who wanted to see Malthus himself as the chief proponent of struggle. Darwin lived in a society in which individual competition was increasingly acknowledged as the driving force of economic progress. Malthus and the laissez-faire economists may have tried to rationalize this situation by claiming that the competition was designed for the good of all, but ordinary people were probably well aware of the true state of affairs. With the success of the industrial revolution, the commercial classes had a new sense of their own achievements and now were prepared to accept their prosperity as a reward for individual initiative—a reward that was deservedly withheld from those less active or able. Such views are certainly apparent in the general literature of the time, if not in the formal tracts on political economy (Gale, 1972). Darwin may have unconsciously absorbed this more ruthless attitude so that his mind instinctively was prepared to see the struggle for existence as the natural way in which stability of the population would be maintained.

The modern study of Darwin's notebooks has revealed the enormous breadth of the influences that affected his thinking, but it will not resolve the debate over the ideological origins of natural selection. What some historians consider to be the most important external influence may have been transmitted in so indirect a manner that we may never be able to confirm its presence. The link between Darwin's theory and capitalism is too pervasive for it to be detected in the notes recording his conscious thoughts. It rests instead on the general congruity between the Darwinian image of struggle and the realities of nineteenth-century society. The historian must choose whether he will accept the logic of this indirect argument or concentrate on the wealth of scientific insights revealed by the notebooks.

DEVELOPMENT OF THE THEORY: 1840–59

After putting together the outline of a naturalistic theory of species transformation, Darwin spent the next twenty years exploring its details and expanding its explanatory powers. In 1842, he wrote out a short sketch, followed two years later by a substantial essay that was meant to be published if he should die (reprinted in Darwin and Wallace, 1958). He had no intention of publishing yet, however, especially as the reaction to Chambers's *Vestiges* had revealed the strength of public and scientific opposition to transmutation (Egerton, 1970*b*). Darwin only gradually let a few friends into the secret, including Lyell and the botanists Joseph Hooker and Asa Gray (Colp, 1986). Informally, he was trying to build up a community of scientists who would speak the new language of evolution (Manier, 1980). Only in the later 1850s did he begin to write a large-scale work with the aim of eventual publication (Darwin, ed. Stauffer, 1975). In 1858, his hand was forced by the arrival of Wallace's paper on natural selection, and he began to write the shorter account that appeared as the *Origin of Species*.

This work was carried out under conditions quite unlike those of the earlier years. Darwin married in 1839 and moved to Down House, in the countryside of Kent. Soon he began to develop the debilitating illness that was to remain with him for the rest of his life. The nature of this illness is uncertain. At one time it was thought that Darwin had picked up a nervous ailment transmitted by a South American insect. More recently, it has been suggested that he was poisoning himself with patent medicine (Winslow, 1971) or that the symptoms were the result of psychological stress (Colp, 1977). The illness limited Darwin to a few hours work a day and prevented him from entering public life except within his local community.

Modern historians have focused attention on two issues in the development of the theory during this period. One concerns the relevance of numerous projects in natural history that Darwin undertook, especially his major study of the barnacles (Darwin, 1851–53). We now can see that these projects were not merely sidelines, taken up so that Darwin could establish his reputation as a competent biologist. On the contrary, they were used directly to test the theory of evolution and work out its consequences for general biology (Ghiselin, 1969; Gale, 1982). Even more crucial were developments taking place within the theory itself. As expounded in the 1844 "Essay" it was *not* complete, and much attention recently has begun to focus on the additions he was forced to make. Darwin began to realize that he would have to relate natural selection to broader trends in the history of life, particularly the constant branching and specialization revealed by many trends in the fossil record. Only when he had explained this through his "principle of divergence" did he feel confident enough to begin his big book on the species question. In the meantime, his thinking on the role of factors such as geographic isolation had changed dramatically.

Darwin's interest in the barnacles was stimulated by the discovery of some unusual specimens while on the *Beagle* voyage. He now saw that the preparation of a complete description of this little-studied subclass would be an ideal way of testing his views on evolution in practice when applied to morphology and classification (Ghiselin, 1969; Ghiselin and Jaffe, 1973). His intentions were not made obvious in the published monograph, but in effect Darwin was exploring the effect of evolution on biologists' understanding of natural relationships. When his theory finally was published, he would have to be able to show that the resemblances defining each taxonomic group are the result not of some mystical archetype but of common descent. Evolution allows the Linnaean system of classification to be interpreted as a representation of groupings produced by branching, in which the basic character of the common ancestor is retained by its superficially modified descendants. Darwin also had to recognize that there could be different degrees of change among the branches stemming from a particular source. Some branches might become modified to such an extent that we no longer wish to include them within the genealogical group for taxonomic purposes. Disagreements over what level of difference is required before a new Linnaean group is defined thus would be inevitable, giving rise to those periodic debates among taxonomists.

The study of barnacles also alerted Darwin to the amount of variation that could exist within a species. One of the scandals of mid-nineteenth-century natural history was the inability of naturalists to agree on whether a group of closely related forms were varieties of a single species or separate species in their own right. Technically, the question should have been decided by whether or not the populations could interbreed; but this was often impossible to decide in practice, and naturalists were forced to decide by the degree of morphological difference. Some lumped together all closely related forms as varieties of a single species, while others tended to make a separate species for each distinct form. Darwin was anxious to show that species normally divide themselves into varieties, because he believed that this was the first stage in speciation (the production of new species). Varieties are, in effect, incipient species. Because the buildup of reproductive isolation that eventually defines the species is gradual, Darwin was able to explain why naturalists so often disagree on the status of varieties. In many cases, the process of divergence will be at an intermediate stage, where it will be extremely difficult to decide the true status of the separate populations.

The sterility criterion had been the cornerstone of the creationists' belief that species are real entities formed by God, while varieties are local and transitory products of nature. It sometimes is suggested that by allowing one species to change gradually into another through an intermediate variety,

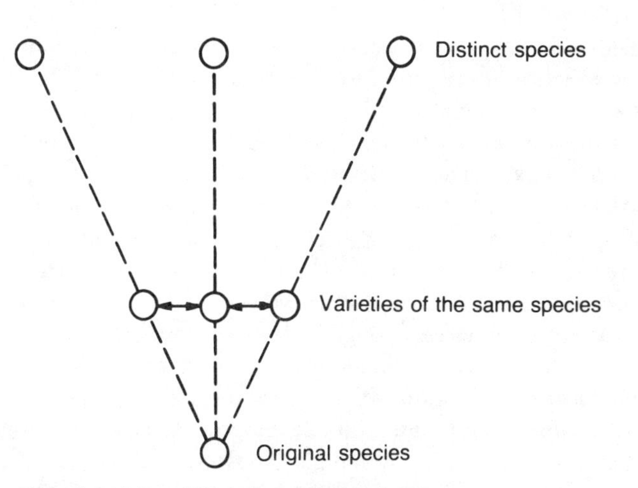

Fig. 19. The Relationship of Species and Varieties.

In Darwinian theory, varieties are "incipient species"—if they survive and continue to change, they eventually may form distinct species of their own. The diagram shows a single original species diverging into three later ones. The first stage of the process forms three different populations that are merely varieties of a single species. They may not normally do so, but members of each group are physically capable of interbreeding successfully. At a later stage, differences between them are so great that interbreeding cannot take place. The populations are now completely distinct and can be counted as separate species. But there is no clear dividing line between species and varieties. The process of divergence is a continuous one in which the possibility of interbreeding gradually diminishes.

Darwin's theory turns the species into an arbitrary figment of the taxonomist's imagination. Indeed, the principle of continuity can have this consequence. Lamarck's theory, for instance, had not so much explained the origin of species as destroyed the species concept altogether. Each apparently distinct form was supposed to grade insensibly into others, so that the taxonomist had to draw the line defining "species" according to mere convenience. This is not a consequence of continuous change in a system of divergent evolution, however, because as the branches separate, a morphological gap opens up between them. Darwin thus could hold that species are real, that is, distinct entities, although there is obviously no underlying essence fixing their character (Kottler, 1978). In modern Darwinism, species are equated with distinct breeding populations, and some commentators think that Darwin himself had gone a long way toward this kind of population thinking. It is agreed more generally, though, that he was never able to shake off completely the traditional method of defining each species by its morphological character (Beatty, 1982).

Darwin's interest in the process of reproduction was stimulated by his discovery that, although most barnacles are hermaphrodites, a few forms have males that are minute parasites upon the female. He also discovered a case in which a hermaphrodite form had dwarf males. He deduced that the whole group originally had been hermaphroditic and that some forms evolved separate males later on (Ghiselin, 1969). The case of the hermaphrodite with separate males illustrated how his theory could explain the well-known problem of vestigial or rudimentary organs. In this case, the male parts of the hermaphrodite had become superfluous and were in a state of degeneration. Many other forms exhibited such vestigial structures; a well-known example is the teeth of certain whales, which are never used for feeding and do not grow large enough to break through the gums. If Darwin's theory were true, one could not argue that these structures were being prepared for *future* use: natural selection can only enlarge a structure if it has immediate value at all points in its development. This point was crucial for the elimination of teleology, although it created a major conceptual problem when Darwin tried to imagine all the intermediate stages through which a complex organ such as the human eye must have passed. In this case, however, he was reassured by the fact that various grades of complexity in eye structure are found in a wide range of animals, showing that each level of development is perfectly capable of functioning to some effect. Vestigial organs, by contrast, had to be those that once had been of some use but which now were degenerating because they served no purpose and were a waste of energy for the organism to produce. It should be added that Darwin was prepared to consider that the degeneration might be produced by the inherited effects of disuse, because he always admitted that Lamarckism might play a subordinate role in evolution.

Turning to the developments that occurred within the theory of natural selection, we must first note a disagreement over the state of Darwin's thoughts in the early 1840s. It normally is assumed that by this time he already had come to see natural selection as an unrelenting force, the struggle for existence constantly promoting the fit at the expense of the unfit, even in a stable environment. But Ospovat (1979, 1981) has argued that in its early form, the theory had not yet outgrown its origins in Paley's natural theology. Even in the 1844 essay, Darwin still implied that species normally exist in a state of perfect adaptation, with little or no individual variation, where the struggle for existence is unnecessary. Only when the environment changed did individual variation appear, providing natural selection with the raw material it needed to change the species until it was once again in a state of perfect adaptation to the new conditions. According to this interpretation, Darwin only gradually over the next decade came to appreciate the full significance of population pressure, which would generate a struggle for existence whatever the external conditions or the degree of individual variation. He thus moved slowly toward the mature version of his theory, in which evolution is not an episodic process separating periods of stability but a constantly active force that will not only adapt species to new conditions but wherever possible also will increase the level of adaptation even in a stable environment. Only then did he begin to realize the difficulty of reconciling a mechanism based on never-ending struggle with the existence of a benevolent Creator.

Whether or not we accept Ospovat's view of Darwin's original theory, it is certainly true that he became aware increasingly of the extent to which natural selection produced a steady trend toward increased specialization of function. He always had known that evolution was a branching process, an inevitable consequence of the discovery that new species arise when small populations derived from an existing form become geographically isolated under new conditions. At first this point simply was taken for granted, but a problem emerged when Darwin began to realize that after the initial branching there was a continued tendency for the branches to diverge ever father away from one another. His own research must have helped him to recognize this point, but Ospovat (1981) argues that the main stimulus was the work on divergence and specialization by other naturalists such as W. B. Carpenter, Richard Owen, and Henri Milne-Edwards. Modeling their concept of development on K. E. von Baer's theory of embryological growth, these biologists argued that not all species are adapted perfectly to their own way of life. On the contrary, they show differing degrees of specialized divergence from the common archetypical form of their group. Carpenter and Owen also showed that the degree of specialization within a group increased in the course of geological time, as shown by the sequence of fossils (see also Bowler, 1976*a*). This forced Darwin to see that evolution was not a process for merely maintaining adaptation. Instead, it would have to be considered a developmental force that acted on the once very generalized types from which each class began, producing first a number of separate branches and then progressively specializing each branch for its own particular way of life.

By reinterpreting adaptive evolution as a process of constant specialization, Darwin was able to link his theory with the work of more conventional naturalists. In the *Origin of Species* he even could refer to Owen's paleontological studies on this theme as evidence in favor of his own theory. At the detailed level, Darwin's opposition to teleology forced him to disagree with the interpretation that most of his contemporaries put on the trends they had discovered. Yet in its general outline, his theory was now able to mesh quite well with the latest developments made by those who adopted a more conventional approach to morphology, paleontology, and embryology. His theory would be even easier to promote now that it required only a reinterpretation of existing knowledge, rather than setting up of a whole new picture of the history of life.

The link between evolution and embryology has frequently been misunderstood. Clearly, Darwin did not follow those who saw the development of the human embryo toward its final goal as a model for the history of life on earth. His theory required something like von Baer's approach to embryology, because this was linked to a branching concept of development, although Darwin was apparently unaware of von Baer's writings (Oppenheimer, 1967). Contrary to popular opinion and the views of some later evolutionists, the Darwinian theory does not require that embryological growth must recapitulate the evolutionary history of the species (Gould, 1977*b*). Variation is not purposeful and must be considered a disturbance of the growth process. There is thus no necessary reason that an adult structure should be reduced to a mere stage in the embryology of a form's evolutionary descendants. This does not mean that embryology is of no value in helping us to understand the relationships between species. Sometimes two quite different adult forms have similar structures as embryos, and this identifies them as evolutionary relatives. Darwin, however, did not expect the growth of the embryo to reveal a speeded-up picture of evolution. The belief that ontogeny (individual growth) recapitulates phylogeny (the course of evolution) was grafted onto the theory later in the century by naturalists with pronounced neo-Lamarckian views.

The concept of branching evolution inevitably undermined the traditional view that man is the preordained goal of organic progress, but did it necessarily destroy the whole idea of progress? In a branching scheme, there can be no unambiguous hierarchy of forms each with its allotted position. Instead, it might be possible to set up an abstract scale of organic complexity for ranking entirely different kinds of living structures. Darwin realized that this would be very difficult in practice, but he admitted that most naturalists have an instinctive feeling that some organisms are "higher" than others. In this case, it would be possible to say that evolution was progressive in the sense that it pushed each form toward a higher level of organization within the context of its own peculiar kind of structure. Darwin was tempted to believe that increasing specialization was indeed a form of progress, because it meant that the descendants were better prepared than their ancestors to cope with a particular way of life. This had the advantage of allowing him to retain his old faith in an overall purpose for natural selection, a comfort in his effort to retain a link with the argument from design. He was forced to admit, however, that some kinds of specialization—parasites, for example—result in actual degeneration. Obviously, evolutionary breakthroughs that lead to the establishment of entirely new classes do *not* arise from the more highly specialized members of the previous class. Only a more generalized form can undergo so drastic a change of structure, while specialization easily can become a trap that prevents a species from adapting to rapid change in its environment and thus leads to extinction. Darwin continued to believe that natural selection could give rise to a form of progress, but he had to concede that it was at best a slow and irregular by-product of the mechanism's chief function of adaptation.

The link between Darwinism and branching evolution created many of the characteristics we associate with the modern form of the theory. Yet it also created a major problem for Darwin, because he had to explain why natural selection gave rise to this constant increase in level of specialization. In the original form of his theory, speciation split a single form into several separate branches, but there was no obvious reason the branches should then continue to move farther apart. It now seems that biogeography played an important role in leading Darwin toward an explanation of this further level of diversity (Browne, 1980). He began to examine the relative size of genera (i.e., the number of species they contain) and the degree of diversity among component species. It became obvious that small genera fell into two types: those containing a few very similar forms and those containing a few very diverse, aberrant forms. Darwin eventually realized that these represented two stages, the beginning and the end of a historical process. When speciation from an original form first creates the genus, all component species still are very similar to one another. In the course of time, divergence increases the number of species *and* the amount of difference between them. At the end of its career, however, when the genus is being replaced by other, more successful forms, only a few highly diverse species will be left before they too are driven to extinction.

By 1854, recognition of this point had greatly altered Darwin's views on how selection operated. It now appeared that the best location to encourage divergence within a genus was a crowded area where there could be *no* geographic isolation. Expansion over a wider area only came later in the genus's history. Thus Darwin began to reconsider the original insight gained in the Galápagos, that geographic isolation is crucial for speciation (Sulloway, 1979). He did not abandon isolation altogether, but he now began to think that it was not a crucial requirement for splitting up the original population. As an alternative, he invoked natural selection acting to promote different life-styles at opposite ends of the original population's geographic range. In other words, selection itself was powerful enough to create a division within the population, pulling apart the two halves of the original group by its adaptive pressure. Thus Darwin converted to what is now called sympatric speciation (speciation without geographic isolation).

Although we now can see why Darwin was led to change his mind on the importance of isolation, his decision was to create major problems for the development of his theory in later years. The majority of modern naturalists believe that sympatric speciation is impossible; the original splitting of the population is always allopatric, that is, it requires at least an initial phase of geographic isolation that prevents interbreeding. Once the populations have built up "isolating mechanisms" that will prevent interbreeding—these can be behavioral rather than genetic, a simple unwillingness to mate with the other form—the geographic barriers can disappear and the split still will remain. The problem with sympatric speciation arises when we consider that in traditional Darwinism all changes must be slow and gradual. It is thus difficult to imagine how natural selection can prevent interbreeding between two halves of the population, if there is an intermediate area where selection pressures do not operate, inhabited by individuals who still can mate with those at either extreme. Constant blending of the divergent characteristics then can take place by interbreeding across intermediate areas, and the crucial separation of breeding populations required for speciation is prevented.

It should be noted that the problem of "blending" mentioned above is *not* that discussed by some historians (e.g., Eiseley, 1958) to illustrate the weakness of Darwin's views on heredity. We are dealing here with the blending of characteristics by interbreeding throughout a population, not the blending of parental characteristics in individual offspring. It is quite true that in the absence of Mendelian genetics, Darwin and most of his contemporaries did believe in "blending inheritance," where the offspring is always a blend or average of the parents' characteristics (Vorzimmer, 1963). This later gave rise to another, quite separate, problem when a critic pointed out to Darwin that blending with unchanged mates soon would dilute the advantage of any individual born with a favorable characteristic (Bowler,

1974*b*, see chap. 7). Darwin sometimes wrote as though favorable variants were extremely rare, and in this case, blending soon would swamp the effect of the favored individual, like a spot of black paint mixed into a bucket of white. Originally, Darwin had avoided this problem by invoking isolation, where the small size of the population would prevent complete swamping. His switch to sympatric speciation eliminated this solution, however, and thus indirectly set up the problem of blending heredity. Darwin's response to this later problem was to move even more firmly toward a gradualist position, emphasizing that favorable variations were not just single individuals but integral parts of a complete range of variability affecting the whole population.

Whatever the merits of his decision, Darwin now had a link between biogeography and the fossil evidence for increasing divergence. He also had convinced himself that selection must be a far more powerful and ruthless force than he once had imagined. This last decision once again helped to drive home the increasing difficulty of reconciling the theory with design. But he still had to explain *why* natural selection constantly acted to diversify life in this way. Selection had to become something more than just a passive force for adapting species to changes in their environment. He found the answer, once again, in an economic analogy: the division of labor (Limoges, 1970; Schweber, 1980; Ospovat, 1981). In the economic sphere, it is more profitable for a number of workmen to specialize in particular aspects of a manufacturing process than for each to try to do the whole thing for himself. The French physiologist Henri Milne-Edwards had applied this analogy to the workings of the individual organism, which would be more efficient if each organ specialized for a single function. Darwin now realized that a similar analogy could apply in ecology and evolution. If two species were very similar, there would be a tendency for them to compete with each another, which would be detrimental to the future prospects of both. It would be to their advantage if they could diverge, each specializing for a different way of life and thus lessening the chances of competition. The same area of territory, furthermore, would support a far larger component of living things if they were divided into a wide range of forms, each capable of exploiting the area's resources in a different way. Because natural selection works for the advantage of each form, it thus will tend to increase the level of diversity, constantly specializing each for its own life-style even when the environment remained stable. Here at last was a principle of diversity that would turn natural selection into an active force with the properties Darwin desired.

When Darwin put the pieces of this puzzle together in 1856, he regarded his theory as complete. With divergence seen as a necessary consequence of natural selection, he could explain a far greater range of biological phenomena than with the original form of the theory. Only then did he feel confident enough to begin writing his projected "big book" on the species question (Darwin, ed. Stauffer, 1975; Hodge, 1977), a project that would soon be interrupted by his decision to write the *Origin of Species*.

WALLACE AND PUBLICATION OF THE THEORY

The stimulus that prompted Darwin to publish was the arrival in 1858 of a paper outlining a similar theory of natural selection. The author was Alfred Russel Wallace, who thereby established his place as an independent codiscoverer of natural selection (Marchant, 1916; George, 1964; Beddall, 1968; Williams-Ellis, 1969; McKinney, 1972; Fichman, 1981). Some accounts of Wallace's work imply that history has been less than just to him. Why should the theory be known as "Darwinism" if Wallace contributed equally to its discovery? There are hints that Darwin himself tried to play down Wallace's role to preserve his own priority. A recent variation on this theme implies that Darwin plagiarized the principle of divergence from Wallace (Brackman, 1980; Brooks, 1983; for a rebuttal, see Kohn, 1981).

There are good reasons for treating these suggestions of a conspiracy of silence as exaggerated. The most obvious fact that counts against giving Wallace a role equal to that of Darwin is that his discovery came two decades later. By this time, Wallace had spent some time working out the implications of branching evolution, but he did not see selection as the actual mechanism until 1858. Even if we accept the idea that his brief paper contains the essence of the whole theory, we know that by itself it had little impact. When published along with an extract from Darwin's work, there was little reaction in the scientific community. It would have taken years for Wallace to prepare a complete account of the theory himself. But Darwin had already done all the groundwork and was in a position—now that the cat was out of the bag—to condense his own thoughts into the single volume that we know as the *Origin of Species*.

Wallace came from a poor background and eventually became a professional collector, supporting himself by the sale of specimens gathered in remote parts of the world. His first expedition was to South America from 1848 to 1852, with his friend Henry Walter Bates (Beddall, 1969). This gave him an insight into just those problems that had first started Darwin thinking about evolution, ecology, and geographic distribution. The ship carrying Wallace's specimens back home was destroyed by fire, but fortunately he was insured; he soon set out again for the Far East and the islands of the Malay archipelago, modern Indonesia. There, during a bout of fever, he conceived the idea of natural selection. He was then on the little-known island of Gilolo, not—as he claimed—on the famous spice island of Ternate

(McKinney, 1972). He wrote the idea up in a short paper and mailed it to the naturalist he thought best able to evaluate it—Darwin.

Like Darwin, Wallace had taken Lyell's *Principles* with him to South America. He adopted the system of gradual geological change and soon found himself questioning Lyell's assumption of the fixity of species. He became interested in how geographic barriers define a species' territory, reaching conclusions similar to those impressed on Darwin by the Galápagos islands. Wallace began to feel that creationism required too many arbitrary assumptions to explain the facts of distribution. He was impressed by Chambers's *Vestiges,* using the book to focus his attention on the problems of transmutation. During the 1850s, Wallace became convinced that branching evolution had occurred, although as yet he did not understand the mechanism of change. He began to plan a book on the species question and in 1855 published an important article stating that "every species comes into existence coincident in time and space with a preexisting closely allied species" (Wallace, 1855; reprinted in Wallace, 1870, 1891). He began to correspond with Darwin, who had expressed high regard for this paper. Darwin did not discuss his own theory, however, leaving Wallace to assume when he first thought of the idea that he had had an important new insight.

The title of the 1858 paper was "On the Tendency of Varieties to Depart Indefinitely from the Original Type" (reprinted Wallace, 1870, 1891; Darwin and Wallace, 1958). It generally has been assumed that it contains the essence of the theory already worked out by Darwin, although there are actually considerable differences between the ways in which the two naturalists presented the idea. Wallace had been impressed with Lyell's discussion of the "war of nature" between species and understood the struggle for existence at this level. Like Darwin, he read Malthus because he too was interested in the problem of human evolution. But Wallace seems to have applied the concept of struggle in a very different way, leading some commentators to ask if he can really be counted as the co-discoverer of Darwinian natural selection (for an evaluation of these claims, see Kottler, 1985).

A. J. Nicholson (1960) pointed out that Wallace tended to think of the environment as setting absolute standards of fitness against which the members of each species are measured. Only those who do not measure up to the standard are eliminated, and there will be no evolution unless the environment changes. This would be a much less severe concept of selection than Darwin's, where the members of a population are always competing against one another, whatever the state of the environment. The possibility of an even more striking difference between the two men's views centers on Wallace's use of the term "variety" (Bowler, 1976*c*). It can be argued that throughout most of the 1858 paper, Wallace stressed competition between established varieties or subspecies, not between individual variants. Darwin always had thought of struggle at an individual level and made this the basis of this theory. Wallace instead presented selection acting between varieties, eliminating those least fitted to cope with the overall conditions of the species' range. He realized that to make the mechanism self-sustaining, the species must constantly be forming new varieties, but he does not argue very clearly that this is caused by a more basic level of natural selection acting on individual variations. The main concern of Wallace's 1858 paper was thus the second level of selection, which Darwin eventually had recognized as the cause of divergence.

These alternative interpretations of Wallace's views can be made plausible by noting one striking difference between the way in which he and Darwin made their discoveries. Darwin became convinced of evolution by his experiences on the *Beagle*, but after returning home, he turned to the study of artificial breeding to throw light on the mechanism of change. This focused his attention on processes occurring within a population and thus emphasized selection acting on individual differences. Wallace was converted to evolution in the same way but did not develop an interest in artificial breeding; indeed, he denied that artificial selection serves as a good analogy of the natural process. Without the model of artificial selection to influence his thinking, Wallace would have been less likely to focus on individual competition. His interests in geographic distribution might then have led him to think in terms of an absolute standard of fitness against which each population is judged.

Whatever the differences of approach, Wallace's paper convinced Darwin that his own views now had been anticipated to a large extent. A less honorable man might have destroyed the paper and put Wallace off. Darwin did not contemplate such a move, yet at the same time he did not want to lose his twenty-year priority. He turned for advice to Lyell and Hooker, who arranged for a reading of Wallace's paper and two short extracts from Darwin's writings, one a letter to Asa Gray confirming his prior recognition of the principle of divergence. The papers were read before the Linnean Society in London and subsequently published in the society's journal. Curiously, there was little debate when the papers were read and no response to the publication. So momentous a subject could not be brought to the public attention by such short expositions. But Darwin now realized that he could hold back no longer and began to write a substantial account of his theory for immediate publication. The *Origin of Species*, published at the end of 1859, sparked off the great debate.

7

Darwinism: The Scientific Debate

When the *Origin of Species* was published by John Murray on November 24, 1859, the 1,250 copies of this edition were snapped up by the booksellers on the first day. Darwin, exhausted by the labor of writing, took the waters at Ilkley while waiting for the storm to break. There was some support from younger scientists, but most of the initial response was indeed negative. There is a story that Darwin was pointed out as the most dangerous man in England by a clergyman. Nor should we be surprised at the extent of the reaction from conservative forces: evolutionism threatened religion and hence the church, still seen as a pillar of established society. A number of scientific arguments also were used against the theory, some of which Darwin and his followers had to take very seriously. Yet despite the storm of opposition, evolutionism began to make headway, although perhaps not quite as rapidly as sometimes has been assumed (Hull et al., 1978).

Given the strong religious beliefs of many naturalists, it would be easy to dismiss all their scientific objections as merely desperate attempts to scrape up anything that might defend the old theological view of nature. There seems little doubt that the motivation behind many of the attacks was a general dislike of the materialistic world view of Darwinism. Yet this does not mean that the arguments themselves were without scientific validity. Darwin had built his theory on foundations provided by the contemporary understanding of nature, and in some areas—particularly the study of variation and heredity—these foundations were very unsound. Thus, it was possible to construct scientific arguments against the theory which would appear valid at the time, although we may be able to see through the objections today because of advances in areas such as genetics. Some problems never have been resolved: modern creationists still use the discontinuity of the fossil record as an argument against any form of evolution. Such objections may well be put forward in defense of natural theology, but in Darwin's time they posed genuine threats to the selection theory.

David Hull (1973*b*) has edited a valuable collection of critical reviews of the *Origin* revealing the wide spread of opinion even among those scientists who objected to the theory. This is hardly surprising, because the old creationism was by no means a unified scientific theory. Obviously, scientists of different intellectual backgrounds would see different objections as more important. By contrast, those naturalists who wholeheartedly supported Darwin tended to be impatient with the limitations of natural theology and often were concerned with those areas of study such as geographic distribution where these limitations were most obvious. But how did this small group of scientists help to influence public opinion? Here the survey of the periodical press by Ellegård (1958) is useful because it covers journals from various religious and political backgrounds. As might be expected, the more radical organs took the theory seriously, while conservative journals stuck with creationism. The survey, however, reveals a general shift of opinion during the 1860s. More periodicals began to appreciate the artificiality of miraculous creation and published articles advocating creation by law. This was Darwin's most immediate effect in changing public opinion: creating not a wide understanding of his theory but a growing willingness on all sides to admit that the world somehow is governed by law rather than divine caprice.

It is thus necessary to distinguish carefully between evolutionism, in general, and the Darwinian selection theory, in particular. The term "Darwinism" has come to be regarded as virtually synonymous with evolutionism, yet understanding of the later history of the topic will be confused if some distinction is not made. It is clear from Ellegård's survey that by 1870 the basic idea of evolution was becoming widely accepted, and because the public associated the idea with Darwin, the name "Darwinism" became popular. But many who accepted evolution had doubts about Darwin's particular explanation of how it occurred. Even among Darwin's close supporters there were differences of opinion over the role of the selection mechanism, although the group was careful to prevent these from appearing crucial. It is seldom realized that even some of Darwin's closest supporters, including T. H. Huxley, made very little use of the selection theory. There is thus a sense in which most of what passed for Darwinism owed little to those aspects of the theory that are singled out for attention by modern biologists. By today's standards, many of these early evolutionists were only "pseudo-Darwinists"—they were unified only by their recognition of Darwin as the figure who had prompted their conversion to the general idea of evolution.

The extent to which non-Darwinian evolutionary ideas flourished in the late nineteenth century will be explored below (chap. 9). But our recognition of this point must be taken into account when evaluating the arguments used against the selection theory. All too often, historians have tended to regard Darwin's theory as the only worthwhile focus of attention in the post-Darwinian era. Even when acknowledging the strength of the opposition, they tell us more about the objections to natural selection than about the alternative theories that were gradually brought in to replace it. This approach is particularly misleading when—as in Eiseley's (1958) account—the problem of heredity is singled out as the chief area of debate. Darwin had a non-Mendelian view of heredity, and it is easy for the historian to present this as a gap in his theory which prevented its full acceptance until filled in by the rediscovery of Mendel's laws in 1900. Modern Darwinism is presented as a jigsaw puzzle in which Darwin left one crucial piece to be inserted by later biologists. In fact, there were many objections to natural selection outside the field of heredity, and the few genuine Darwinians were able to create a viable foundation for their theory even without Mendel's work. The decision to concentrate on heredity is unhistorical: it reflects our modern preconceptions about the direction in which evolution theory ought to have been moving. A truly historical perspective must look at the whole range of scientific objections, including those that eventually became the source of non-Darwinian theories about the evolutionary mechanism.

THE GROWTH OF DARWINISM

From the start, a few naturalists saw that Darwin had at last opened up the question of the origin of species to scientific investigation. Without the support of this group, Darwinism would have died an early death. But because there were enough competent naturalists to defend the theory, it could not be swamped by the opposition. By 1870, many of those scientists concerned with the most relevant areas of biology had conceded that evolution was preferable to special creation, although the mechanism of change remained a subject for debate. Our survey of this revolution begins with an outline of the evidence that initially helped to convert naturalists to evolution. The *Origin* showed that a number of well-known phenomena could be explained for the first time, if it were assumed that a mechanism such as natural selection was responsible for the production of new species. An ever-increasing number of naturalists accepted this explanatory power as a valid reason for adopting Darwinism as a working hypothesis. In Britain, especially, the number of converts expanded until by the 1880s, it had become a dominant force in the scientific establishment. Yet because natural selection remained to some extent controversial, we also shall have to take a wider look at the tactics that ensured the success of Darwinism within the scientific community.

The *Origin of Species* does not begin, as one might expect, by making the case for evolution. Instead, it launches straight into a detailed discussion of natural selection, for the reason that Darwin had to convince his fellow naturalists he had found a new and more plausible mechanism of change. Many were aware of the general arguments for evolution but had refused to commit themselves until a scientifically testable hypothesis for the means of change became available. Lamarckism had been discredited earlier in the century, while the vague "laws of development" advocated in *Vestiges* were seen to have little scientific value. It was essential for Darwin to show that he had come up with a new approach to the problem, and thus he had to explain natural selection before making the general case for evolution. It is important to note that Darwin was not dogmatic on the role of selection. He certainly regarded it as the most important mechanism, yet from the beginning he conceded that it might be supplemented by others such as the inherited effects of use and disuse, that is, Lamarckism. In response to criticisms of selection, he altered the details of his theory and gradually admitted a greater role for the alternatives. Each edition of the *Origin* thus differed from its predecessors, including eventually an additional chapter on the difficulties faced by the theory (variorum text, ed. Peckham; Darwin, 1959*b*).

Darwin's response to his critics provides a touchstone by which we can test the attitude of historians. Those who regard selection as a great mistake assume that his concessions indicate a growing awareness of just how weak his theory was. They picture him as full of inconsistencies, frantically rushing from one point to another in a desperate effort to shore up a fundamentally unsound structure (Barzun, 1958; Himmelfarb, 1959). Peter Vorzimmer's account (1970) appreciates the problems caused by Darwin's inadequate knowledge of heredity but presents a similar picture of him tackling difficulties piecemeal, often weakening one point in order to strengthen another. Some writers have implied that by the end of his career, Darwin had abandoned selection in favor of Lamarckism (Eiseley, 1958; Himmelfarb, 1959). This certainly is going too far: Darwin did concede a greater role for the alternative mechanisms, but the *Origin* remained basically a description of the selection theory. It is possible to take a far more positive view of Darwin's efforts to solve the problems confronting him (Ghiselin, 1969; Hull, 1973*b*; Ruse, 1979*a*). Given the limitations imposed by what was available to him, his efforts to maintain the integrity of his original view of evolution can be seen as heroic rather than pathetic. The fact that historians can disagree to such an extent over Darwin's success indicates that we cannot evaluate the reception of his theory by simply counting the arguments for and against selection. The emergence of Darwinism represents a change within the scientific community which only can be accounted for in terms of social factors within that community and a wider revolution in the values accepted by scientists.

The confusion over the status of Darwinism arises because so many of the best arguments in favor of the theory were of an indirect nature. They supported Darwin's overall vision of branching, adaptive evolution, but did not prove that selection was the actual mechanism of change. It proved extremely difficult to confirm the effectiveness of natural selection directly, because the artificial form of selection was confined to the production of varieties within existing species. Many critics argued that this was a major weakness of the theory, but Darwin insisted that his idea should be taken seriously because it helped to explain such a wide range of otherwise incomprehensible biological facts. One new discovery did help to lend direct support to natural selection in the confused years of the early 1860s: Henry Walter Bates's investigation of mimicry in insects. Bates had been Wallace's companion on his first trip to South America and had devoted himself to a study of the brightly colored insects of the Amazonian forests (Bates, 1863; Beddall, 1969; Woodcock, 1969; for a different interpretation of Bates's discovery, see Blaisdell, 1982). His work concerned the relationship between mimicry and warning coloration. The most obvious form of mimicry occurs when an insect camouflages itself as an inanimate object to escape the attention of predators. Insects that are unpalatable to birds go to the opposite extreme of developing bright warning colors to protect themselves from being seized accidentally. Bates discovered that there was a second kind of mimicry in which an edible insect had gained protection by copying the warning colors of an inedible form (Bates, 1862). Selection was by far the most plausible mechanism to account for these peculiarities of insect coloration. Because insects cannot control the color of their wings, the effects could not have been developed by Lamarckian use-inheritance. Selection, however, would favor those individuals who happened to vary in such a way as to gain some protection: they would live longer and leave more offspring, so that eventually the whole species would take on the protective coloration.

In the later chapters of the *Origin,* Darwin discussed those areas of existing biological knowledge that would be illuminated by his theory. Although he had to deal at length with the difficulties created by the imperfection of the fossil record, a whole chapter was devoted to showing that the available outline of the history of life was consistent with his theory. Natural selection was not a mechanism of inevitable development: species changed when it was to their advantage but otherwise remained in their original form. The fossil record showed that there was no fixed rate of change: some forms changed rapidly, while others remained stable over long periods of geological time. In some cases, very ancient forms are still alive today, virtually unchanged—the so-called living fossils. This lack of change was perfectly understandable within Darwin's theory, because once a species was well adapted to a stable environment there would be no reason for it to change. Nor was extinction inevitable, so long as a better-adapted competitor did not emerge. There is no preordained pattern of development, so that no two branches of evolution can produce the same result (although superficial resemblances may result from convergence, when two branches adapt for similar life-styles). Evolution is, in effect, irreversible—once a particular form has been eliminated by extinction, it can never be re-created. This generalization was sufficiently well known to paleontologists that it was subsequently given the name "Dollo's law" (Gould, 1970). When a new class appears, natural selection will produce a general trend toward divergence and specialization, and here Darwin was able to cite the paleontological work of Richard Owen and others to show that such trends indeed are found in the fossil record. Finally, Darwin could point to the law of succession of types he had noted in South America. There was a basic similarity in the succession of forms within a continental area, which is exactly what would be expected if new species were produced by the transmutation of old ones. Darwin did not believe that the giant armadillos and sloths whose remains he had discovered were direct ancestors of their smaller modern counterparts. The giant forms were extinct branches of the groups to which they belonged, but the record also showed that they had been accompanied by smaller relatives that had survived to become the ancestors of the modern South American animals.

Two chapters of the *Origin* were devoted to geographic distribution of living forms, the original stimulus for Darwin's conversion to evolution. He insisted that it was impossible to explain the distribution by claiming that each species occupied the territory for which it was perfectly adapted. Areas with very similar climates sometimes were inhabited by quite different animals and plants, while elsewhere very similar forms were spread over a wide geographic range. The real clue to understanding distribution lay in the study of geographic barriers that limited the spread of each form of life outward from its original location. This confirmed the idea that the present distribution was the result of a historical process, each successful form expanding as far as it could and adapting to the various environments it encountered. Apparently, impenetrable barriers could sometimes be crossed by unusual means, and Darwin reported a long series of experiments into dispersal mechanisms, such as transportation of seeds on the feet of birds. Isolated islands such as the Galápagos received their inhabitants by such haphazard processes, usually from the nearest continental area. On the continents themselves, climatic fluctuations on a geologic time scale could explain many anomalies in the present distribution. Darwin emphasized how recent ice ages had permitted the southward spread of many arctic forms, which had been left marooned in mountainous areas as the climate warmed up again. The present distribution of animals and plants was entirely consistent with the theory of branching, adaptive evolution but virtually inexplicable in terms of supernatural design.

It is significant that two of Darwin's earliest converts were botanists with a strong interest in the geographic distribution of plants. Joseph Dalton Hooker played a leading role in promoting the theory in Britain (L. Huxley, 1918; Turrill, 1963; Allan, 1967). He had explored many parts of the world and had long been puzzled by the difficulty of providing a rational explanation of plant distribution. As one of Darwin's early confidants, he soon came to appreciate the value of the selection theory. Hooker's introductory essay to his flora of Tasmania (1860) was one of the first scientific works to offer open support for Darwin. In America, Asa Gray played a similar role, exploring the distribution of plants on that continent in Darwinian terms (collected papers, reprinted 1963; Dupree, 1959). Alfred Russel Wallace continued to provide geographic evidence for the theory from his zoological studies (Marchant, 1916; George, 1964; Williams-Ellis, 1969; McKinney, 1972; Fichman, 1981). An 1864 paper on the Malayan butterflies (reprinted Wallace, 1870) showed how local varieties often were found on islands and pointed out the impossibility of drawing a sharp line between varieties and species. More extensive works (1869, 1876, 1880) showed how the worldwide problems of geographic distribution could be resolved by combining evolution with a study of the barriers limiting migration. Southeast Asia continued to supply some of Wallace's best illustrations, and the line dividing the Asian from the Australian faunas among the Indonesian islands still is known as "Wallace's line." Here he showed how the line of separation among the islands could be explained by taking into account the possibilities of migration not only under present conditions but also in the past, when geological changes may have raised or lowered the sea level (Beddall, 1969; Mayr, 1954; Fichman, 1977).

In addition to paleontology and geographic distribution, the *Origin* also showed how existing techniques in classification and morphology could be seen as natural consequences of Darwinian evolution. The principle of divergence explained the subordination of groups within groups on which the Linnaean taxonomic hierarchy depended. The natural system of classification for which naturalists had searched so long was an expression of the relationships created by common descent. The "unity of type" within each group was the result not of some mysterious archetypical model but of the preservation of the most basic characteristics of a common ancestor in all its descendants. The present relationships between living forms could only be understood as the result of a historical process. It was no good trying to pretend that each species is perfectly adapted to its way of life, because in many cases of homology the structure's use seemed highly inefficient. A competent Designer would never have modified the same basic structure to serve such highly diverse functions as the wing of a bat and the paddle of a whale. But evolution could only work by modifying the structures available to it, and the wide range of homologies was the inevitable consequence of this opportunism. The specialized nature of the adult structure appears gradually in the course of embryological growth, because the variations that natural selection feeds on are small and generally affect only later stages of the growth process. Earlier stages of growth are preserved intact, allowing embryos to reveal the common evolutionary origin of widely divergent adult forms. Finally, the theory explains the frequent occurrence of rudimentary organs that never become large enough to function. These are always organs that were once of some use to an ancestral species but have lost their function through a change in circumstances and are being gradually eliminated.

The arguments advanced by Darwin and his early followers were enough to convince the majority of biologists that evolution would have to be taken seriously. Over the next decade, approximately three-quarters of the working biologists in Britain seem to have accepted evolution—although, curiously, there is no evidence that younger scientists were more willing to make the transition than their older, more established colleagues (Hull et al., 1978). By the 1880s, a well-entrenched school of Darwinism had become a dominant feature of the scientific establishment. Yet the rise of this school cannot be explained simply in terms of the theory's obvious technical advantages. Natural selection itself continued to be a highly controversial topic, and later in this chapter is an account of the more prominent objections to the theory. Many of the more conservative naturalists would clearly have preferred a theory based on sudden transitions, linked into a more orderly process with a purposeful goal. The Darwinian mechanism of evolution actually suffered an extensive loss of popularity in the later decades of the century (see chap. 9). Thus, the original rise to prominence of a Darwinian school of thought requires an explanation in wider terms than a mere demonstration of the theory's scientific powers. Somehow, Darwin's supporters were able to gain a dominant role in the scientific community despite the existence of many factors that could serve as a source of opposition to their theory.

When a theory has many advantages but also serious limitations, its success or failure may depend on the skill with which its supporters argue their case before the scientific community (Hull, 1978). The outcome of the debate is determined, in effect, by the public relations skills of its supporters and their opponents. The original group of Darwinists possessed a number of skills that were extremely useful in this situation. They were quite flexible in their approach to the theory, so that Darwinism was not presented as a dogmatic, all-or-nothing commitment to natural selection. Their group thus could be joined by anyone who accepted the most general principles of the

Darwinian world view: evolution by common descent, through an adaptive process that probably owed a good deal to natural selection. Within these principles, the chief Darwinists themselves disagreed considerably over details. Darwin accepted a small element of Lamarckism; T. H. Huxley was a saltationist; while Wallace even advocated divine intervention in the evolution of man. This flexibility helped to disarm the critics, because objections to natural selection could be sidestepped by appealing to the possibility of supplementary mechanisms. At the same time, the Darwinists never fought openly among themselves. They agreed to differ in the hope that future research would solve their problems and thus were able to present a united front to the world, confident that their basic ideas were sound. It is significant that when a dogmatic "neo-Darwinism" emerged later in the century, insisting that selection was the only mechanism of evolution, it promptly began to lose ground to supporters of alternative theories, such as neo-Lamarckism.

Darwin himself could not stand the excitement of public debate, nor was he active within the emerging community of professional scientists. A second-in-command was needed, who could look after the practical problems of establishing the theory. One man, Thomas Henry Huxley, filled this position so effectively that he became known as "Darwin's bulldog" (L. Huxley, 1900; Irvine, 1955; Bibby, 1959; Ashforth, 1969; Di Gregorio, 1984). Huxley had exactly the kind of character required; he was willing to take on any opponent in defense of free thought—and in the course of his career, opponents ranged from Mr. Gladstone to the Salvation Army. Like Darwin, Huxley gained his early experience as a naturalist aboard a British naval vessel, but his interests lay more in anatomy and paleontology. During the 1850s, he became dissatisfied with creationism and convinced that science must proclaim its independence from theology. But he felt unable to support transmutation because there was no satisfactory mechanism to account for it. He ridiculed *Vestiges* and the idea of "creation by law," but on reading the *Origin,* he immediately saw that here at last was a plausible hypothesis that opened up the subject to scientific investigation. Luckily, the London *Times* asked him to review Darwin's book. His favorable account in that influential newspaper on December 26, 1859, helped to ensure that the theory was not howled down by the opposition. He also wrote a longer account for the *Westminster Review* (reprinted in Huxley, 1893).

Huxley conceded that selection would not be proved as a valid evolutionary mechanism until an experimental test with artificial breeding had produced a totally new species. He also criticized Darwin's commitment to gradual evolution and suggested instead that large mutations sometimes might produce new forms directly. To a paleontologist, this would have seemed quite reasonable, and Huxley does not seem to have been aware of the problems such an assumption might lead to in other fields. His own detailed anatomical work was changed very little by his conversion to evolutionism (Bartholomew, 1975; Desmond, 1982; Di Gregorio, 1982, 1984). Huxley was determined, nevertheless, that the theory should be given a fair hearing, and he was in a position to ensure that its supporters would be helped in their professional careers.

It was Huxley who was most visible in the famous confrontation between the Darwinians and Bishop "Soapy Sam" Wilberforce at the Oxford meeting of the British Association for the Advancement of Science in 1860. The meeting degenerated into uproar when Huxley declared that he would rather be descended from an ape than from a man who misused his high position to attack a theory he did not understand. Modern research has suggested that Huxley's defense of Darwinism at this meeting may not have been as effective as the popular legend implies (Lucas, 1979). His real triumph was in the far subtler process of ensuring that the evolutionists gained control of the scientific community.

Huxley was typical of a new generation of scientists determined to wrest intellectual authority away from its traditional sources. Evolutionism was useful to them precisely because it demonstrated that science could now determine the truth in areas once claimed by theology (Fichman, 1984). Huxley went on to become a leading public figure, serving as a scientific expert on numerous government commissions. He was also a member of the "X-club," an informal but extremely influential group of men whose behind-the-scenes activity shaped much of late Victorian science. It was by exploiting their position within this network that Huxley and his fellow converts ensured that Darwinism had come to stay (Ruse, 1979*a*). They avoided open conflict in scientific journals but used their editorial influence to ensure that Darwinian values were incorporated gradually into the literature. The journal *Nature* was founded at least in part as a vehicle for promoting Darwinism. Academic appointments were also manipulated to favor younger scientists with Darwinian sympathies, who would ensure that the next generation was educated to take the theory for granted. So successful was this takeover of the British scientific community that by the 1880s, its remaining opponents were claiming that Darwinism had become a blindly accepted dogma carefully shielded from any serious challenge.

The Darwinists' success clearly points to a change of attitude within the scientific community. Whatever the debates over the actual mechanism of evolution, the new movement was committed to a causal interpretation of the development of life, repudiating not only divine creation but any teleological explanation in which evolution was drawn toward predetermined goals. The permanent success of Darwinism lay in the triumph of this attitude, because the arguments over natural selection itself did not diminish as the century drew to a close. One might still hope—as Darwin himself did at times—that the laws of evolution would produce a gradual ascent of life toward higher forms, but it was no longer legitimate to use future goals to explain evolutionary trends. If the Creator designed the laws of evolution, His actions were no longer demonstrable by science and His role in the universe had to be accepted as a matter of faith. The rise of Darwinism corresponds to the emergence of a new generation of biologists determined to allow the scientific method complete access to the question of the origin of species. Chapter 9 will show that many of the old idealist and teleological attitudes resurfaced later in the century as the intellectual foundation of theories such as neo-Lamarckism; but even the supporters of these alternatives were determined to show that the evolutionary trends they postulated were controlled by causal factors, not by mysterious forces outside the scope of scientific investigation.

The fact remains, however, that in the 1860s a large number of naturalists still were attracted to the argument from design and were suspicious of natural selection precisely because of its apparent materialism. Sir J. F. W. Herschel's dismissal of selection as the "law of higgeldy-piggeldy" expresses his contempt for a mechanism that left the history of life to such a haphazard and undirected combination of circumstances. The success of Darwinism may be explained at least in part by the difficulty these opponents encountered in formulating a coherent alternative to selectionism and by the clumsiness of their tactics (Desmond, 1982; Bowler, 1985). Later in the century, Lamarckism offered an alternative that avoided the harsher implications of selection and was appreciated for this reason, but in the 1860s this mechanism still was thought to have been effectively discredited by earlier critics, such as Cuvier and Lyell. Opponents of Darwinism, such as Richard Owen and St. George Jackson Mivart, were prepared to accept transmutation if it were supernaturally guided in certain directions, but this seemed all too obvious a compromise with the old argument from design. It was difficult to pinpoint any characteristic of the supposed guidance that could be used to distinguish its effect from that of natural selection, especially as Owen had helped to popularize the concept of specialization as a key to understanding the fossil record. The opponents of Darwinism thus were reduced to listing the arguments *against* natural selection, without having any clear-cut alternative they could argue *for*. In addition, they seemed unable to organize their movement effectively. Owen was a respected but very isolated figure, while Mivart allowed himself to be ostracized from the scientific community. Thus, no rival school emerged to challenge the growth of Darwinism in Britain.

The reception of Darwinism by naturalists outside Britain varied considerably (Glick, 1974; Kohn, ed., 1985). A great deal has been written on American reactions, although most works have concentrated on the intellectual and social issues (Hofstadter, 1959; Persons, 1956; Daniels, 1968; Loewenberg, 1969; Pfeifer, 1974; Russett, 1976). Among scientists the response was generally negative from older naturalists committed to idealism but far more positive among their younger pupils. One man stood out as the leading opponent—Louis Agassiz, still the most widely known American naturalist. Agassiz's idealist world view was totally incompatible with any theory of natural evolution (Lurie, 1960; Mayr, 1959*a*; Winsor, 1979). For him, species were immutable elements in God's plan of creation whose forms could never be changed by any amount of natural variation. In the end, Agassiz rendered his whole position ridiculous. In his efforts to minimize the amount of natural variation, he insisted that every form with the slightest distinct characteristic must be a separately created species, even those considered to be merely local varieties by most naturalists. If this was the only way that creationism could be defended, younger naturalists would have none of it, and in his later years Agassiz was disappointed to see many of his own students turning to some form of evolutionism.

The brunt of Agassiz's attack was borne by the botanist Asa Gray (Dupree, 1959). Like Hooker, Gray was ideally situated to appreciate the arguments drawn from geographic distribution of plants and knew of natural selection before its official publication. His collected papers (reprinted 1963) explore the evidence provided by distribution of American plants. Although he was a convinced evolutionist and called himself a Darwinist, Gray had problems with the selection mechanism itself. A deeply religious man, he was concerned to show that selection still could be reconciled with some form of natural theology. More straightforward support came from the paleontologist Othniel C. Marsh (Schuchert and Levene, 1940). Like Huxley, Marsh accepted a simplified version of selectionism and got on with the job of trying to understand the fossil record in terms of evolution, and he made important discoveries that helped to fill gaps in the fossil record.

Although many American naturalists converted to evolution, there was a more deliberate effort among some of them to search out alternatives to natural selection. In a sense, Agassiz's influence was still powerful, because his students who accepted evolution found it difficult to assimilate the Darwinian scheme of random variation and selection. Their position within the scientific community prevented the Darwinians from engineering the kind of takeover that was so successful in Britain. Before the 1860s were over, paleontologists, such as Alpheus Hyatt and Edward Drinker Cope, were laying the foundations of what later would be called the American school of neo-Lamarckism.

On the European continent, Darwinism provoked very different responses. In France, the theory aroused little excitement (Conry, 1974; Farley, 1974; Stebbins, 1974). French scientists gradually converted to evolutionism later in the century, but the selection mechanism held no attraction for them and was not the stimulus responsible for the conversion. This negative response to Darwinism was conditioned by the legacy of Cuvier.

French naturalists saw evolution through the eyes of the anatomist, as a process of purely structural change, and found it easier to revive Lamarck's belief that new habits could influence structure directly. There was nothing like the British tradition of natural history through which Darwin and Wallace had come to study the problems of geographic distribution. In addition, the rationalist way of thought dating back to Descartes prevented the French from taking seriously anything so untidy as a mechanism based on random variation. Thus, for the first time France begins to drop out of our story, because the great debates over the validity of Darwinism did not trouble the body of French science. Even the modern synthesis of selection and genetics has had less impact in France than in most other countries.

Germany, too, had a traditional interest in comparative anatomy dating from the idealist period. Much of the German contribution to the growth of evolutionism consisted of an effort to translate the idealist view of anatomical relationships into an evolutionary one. Yet, superficially at least, certain social factors ensured that selectionism would attract more attention here than in France (Gasman, 1971; Montgomery, 1974). Some German scientists, of whom Ernst Haeckel is the best known, were political radicals who saw Darwin's rejection of design as a weapon in their fight against conservatism. It has been said that although Darwinism was born in England, it found its true home in Germany (Radl, 1930; Nordenskiöld, 1946). There can be no doubt that by the 1870s, *Darwinismus* had become widely popular, but this was not based altogether on a full appreciation of the selection mechanism. Haeckel openly proclaimed his intention of synthesizing the evolutionary theories of Darwin, Lamarck, and Goethe (translation, 1876). Darwinism thus gained popularity only as a symbol opposed to traditional religion and as a promise of progress in human affairs. When August Weismann later tried to insist on selection as the sole cause of evolution, the German holiday with *Darwinismus* came to an end.

THE HISTORY OF LIFE AND THE FOSSIL RECORD

Although Darwin argued that available knowledge of the fossil record was compatible with his theory, he knew that paleontology harbored a major problem. The record certainly did not reveal gradual transformations of structure in the course of time. On the contrary, it showed that species generally remained constant throughout their history and were replaced quite suddenly by significantly different forms. New types or classes seemed to appear fully formed, with no sign of an evolutionary trend by which they could have emerged from an earlier type. Darwin devoted a chapter of the *Origin* to explaining the "imperfection of the geological record," arguing that the fossils we discover represent only a tiny fraction of the species that actually have lived. Many species, and many whole episodes in evolution, will have left no fossils at all, because they occurred in areas where conditions were not suitable for fossilization. Apparently sudden leaps in the development of life are thus illusions created by gaps in the evidence available to us. Future discoveries may help to fill in some of the gaps, but we can never hope to build up a complete outline of the history of life.

Darwin himself was reluctant to speculate about the missing episodes, but many of his followers felt it necessary to attempt a complete reconstruction of the history of life. Where fossils were lacking, comparative anatomy and embryology could show what might have happened. This evolutionary morphology went far beyond hard evidence and eventually got bogged down in a morass of untestable hypotheses, but its popularity is a clear indication of the faith inspired by Darwinism. It normally is assumed that the movement was made possible by the fact that Darwin did not challenge the basis of existing morphology—rather, he enabled naturalists to reinterpret existing knowledge in more realistic terms. In a sense, it made little difference to the description or classification of a form whether its affinities were attributed to archetypes or to common descent. Evolution merely focused attention more strongly on those forms, living or extinct, that seemed most likely to yield information on crucial developments. The Darwinist, however, was forced to create sequences that could be explained as the result of causal processes. His views on the nature of the adaptive pressures involved might be highly speculative, but he could not in principle suggest a purely formal pattern of development.

To opponents of evolution, this speculation went much too far. They refused to admit that the fossil record could be as incomplete as Darwin claimed, regarding the gaps as genuine indications of discontinuous steps in the advance of life, best explained by divine creation. Arguments that had been valid against *Vestiges* were still valid against the *Origin,* however much Darwin might try to evade the issue by invoking the imperfection of the record. The evolutionists must produce at least some evidence of continuous change within a group and at least some evidence of intermediates linking the major groups that are now so distinct. Despite Darwin's efforts to defuse the issue, the fossil record thus became a major debating point—and remains so even today.

The most ambitious efforts to reconstruct the course of evolution took place in Germany. There, the introduction of Darwinism allowed naturalists to pin down their search for an alternative to idealism. Instead of seeking archetypical relations between species, they would try to work out real evolutionary connections. Carl Gegenbaur was a leading exponent of this idealist-turned-evolutionist approach, little concerned with the details of the Darwinian mechanism but absorbed by the morphological problems of working out the most likely connections between known forms (Russell, 1916; Nordenskiöld, 1946; Coleman, 1976). He made important suggestions concerning how the air-breathing vertebrates might have evolved from the fish, although later naturalists were forced to abandon some of his pioneering insights.

By far the best-known exponent of this approach was Ernst Haeckel (Bölsche, 1906). Selection played only a secondary role in Haeckel's evolutionary philosophy, because he combined it with an idealist desire to search out the structural relationships between living things and a revived version of Lamarck's mechanism of the inheritance of acquired characters. Biological races were formed by direct response to the environment and were then screened by a struggle for existence at the racial level which allowed only the fittest to survive. The ultimate direction of evolution was always toward progress; man was the highest form produced so far and contained within him seeds of an even greater development. Beginning as a materialist, Haeckel eventually expounded a philosophy of "monism" according to which mind and matter are aspects of the same universal substance. These views had a strong ideological slant, and Haeckel's "Monist League" has been seen as an influence on the later foundation of Nazi thought (Gasman, 1971).

Haeckel's reconstructions of the history of life were treated as an important part of the new evolutionism, and his books reached a wide audience even in English translation (1876, 1879). Even where fossil evidence was lacking, Haeckel confidently predicted the course evolution must have taken to reach man. While aware of the complex nature of the process, he tended to stress the constant production of ever-higher forms, giving an impression of inevitable progress. One of the chief problems was his willingness to see modern forms as almost unchanged relics of principal stages in the advance toward man, forgetting that any kind of life may continue to change even though it is no longer at the head of what we so anthropomorphically designate as the key line of progress.

As a microscopist, Haeckel described the earliest stages of life, in which a single-celled organism evolved into a multicellular form with a primitive intestinal canal. This hypothetical ancestor of the whole animal kingdom he called the "gastraea." From it evolved the various invertebrate types, including the Tunicates, which Haeckel regarded as closely related to the ancestors of the vertebrates. The work of A. Kowalewsky on the Amphioxus or lancelet was used to suggest that this primitive modern creature, lacking a bony skeleton, represents the ancestor of the whole vertebrate or chordate phylum. From it the true fish evolved, and in particular the lung fish or Dipneustra, capable of breathing air for a short period of time. By acquiring limbs and improved lungs, these evolved into the amphibians, which in turn gave rise to the reptiles when they became able to lay their eggs on dry land. Haeckel suspected that a number of different kinds of reptiles had evolved separately into mammals and that birds were another offshoot in a different direction. The first mammals were similar to the modern monotremes (e.g., the duckbilled platypus), after which the marsupials evolved and then the true mammals. The class branched out to give the various orders of modern mammals, including the Primates—the apes and man. Haeckel even coined the name *Pithecanthropus* to denote his purely hypothetical intermediate by which man had evolved from the apes.

Only a few of the connections postulated by Haeckel's reconstructions rested on fossil evidence; therefore, he made great use of indirect evidence. He assumed that anatomical similarity indicated community of descent. More controversial was his revival of the recapitulation theory, the claim that ontogeny recapitulates phylogeny, that is, that the growth of the modern embryo repeats key stages in the past evolution of the race. Haeckel did not accept Agassiz's belief that a divine plan of creation is repeated both in the succession of animal classes and in the growth of the human embryo. He argued that a natural mechanism would preserve the adult forms of ancestors as stages in the embryonic growth of their modern descendants. Although some types of evolution interfered with the process, by exercising due caution, the naturalist could pick out those embryological structures that do represent stages in past evolution. It is probable that Haeckel's exaggerated use of this "biogenetic law" created the popular belief that evolutionism entails such an embryological parallel. In fact, modern biologists do not accept a simpleminded version of the recapitulation theory and their evolutionism does not rest upon such shaky foundations. Haeckel's use of the embryological analogy illustrates the extent to which his thought was influenced by Lamarckian, rather than Darwinian, factors (Gould, 1977*b*).

In the 1870s and 1880s, much work was done in anatomy and embryology by biologists trying to follow up Haeckel's lead. Alternative interpretations were proposed for the crucial stages of evolution, and the evidence for each argued backward and forward. Eventually, scientists more concerned with experimental studies began to question the value of the whole exercise. Because new fossil evidence had allowed the theories to be tested directly only in a few cases, they argued that the project should be abandoned in favor of more productive studies. Yet the reaction against building up hypothetical genealogies did not represent a rejection of the basic idea of evolution, and paleontologists still tried to fit those fossils that did turn up into an evolutionary scheme. By the end of the century, Sir J. W. Dawson of Montreal was almost the only paleontologist of any reputation who still defended creationism (Dawson, 1890; O'Brien, 1971). Why, then, did the paleontologists accept evolution? The answer seems to be that enough fossil evidence turned up to convince them of the futility of creationism, even if it did not prove evolution. The evolutionists could predict that if, for in-

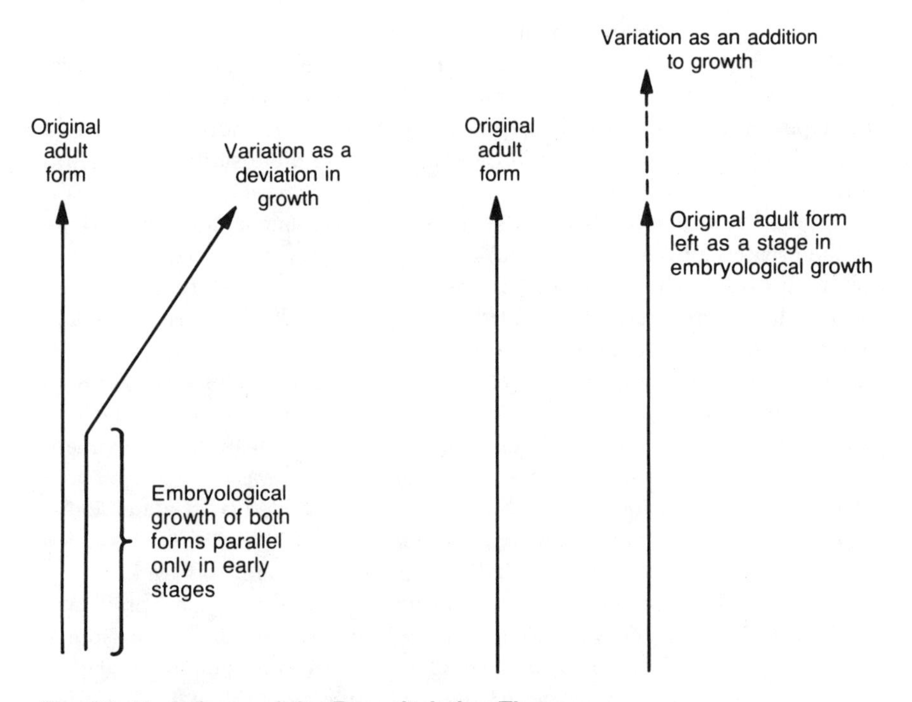

Fig. 20. Variation and the Recapitulation Theory.

Fritz Müller first outlined the circumstances under which recapitulation can take place (translation, 1869). Whatever the mechanism by which it is produced, evolution consists of the accumulation of individual variations. There are two ways in which an individual organism can be thought to acquire some variation from the structure normal to its species. One (left) is by a deviation in the growth process so that the adult form is different from that produced as the end product of normal growth. In this case, there is no possibility of the original adult form being preserved in the embryological growth of the variant. There can be no recapitulation of evolution in embryology, only a similarity of embryological development up to a certain point. In the second mode of variation (right), the new character appears by *adding on* an extra stage to the original growth process. The old adult form then becomes a stage in the embryological growth of the variant, and embryology recapitulates the past stage of evolution. But such a concept of variation is derived more easily from Lamarckism, where evolution proceeds from efforts of the adult organism, than from Darwinism, where random variation more easily can be seen as a deviation in the growth process.

stance, birds had evolved from reptiles, then one might hope to find a fossil with characteristics intermediate between the two classes. Creationists, by contrast, had a vested interest in maintaining the absolutely distinct characteristics of existing groups and were unwilling to postulate intermediates in the past. When such intermediates *were* discovered, they could be fitted into a creationist scheme by adding yet another miraculous event, but creationists would make this concession only after the discovery, depriving their position of any predictive power. A single intermediate thus would support the evolutionists' position, even if the complete sequence was still missing.

Not all fossil discoveries turned out to be of permanent significance. One of the greatest problems for evolutionists was lack of any fossils in the Precambrian rocks, which gave the impression that the earliest forms of invertebrate life appeared suddenly at the beginning of the Cambrian. Darwin and his followers at first were cheered by the discovery in 1865 of a strange "fossil" in the Precambrian rocks of the Canadian shield. This was described as the remains of a giant Foraminifera (a primitive marine organism) and was named *Eozoön canadense* (O'Brien, 1970). To evolutionists, this "dawn animal of Canada" was a sign that the Precambrian period was not totally devoid of evolving life. Curiously, the discovery also was hailed by the creationist J. W. Dawson, who believed that it supported his position; more advanced than its closest modern relatives, the organism indicated no evolutionary progression. Unfortunately, it was soon demonstrated that the "fossil" was the result of purely mineral action. Only in recent decades has the existence of genuine microfossils been confirmed in the Precambrian rocks, with a number of more advanced types immediately preceding the main Cambrian "explosion."

If Darwinists could draw no permanent support from *Eozoön*, there were other areas in which more promising discoveries were made. Attention soon began to focus on the link between the reptiles and the birds. T. H. Huxley first suggested that certain dinosaurs have legs and feet almost indistinguishable from those of birds (Rudwick, 1972; Bowler, 1976*a*; Desmond, 1976, 1982). This, he claimed, was the first sign of a "missing link" between two important classes. The subsequent discovery of *Archaeopteryx* added further confirmation: a creature with feathers like a bird, a body with many reptilian characters, and a mouth with teeth rather than a beak. O. C. Marsh discovered more toothed birds in America and named them the Odontornithes (Marsh, 1880). The exact course by which the birds had evolved from dinosaurs was still a mystery, but such intermediates would not be expected if, as creationists claimed, the later class had a completely separate origin. The possibility of links between other classes also was recognized, as in the case of the "mammallike reptiles" (Desmond, 1982). In this case, however, it was the anti-Darwinian concept of evolution promoted by Richard Owen and his followers that encouraged recognition of the fossils' significance. The idealist concept of orderly development was modified to serve as the basis for a claim that the mammals were not the descendants of a single evolutionary breakthrough but a "grade" of organization reached by a number of separate lines of reptilian evolution. Some more purposeful force than natural selection thus was supposed to be at work in evolution.

On a smaller scale, a number of almost continuous evolutionary sequences were also discovered. The example that attracted most attention was the evolution of the horse. The modern horse is highly specialized for running on open plains, with its broad hoof derived from a single toe. But if evolutionists were correct, it must have evolved from an ancestor with five toes like any normal mammal. Huxley at first had tried to link the horse with certain European fossils—naturally enough, because the modern horse was introduced into America by the white man. But Marsh soon convinced him that the horse must have evolved in America before migrating to Eurasia. Marsh had uncovered a whole series of fossils tracing the modern horse back to a small, multiple-toed ancestor he named *Eohippus*, the "dawn horse" from the Eocene. This series Huxley (1888) proclaimed as "demonstrative evidence of evolution." The true story of the horse's evolution has turned out to be rather more complicated, with many of the fossils representing extinct side branches rather than steps on the way toward the modern horse. The general outline of the sequence, nevertheless, corresponded closely to the Darwinists' predictions. Less spectacular, but even more obviously continuous, sequences of fossils were discovered among the invertebrates, leading the majority of paleontologists to agree that remaining gaps in the record were indeed a result of its imperfection. It must be emphasized, however, that some paleontologists thought the lines of evolution were too regular to be explained by the haphazard process of natural selection. Americans such as Edward Drinker Cope and Alpheus Hyatt portrayed adaptive evolution as a sequence of linear developments aimed toward a purposeful goal by a Lamarckian mechanism (Bowler, 1977*c*, 1983). These views would become the basis for a major outburst of anti-Darwinian feeling later in the nineteenth century (chap. 9).

THE AGE OF THE EARTH

If paleontology offered less of a problem than Darwin had feared, geology became one of his greatest headaches. As a follower of Lyell, Darwin had assumed that he had virtually unlimited amounts of time to play with in explaining the history of life. This assumption was crucial because of his insistence that evolution has been a very slow process. The accumulation of minute variations by selection would take hundreds, perhaps thousands of millions, of years to produce the diversity of life as we see it today. Lyell's geology allowed time for this slow development of life to occur, but now the uniformitarian theory was to face renewed opposition. An attempt was made to alter the whole time scale of the earth's history and to challenge, by implication, the Darwinian theory of evolution.

The challenger was the physicist William Thomson, Lord Kelvin (Burchfield, 1975). Kelvin's basic point was that the kind of steady-state geology advocated by Lyell was incompatible with the laws of physics. This argument should have been obvious from the start—probably the only reason it had not been raised against Lyell in the 1830s was the primitive state of thermodynamics at that time. Now, Kelvin could incorporate what was essentially a commonsense argument into an established scientific framework. All geologists accepted the fact that volcanic activity resulted from high temperatures in the earth's interior. But if the earth is hot, it must cool down like any other hot body, and the heat must be conducted to the surface and radiated off into space. Chemical reactions might generate some heat to offset this cooling, but Kelvin showed that this could have only a trivial effect on the final outcome. If the earth is hot now, it must have been hotter in the past and have gradually cooled down to its present state. If we trace the cooling far enough back in time, we must come to a point where the whole planet was a mass of molten rock, as one would expect it to have begun according to the nebular hypothesis. In 1868, Kelvin showed by detailed calculations that the amount of time involved was far less than predicted by the uniformitarian method in geology. Thus, he claimed, we are not justified in assuming that earlier geological processes took place at the same rate as their modern equivalents. The catastrophists had been right not only to adopt a directionalist philosophy of the earth's history but also to assume that past causes operated more vigorously than those of today.

The strength of Kelvin's position lay in his appeal to the fundamental laws of physics, which to most scientists seemed far more reliable than the flimsy theories of the geologists. If the earth is hot, the basic laws of thermodynamics dictate that it must cool down in a definite period of time. The planet cannot have been maintained indefinitely in a steady state unless some factor has continually compensated for the energy radiated into space. Kelvin knew of no such factor and simply assumed that his basic theoretical framework was valid, as did most of his contemporaries. Even allowing a margin of error in his estimates of internal temperature and rate of cooling, he believed that the total age of the earth could not be more than about a hundred million years, far less than the time required for Darwinian evolution. So great were Kelvin's prestige and the apparent strength of his argument that most geologists began to rework their theories to incorporate much faster rates of change than those postulated by Lyell.

The retreat of the geologists placed Darwin in a difficult position, because he simply could not accept Kelvin's estimate of a hundred million years (Burchfield, 1974). He had no positive arguments to offer in the realm of physics but insisted that the estimates he and Lyell had been making could not be in error to the extent implied. To the end of his life, Darwin believed that there was something wrong with Kelvin's calculations, but he had nothing to back up his belief. Even other evolutionists such as Wallace and Huxley began to accept the idea that evolution must have taken place more rapidly than they had at first thought possible. They argued that biology must take its time scale from geology, so that if geologists were prepared to accept Kelvin's estimates, evolution must have occurred in the same time span. By the end of the century, a number of alternative evolutionary mechanisms had been suggested, at least in part to show how the process could have advanced more quickly than by natural selection. Kelvin's arguments did not impede the growth of evolutionism, but they certainly contributed to the declining popularity of the selection mechanism in the late nineteenth century.

In the end, Kelvin's estimate of the earth's age became a casualty of the revolution in physics that took place after 1900. The discovery of radioactivity introduced a totally new factor that Kelvin had not been able to take into his calculations. In 1903, Pierre Curie announced that the radioactive decay of elements such as radium liberated a slow but steady supply of energy in the form of heat. These elements are present in small quantities throughout the earth. By 1906, Lord Rayleigh had shown that the heat produced by radioactivity deep in the earth would balance the cooling effect postulated by Kelvin. The rate of decay of these radioactive elements, furthermore, is so slow that this balance can be maintained over a vast period of time, allowing the earth to remain in a steady state. It took the geologists a little while to realize that they could now go back to using vast amounts of time just as Lyell had done; but by the 1930s, techniques of radioactive dating were developed to show that the Cambrian period must be placed over half a billion years in the past. The modern synthesis of Darwinism and genetics would not be troubled by the problem of time.

We tend to think of the nineteenth century as a period of optimism and of evolution as a symbol of Victorian faith in a progressive universe. Yet Kelvin's ideas point to a contradictory tendency, a pessimistic view in which the earth and ultimately the whole universe are running downhill toward degeneration and death. When Kelvin suggested that the energy supplies of the earth and the sun are limited, in effect, he was predicting that eventually the cold will kill off all life on the earth. On a wider scale, the physicist Rudolph Clausius predicted the "heat death" of the whole universe. All natural processes make energy less available by equalizing temperature differences, so that in the end the universe will contain nothing but matter at the same uniform temperature in which no natural processes can occur (Gillispie, 1960; Brush 1978). In this totally pessimistic philosophy, all natural—and human—activity ultimately must be brought to nothing by the inexorable laws of physics. The existence of such an alternative to the optimism of many evolutionists should not be ignored in any attempt to characterize late-nineteenth-century thought.

UTILITY, HEREDITY, AND VARIATION

Kelvin's calculations struck indirectly at Darwin's theory via the assumption that natural selection was a very slow process. But there were other arguments directed more specifically against the selection mechanism which required all Darwin's ingenuity in defense of his brainchild. There were fundamental problems, particularly with his understanding of variation and heredity, and it is not surprising that he found it difficult to counter the attacks launched in this area. This does not mean that the basic idea of selection was without value, only that there were problems caused by Darwin's inability to develop the idea properly within the limitations of contemporary understanding. The modern combination of selection and genetics has resolved many of these earlier difficulties.

Natural selection is a strictly utilitarian mechanism: it can develop only those characteristics that are useful to the individual in the struggle for existence. If it were the sole mechanism of evolution, we should expect that every characteristic of every species, even the most trivial, serves some adaptive purpose. But is this always the case—are there not examples of characteristics that are useless or even harmful to the species? How can one explain the bright colors of, say, the hummingbirds, which must make them extremely visible to predators? Darwin made one major addition to his theory to account for such cases. This was the mechanism of "sexual selection," outlined in the *Origin* and discussed at length in the *Descent of Man* (1871). Some characteristics, he argued, are useful not in the struggle for existence but in allowing the individual to mate. If a particular color scheme on the male bird, for instance, serves to attract the female in the courtship ritual, any male in which those colors are particularly well developed will mate more often and leave more offspring who will inherit the characteristic. Over many generations, female choice of the most brightly colored males would greatly enhance the degree of coloration. On a similar basis, organs such as the horns of male deer would be developed, because they are used in the competition by which males gather a "harem" of females for mating. In each generation, only those stags with large antlers would succeed in mating, and the average size of the horns gradually would increase.

Sexual selection was not without its problems. Darwin and Wallace engaged in a substantial debate on the topic, and eventually Wallace rejected the concept of female choice altogether (Kottler, 1980). Other naturalists believed that it would be impossible to explain the whole range of animal coloration in terms of the colors being useful to the individual organisms.

On a more general level, the botanist Carl von Nägeli argued in 1864 that many kinds of characteristics are of no value and cannot have been developed by selection. In particular, he claimed that the trivial characteristics used by naturalists to distinguish between closely related species could be assigned no utilitarian value. By the end of the century, the belief that nature is not built along lines dictated solely by adaptation had swelled into a major current of opposition to Darwinism.

Darwin took this objection very seriously and conceded that selection was unable to explain all the examples of nonadaptive characteristics. He did insist that Nägeli had exaggerated the prevalence of such characteristics. It was difficult for the naturalist to be sure that a particular characteristic was without purpose, because often he knew little about the life-style of the species in the wild. But Darwin was forced to admit that in some cases he could see no possibility of a useful purpose. The only hope of explaining such a characteristic within the context of the selection theory was by use of an additional concept that Darwin called "correlation of growth." If two characteristics are linked in the growth process of the organism, then the action of selection in favoring one of them automatically would promote the other even if it had no value in itself. Other naturalists, however, tended to assume that processes within the organism actively forced variation in nonadaptive directions.

This point was taken up by St. George Jackson Mivart in his *Genesis of Species* (1871). Mivart, a Catholic who was deeply concerned over the implications of human evolution, was one of Darwin's most persistent critics (Gruber, 1960). He pointed out that even those characteristics that are useful when fully developed must have passed through an incipient stage when they would have been of no real value. The Darwinists later countered this by suggesting that an organ frequently does not grow from nothing but is switched from one function to another after it is fully developed. Mivart also pointed to a number of strange coincidences linking quite different branches of evolution. Why does the octopus, for instance, have eyes that are very similar to our own? Surely selection of random variation cannot have produced this similarity between two such widely differing types. Mivart suggested instead that there must be some kind of predisposition for life to evolve in certain ways, whatever the action of selection. Darwin regarded this as unnecessary, because the general structure of an efficient eye would be determined by the laws of optics, while the detailed form of the eye in cephalopods and vertebrates is actually quite different.

An equally persistent line of objections centered on the internal workings of the selection mechanism. We have noted (chap. 6) that Darwin shared the common belief of his time in "blending heredity," which states that any characteristic in the offspring is a blend or average of the parents' characteristics. It was known that some characteristics are inherited on an all-or-nothing basis, but these were regarded as exceptions to the general rule. Darwin explained blending through his theory of pangenesis (published in Darwin, 1868; see Geison, 1969; Robinson, 1979; Farley, 1982; Hodge, 1985). This theory originated from his earliest thoughts on the role played by reproduction and growth in the evolutionary process Each part of the body was supposed to bud off particles called "gemmules" responsible for reproducing that part in the offspring. The gemmules are carried through the bloodstream to the reproductive organs, and fertilization consists of a mixing of gemmules from both parents. Because each will normally contribute some gemmules for each characteristic, blending will be the general rule, although there may be exceptional cases in which gemmules from one parent predominate. Today we regard the all-or-nothing cases as better illustrations of the fundamental process of heredity as described by Mendelian genetics. Thus, Darwin's theory of pangenesis and his whole understanding of heredity were, by modern standards, seriously in error. It must be remembered, however, that pangenesis reflected typical mid-nineteenth-century attitudes toward growth and reproduction. Despite the apparently modern character of his biogeographical approach, Darwin's thinking on the link between growth, heredity, and evolution remained grounded in a conceptual system that would only be overthrown with the emergence of Mendelian genetics.

To what extent did this error affect the validity of the selection mechanism? The fact that there was a problem is illustrated by Darwin's reaction to a review of the *Origin* by the engineer, Fleeming Jenkin (1867; reprinted in Hull, 1973*b*). Eiseley (1958) writes as though this review convinced Darwin that selection fundamentally was unsound and forced him to accept Lamarckism instead. He points out (quite correctly) that pangenesis will allow Lamarckism: because the parts of the body manufacture their own hereditary material, changes in structure caused by use and disuse will be reflected in the production of gemmules and can be inherited. But it would be wrong to imply that Darwin proposed pangenesis as a means of converting to Lamarckism. Vorzimmer (1970) goes to the opposite extreme of minimizing the effect of Jenkin's review, suggesting that it did no more than confirm certain points already in Darwin's mind. A more reasonable interpretation of the obvious concern shown in Darwin's letters about the review is that he was forced to change his way of describing selection without abandoning it altogether (Bowler, 1974*b*).

Jenkin conceded that natural selection will work if we imagine it acting on a complete range of variation within a population. If extra height, for instance, turns out to be an advantage, there always will be a proportion of the population with above average height who will benefit in the struggle for existence. Similarly, there will be many with below average height who will be at a disadvantage, and the next generation will be bred more from taller than from shorter parents. In these circumstances, blending heredity would not interfere with natural selection but would merely spread its effect more evenly through later generations.

Jenkin, however, then went on to argue that selection acting on the ordinary range of trivial variation within the population cannot be the cause of large-scale evolution. He accepted the common belief that ordinary variation is constrained within fixed limits that define the species. Selection at this level could certainly produce varieties within a species, but it could not produce a new species because the variations themselves could not be pushed beyond the limit. Evidence for such a barrier came from domesticated species. The dog, for instance, has been selected for centuries to produce a host of widely differing breeds, yet all dogs are still members of the same species. Darwin, of course, did not accept such a barrier, believing instead that variation would be able to overstep apparent limits if given enough time. On this assumption, there was thus no reason for him to abandon selection, and even Jenkin conceded that it *could* work. Why, then, did Darwin admit that the review had made him think so carefully about the problem of variation and heredity?

The answer lies in the second stage of Jenkin's argument. He had shown to his own satisfaction that selection acting on normal variation could have only a limited effect, and he went on to suggest that the only way of bypassing the limit was for it to work with those rare individuals who are born with some gross abnormality of structure, that is, with monstrosities or what were then called "sports of nature." Such sports could outstep the bounds of normal variation, and it was not impossible that occasionally one might come along whose deformity was useful in the struggle for existence. Could natural selection make use of such "hopeful monsters"? It was known that artificial selection could create a new breed from a single individual born with the required characteristic. In the case of the "Ancon sheep," a farmer had noticed that one of his male lambs had been born with abnormally short legs and realized that if he could spread the characteristic through his whole flock, it would be useful in preventing the sheep from jumping over fences. By careful selection among the offspring of the original ram, he was able to create such a breed, and it had become quite popular. Jenkin's point was that this could not occur in nature. Nothing there ensured that offspring of the sport bred together, and in a few generations its effect would be swamped out through mingling with the mass of the unchanged population. Even if the sport bred more often than normal individuals because of some advantage in the struggle for existence, it would have no effect on the population in the long run. It would be like putting a drop of black paint into a bucket of white—after stirring, there would be no trace of it left.

If Darwin had based his idea of selection on such hopeful monsters, it would be clear why Jenkin's review came as such a shock. Instead, he was committed to the view that "nature takes no leaps" and that evolution proceeded through the accumulation of minute changes. The problem was that he had sometimes written as though even *small* favorable variants occurred as rare individuals. Instead of promoting the idea of a complete range of variation across the whole population, he had pictured as very rare individuals with even a slightly favorable characteristic. When there is a range of variation for a characteristic such as height, then by definition half the population will be above and half below average, and selection will be able to work even with blending heredity. But if favorable variants (small or large) are very rare individuals, then Jenkin's swamping argument will apply.

The effect of Jenkin's review thus was not to make Darwin abandon selection but to require a different way of describing the variation within a population. Darwin would have to stop thinking of variants as single individuals and adopt the concept of a range of variation instead. Selection could work only if the full implications of "population thinking" were exploited. A. R. Wallace was particularly insistent that Darwin recognize this point and in his own later works devoted a great deal of space to showing that there is a significant range of variation for any characteristic within all wild populations (e.g., Wallace, 1889). The selection mechanism thus was preserved by adopting a thorough populational description of variation, which Darwin had tended to abandon because of his pessimism about the extent of variation in the wild. Toward the end of the century, the science of "biometry" emerged to apply statistical techniques to the measurement of variation in wild populations.

A rather different problem of blending centered on the process of speciation, the division of one species into a number of distinct later ones. The example of the Galápagos islands originally had convinced Darwin that geographic isolation was essential for this splitting to occur. Without a physical separation of the populations, constant interbreeding would blend their characteristics and prevent any permanent division of the species. The problem raised by Jenkin would be circumvented, because in a small island population a single variant would not be swamped completely. As we saw in chapter 6, Darwin had abandoned this view by the time he wrote the *Origin* and now believed that natural selection was powerful enough to break up an originally continuous population by adapting its extremes to different life-styles. He assumed that the sheer size of the population would guarantee enough favorable variants, even if the rate at which such individuals were produced was very low. Extreme isolation would hinder the process, because in a small population the appearance of a well-favored individual would be a very rare event. Jenkin convinced him that he must not think of favorable variants as single individuals: *any* population, even a small one confined to an island, must exhibit a significant range of variation in all its characteristics. Darwin, nevertheless, did not reconsider his position on isolation, because he still felt that the more vigorous competition occurring in a continuous geographic area would best promote divergence. He rejected the claim of Moritz Wagner (translation, 1873) that geographic isolation is essential for speciation (Mayr, 1959*b*; Sulloway, 1979). But this left him with the second problem of blending: what prevented the incipient varieties from interbreeding and losing their distinct characteristics? The search for a mechanism that would allow sympatric speciation (speciation without geographic isolation) continued throughout the rest of the century, and its lack of success helped to turn many naturalists against the selection theory. Eventually, Wagner was vindicated, as most naturalists have come to accept the idea that geographic isolation is indeed essential for the first stages in separation.

The difficulty of explaining speciation was compounded by the failure of most early Darwinists to distinguish between geographic and reproductive isolation. This was, in turn, a result of their unwillingness to adopt a completely populational concept of the nature of species. No intermediate stage was defined between "varieties" that still are capable of interbreeding and true species, which are genetically incompatible. Modern biologists recognize the importance of "isolating mechanisms" that may be behavioral rather than genetic: individuals from two populations may be genetically capable of interbreeding but will not do so because their mating behavior is different. Once such an isolating mechanism has appeared, two populations may occupy the same territory without intermingling and can go on separating until they are true species with genetic incompatibility. The crucial question is how the separation is maintained until the isolating mechanism has become established. Current thinking favors the belief that a preliminary stage of geographic separation is essential, even if barriers are subsequently removed. Darwin did not take this view, but he was unable to explain how natural selection can divide an originally continuous population until genetic sterility is achieved.

This problem was related to the whole question of interspecific sterility. Why is it that the crossbreeding of varieties within a species gives normal offspring, while attempts to cross distinct species either fail altogether or produce sterile hybrids like the mule? Opponents of evolution saw this sterility as proof of the distinct nature of species. Darwin argued that there was no sharp dividing line, only a gradual increase in degree of sterility as two forms become more distinct in their physical characteristics. The mule illustrates that in this case the distinction between the two parent species is not as complete as it might be. This still leaves the question, though, of how evolution builds up the level of sterility. Mutual sterility is valuable to the two populations, because it encourages divergence and thus lessens the degree of competition between them. It is of no value, however, to the *individual*, who seeks to maximize the number of his offspring by pairing with as wide a range of mates as possible. Darwin believed that sterility was only a by-product of divergence, which is the essence of the modern position. Wallace, however, insisted that natural selection actually causes sterility, although he seems to have been thinking of selection acting between groups rather than between individuals. Darwin would not accept "group selection" (Ruse, 1980), and most modern biologists agree that selection is a purely individual affair. The controversy over sociobiology has been generated essentially by the attempt to explain characteristics that *seem* to benefit the whole population as the result of selection based purely on the reproductive success of individuals (chap. 11). In the case of sterility, there is no individual benefit, so that this characteristic must emerge only as a by-product of divergence, even though it is of benefit to the populations as a whole.

THE PROBLEM OF METHOD

In addition to the biological objections, Darwin also found that the validity of his scientific method was being questioned. He had hoped that his work would be received as a truly scientific contribution and was disappointed to find that the *Origin* was criticized as a mass of speculations, unsubstantiated by the evidence he offered. It was claimed that he had deserted the path of inductive science to indulge in wild hypothesizing, moving outside the pale of science altogether (Ellegård, 1957, 1958; Hull, 1973*a*, 1973*b*).

The problem of deciding the role of hypotheses in science confused most philosophers at the time. Important figures, such as Sir J. F. W. Herschel, whose writings had created Darwin's early enthusiasm for science, conceded that there was a role for hypotheses or controlled speculations in the process of discovery. Simple Baconian induction, collecting facts at random in the hope that eventually a pattern would be discerned, was often inadequate to cope with vast amounts of available data. The scientist would have to invent a plausible hypothesis based on facts available to him and then use this as a guide to further research. Today this is called the hypothetico-deductive method (Hempel, 1966). From the hypothesis, empirical consequences are deduced and tested against the facts. If the test fails, the hypothesis is rejected and a new one sought. If the first tests are successful, testing continues, either to build up more support or until, eventually, the weakness of the hypothesis is exposed.

There is no doubt that Darwin used such a method in establishing his theory (Ghiselin, 1969). Why, then, did philosophers of science regard his theory as unfounded? It was partly because of a general feeling that the scientist should not go too far beyond accepted norms in conceiving his hypothesis. Darwin had flown in the face of all received doctrines of natural history and thus had engaged in uncontrolled speculation. Nineteenth-century philosophers, furthermore, believed that a really fruitful hypothesis in science must represent a truth about nature. In the end it should be capable of demonstrative proof, of deduction by a logical chain of reasoning from observed facts. This is something no modern philosopher of science could accept, because by such a standard almost every theory would have to be judged inadequate. One cannot *prove* a theory to be true in most cases, because the complexity of the situation will not allow such complete deductions to be made. The scientist can only provide evidence to *support* his theory, recognizing at all times that this falls short of absolute proof and thus the theory only can be accepted provisionally.

Darwin certainly had provided support for his hypothesis, showing that in a number of areas there were facts that were inexplicable except on the basis of evolution. He realized that to demand demonstrative proof of his theory was, in effect, to bar the whole subject to scientific investigation: neither his nor any other conceivable theory could hope to provide such proof. But many of his contemporaries believed that a theory must be proved before it could be incorporated into the body of accepted scientific knowledge. How controversial even some of the most important theories had been when they were first introduced was forgotten. Huxley fell into this trap when he admitted that Darwin's theory would not be proved until two infertile species had been bred from a single source by artificial selection—an unreasonable demand given Darwin's belief that evolution was an immensely slow process. At least, Huxley realized that the theory was truly scientific in that it could be tested against the facts and thus could be *supported* while one waited for proof. Those who rejected the theory accused Darwin of offering them nothing but unsupported speculation because he could not provide them with absolute proof. It might be added that this technique is still used by modern creationists.

A related problem concerned the value of selection as a *vera causa,* a "true cause" of the kind acceptable to science. This arose from debates over the nature of scientific laws and the concept of causation. The chief protagonists were William Whewell, whose *Philosophy of the Inductive Sciences* had appeared in 1837, and John Stuart Mill, whose classic *System of Logic* was published in 1843. Mill upheld the tradition of British empiricism, according to which all knowledge is derived from the senses and a "cause" simply denoted constant conjunction between one kind of event and another. Whewell had tried to adapt Kant's idealism to British tastes, arguing that intuition provided a source of knowledge independent of the senses which could enable the scientist to intuit the laws of nature and then relate the facts to the law. Intuition, furthermore, recognized a cause as an active power in nature responsible for producing the sequence of events in accordance with the law.

Darwin's supporters were members of the empiricist camp. Mill himself fell into the trap of failing to distinguish between the process of discovery and the logic of proof, suggesting that the theory required absolute proof before it could be firmly accepted. At least, he confirmed the fact that selection was, by empiricist standards, a vera causa. All the elements built into it (random variation, the struggle for existence, etc.) were empirically verifiable effects. Whether or not they fitted together to give the consequences claimed by Darwin would have to be proved, but at least the foundations were sound. Darwin was disappointed that both Huxley and Mill required a kind of proof that this theory by definition could not provide, but he knew better than anyone else how firmly he had based his idea on purely observable causes.

For the idealists, however, Darwin had not fulfilled the requirements for providing a true cause. They believed that they could intuit certain truths about nature that were far sounder than Darwin's flimsy theoretical framework. These truths included the fixed character of species and the existence of a *purpose* governing their structure and relationships. The secondary or natural causes with which the scientist normally deals rest ultimately on the power of the First Cause, God Himself. Because the idealist was intuitively certain of the First Cause and its direct control over nature, he held that it was permissible to call upon this higher cause whenever secondary causes were shown to be inadequate. Thus, Whewell could argue that natural causes are known to be incapable of explaining the origin of species, at which point the scientist had to give up the search, secure in the knowledge that the First Cause must be involved directly. Even if one accepted the possibility of a "law of creation" based on evolution, this would have to account for the purposefulness of nature and still would require the guiding intelligence of the First Cause.

The problem with Darwin's mechanism was that it did not include this element of purpose. Variations were produced by an effect that Darwin could not describe completely and that had no relevance to the requirements of the species. Selection was left to make the best of a bad job by eliminating all but those who accidentally were favored in the struggle for existence. The fact that Darwin could not pinpoint the cause of variation indicated the weakness of his theory: how could he claim to have explained evolution if he could not account for the origin of the variations on which it is based? For the idealists, a true cause of evolution would have to account for production of variation in a way that would incorporate the naturalist's intuitive sense of the purpose built into the history of life. Herschel called natural selection the law of higgeldy-piggeldy to express his contempt for a mechanism that attributed the development of life to a haphazard combination of accidents. The truth is that more conservative thinkers wanted to retain the old ideas of fixed species and the argument from design. Idealism allowed them to claim intuitive knowledge of the purpose of nature and hence to dismiss Darwin's theory as inadequate because it did not support this intuition. If evolution occurred, it would have to be as the unfolding of a divine plan, not as a chapter of accidents. Thus, the idealists' criticisms of selection lead into the religious and moral problems created by the theory.

To sum up the various scientific debates, Darwin converted most of his fellow naturalists to a belief in evolution, but he was less successful with his selection mechanism. Some conservative thinkers rejected any form of natural evolution, because it contradicted their belief in the divine orderliness of nature. Darwin, however, had shown that in a number of areas one could sketch in at least the outline of a natural explanation of facts that hitherto had been ascribed to the Creator's whim. Increasingly, biologists began to accept natural evolution, simply because they saw that the origin of species was not a subject that must remain forever beyond the scope of scientific investigation. The selection mechanism certainly played a role in this: it showed that natural explanations of evolution were not totally inconceivable. But whether or not it was an adequate explanation remained open to question. Darwin and his followers exploited the selection theory where possible but were willing to accept additional mechanisms to cover difficult cases. For a while, they were able to present selectionism as the central, if not the exclusive, principle of scientific evolutionism. There were apparently sound objections, however, first pointed out by those who rejected any kind of natural evolution. As time went on and these objections showed little sign of yielding to the Darwinists' efforts, more naturalists came to have second thoughts about selection. Although no longer tempted to reject natural evolution altogether, they continued the search for alternative mechanisms to selection and generated a mood of increasing hostility toward selectionism at the end of the century (see chap. 9). Since many of these alternatives preserved the almost teleological progressionism typical of pre-Darwinian views on natural development, there is a sense in which the Darwinian revolution was less effective than has often been supposed. The triumph of a theory of natural development did not have the disastrous consequences predicted by Darwin's critics because the most radical aspects of his theory were not accepted in his own time.

8

Darwinism: Religious and Moral Problems

The clash between Huxley and Wilberforce at the Oxford meeting of the British Association is often seen as a symbol of the hostility between Darwinism and religion. Disraeli's famous remark that, given the choice between treating man as an ape or an angel, he was "on the side of the angels" pinpoints one of the chief fears expressed by conservative thinkers. A host of cartoons in the popular press brought home to all the incongruity of treating man as merely an improved ape. Yet the problem created by the evolutionary origin of man must be seen in a wider perspective. In the end, the vast majority of people did accept evolution and hence that man himself had evolved. Nor was the problem solely a product of evolutionism, because psychologists also were moving toward a materialistic explanation of the mind in terms of brain functions (Young, 1970*a*). Evaluating man's place in nature now became the central task (Young, 1973). To ask what *kind* of natural system he was supposed to be the product of was more crucial than ever. Exactly how did the laws of organic development work, and did they offer any hope of seeing the human mind as a necessary goal of nature's activity? Reconciliation of evolution with traditional concepts of teleology and design became the chief aim of religious thinkers who tried to grapple with Darwin's theory.

Surprising as it may seem in today's world of revitalized biblical fundamentalism, there was little opposition to Darwinism on the grounds that it challenged the literal word of the Genesis creation story. The great controversies over geology and paleontology earlier in the century had convinced almost everyone that the text of Genesis must be understood in a less rigorous way that would allow the earth and its inhabitants to change over a vast period of time. The challenge of naturalistic science went much deeper than this and constituted but one aspect of the wider process by which nineteenth-century thought became increasingly secularized (Young, 1970*b*; see Mandelbaum, 1971; Chadwick, 1975). The Darwinian debate came at a delicate moment when the Victorian church was engaged in a number of internal problems (Chadwick, 1966; Symondson, 1970). The Evangelical movement was urging a return to a religion of the heart, while the Oxford movement's concern for ancient tradition led John Henry Newman, at least, back to Rome. Even more disturbing was the debate over the "higher criticism" sparked by the translation of D. F. Straus's *Life of Jesus* in 1846. Many found the prospect of treating the Bible as a historical document rather than the word of God profoundly disturbing. When *Essays and Reviews* offered some British contributions to this new approach in 1860, a heated controversy ensued (Willey, 1956; Brock and MacLeod, 1976). Darwinism fed into an existing debate over whether the divine origin of the sacred texts guaranteed certain knowledge of spiritual truths that could be accepted on faith.

In its most extreme form, the challenge to traditional thought manifested itself not in the details of Darwin's theory but in the more comprehensive world view into which evolution was incorporated by more radical thinkers. Darwinism came to symbolize a new system of values, derived more from the evolutionary philosophy of Herbert Spencer. Universal progress was seen as a necessary outcome of the mechanical operations of the laws of nature. Man was a product of the evolutionary process, and, in the absence of any transcendental source of moral values, he would have to create a new ethics based on the guidance of nature itself (Greene, 1981). An increasing confidence in the power of science to generate a complete explanation of the universe was an important part of this new materialism, proclaimed by the physicist John Tyndall in his notorious "Belfast Address" of 1874 (reprinted in Tyndall, 1902). Darwinian evolutionism played its role in promoting this confidence, of course, although a number of Darwin's immediate followers were unwilling to accept the materialist program. Scientists, like many other thinkers, often found the complete extension of evolutionary naturalism into a way of life distasteful (Turner, 1974). Wallace, for instance, came to believe that the last stages of human evolution were supernaturally guided, while Huxley eventually became a severe critic of evolutionary ethics.

The fact that not all scientists went along with the new materialism alerts us to the need for a flexible approach to interaction between Darwinism and religion. All too often, the complexity of the situation has been obscured by zealots on either side who have insisted that science and religion must necessarily come into conflict. While it is true that a radical group of materialists did try to exploit Darwinism's antireligious implications to the full, this was an extreme position that did not reflect the opinions of many scientists and theologians (Greene, 1961; Moore, 1979; Durant, 1985). Just as biologists differed over their willingness to accept a naturalistic world view, so theologians showed varying degrees of willingness to accommodate evolution into their thinking. A few resisted any compromise with the new theory. The absolute fixity of miraculously created species allowed them to preserve the traditional view that there are divinely ordained truths to which the human mind can aspire. This position continued to be upheld by a few extremely conservative biologists, including Louis Agassiz and Sir J. W. Dawson (O'Brien, 1971; Cornell, 1983). Paradoxically, a few strict Calvinists expressed a willingness to adopt the complete Darwinian scheme, including natural selection (Moore, 1979). For them, the potentially disturbing implications of a theory in which the rigid operations of law do *not* guarantee progress could be reconciled with the Christian view that man is a fallen creature whose salvation must come from a source external to this world.

The majority of scientists and theologians tried to find a middle ground by modifying the structure of Darwinism so that evolution itself could be seen as an essentially purposeful process. Darwin's own theory had its origins in the utilitarian natural theology of Paley and Malthus, where apparent evils were assumed to work for the eventual good of all in accordance with God's design. By now, Darwin himself was finding it difficult to reconcile natural selection with design, and most people found the mechanism too harsh and too selfish to be portrayed as the means chosen by a benevolent God to achieve His ends. Evolution could be accepted as the mechanism of creation only if selection could be replaced by a process showing clearer evidence of its Designer's intentions. In the early debates over the *Origin*, many scientists and theologians opted for a compromise position in which evolution is guided toward adaptation and progress by some kind of supernatural power. This concept of theistic evolutionism proved unsatisfactory in science, however, because it still prevented the biologist from providing a completely natural explanation of the phenomena.

As the century progressed, the Lamarckian mechanism of inheritance of acquired characteristics was increasingly seized on as a solution to this problem, a means of adapting the whole species to a goal chosen by the individual animals. To many religious thinkers, a universal system of evolution in which life directed its own development seemed the best way of salvaging a remnant of the argument from design. This romanticized version of evolution has been called "Darwinisticism" to distinguish it from the "Darwinism" that more closely follows the spirit of the selection theory (Peckham, 1959). It is true that Darwin's name all too often was linked with any form of evolutionism, but this clumsy term seems redundant given that most scientists who opted for a Lamarckian viewpoint were well aware of the gulf separating them from the Darwinians. Theistic evolution and Lamarckism can be represented best as consecutive stages in the emergence of a self-consciously anti-Darwinian philosophy of evolution, aimed at retaining a role for divine providence in nature. In many respects, this philosophy was a direct continuation of the developmental view of nature already outlined in Chambers's *Vestiges*.

This confusion over the extent of Darwin's influence arose because both the materialists and their opponents modified his theory of evolution in a similar way, however different their intentions. Few Victorian thinkers of any persuasion could tolerate the idea of evolution as anything but an essentially progressive system: evolution *had* to have a purpose in which the emergence of man played a key role, whether or not one traced that purpose back to a supernatural Creator. More radical thinkers, such as Spencer, denied any transcendental source of moral values but insisted that it was worthwhile to follow nature's own rules. This would ensure not only individual happiness but also the progress of the race; and progress was guaranteed by the fact that natural evolution is itself progressive. Huxley criticized Spencer's evolutionary ethics precisely because he saw that in a truly Darwinian universe there was no guarantee of progress and hence no point in following nature's harsh methods. By advocating a laissez-faire system in which each individual had to look after his own interests, Spencer had given full rein to the worst aspects of Victorian capitalism. In arguing that moral values would have to be defined by the human conscience, not by nature, Huxley was returning to a more traditional source of ethics, even though he could find no assurance that the conscience reflected divine moral guidance.

Liberal Christians, instead of insisting that moral values must have a transcendental source, assumed that the progressive nature of the material world was a deliberate part of God's plan, intended to teach us how to behave. They might disagree with the character of Spencer's ethics, but they too wanted the reassurance of seeing their own values reinforced by nature (Moore 1985*a*, 1985*b*; Livingstone, 1987). Some managed to rescue traditional values by arguing that the driving force of evolution itself was not struggle but the urge for living things to cooperate. All too often, though, the values adopted by anti-Darwinians were not so different from those of Spencer. The Lamarckian belief that the initiative of individual organisms guided evolution along successful paths was very similar to Spencer's position, and, indeed, it is no coincidence that Spencer himself thought that the inheritance of acquired characters was the primary mechanism of organic evolution. Admittedly, the majority of Lamarckians distrusted the harshness of laissez-faire morality and opted for state-controlled charity; but they were just as quick to portray the "lower" races as failures in nature's upward march (chap. 10). Both sides read their own values into nature and then took their self-created vision of the universe as an assurance that those values were sound. The parallels that can be drawn between the approaches of materialists and liberal theologians can only reinforce the fears of some Christians that the latter had betrayed the cause of true religion.

The intensity of the debate over evolutionism's religious consequences should not blind us to the fact that the basic idea of natural development was beginning to influence many aspects of Western culture (Himmelfarb, 1959; Russett, 1976; Oldroyd and Langham, 1983). Novelists, in particular, were quick to absorb the implications of the new scientific ideas, and their work has been used by a number of modern scholars to illustrate the impact of evolutionism on Victorian life. At first, many writers of fiction portrayed the advent of Darwinism as a source of tension within their society, although this critical attitude waned as the concept of progressive evolution grew in popularity (Henkin, 1963). Minor novelists thus serve as a barometer recording the late nineteenth century's absorption of evolutionism into its optimistic sense of universal progress. More creative writers made use of the tensions engendered by evolutionism in many different ways (Beer, 1983; Morton, 1984). George Eliot's novels show a clear appreciation of Darwin's image of nature as a complex and unpredictable system, while Thomas Hardy reflected a pessimistic sense of biology's control over human destiny. Samuel Butler also came to recognize the harsher side of the Darwinian world view and eventually began to write openly in support of neo-Lamarckism.

THE PROBLEM OF DESIGN

In traditional creationism, God was responsible for introducing new species and was the designer of all their characteristics. Many naturalists were now beginning to agree that a vast series of such creations seemed an unlikely prospect, and they were beginning to explore the notion of creation by law. Chambers's *Vestiges* had argued that we gain a nobler conception of the Deity by assuming that He created life through a process of law. Baden Powell (1855) also had suggested that God's power was demonstrated more clearly by the unbroken operations of law than by isolated miracles. Now that evolution was becoming more widely accepted, the more liberal followers of natural theology were prepared to accept creation by law. Chambers's concept of designed evolution, which had been rejected as too radical when *Vestiges* had presented it in the 1840s, now came to be seen as the only way of saving the connection between God and nature.

Would the particular kind of law associated with Darwin's selection mechanism allow one to retain the design concept? Most religious thinkers saw that natural selection threatened all aspects of the argument from design and turned to alternative kinds of evolution. But at least one notable effort was made to argue that Darwin's approach was not necessarily atheistic. It certainly destroyed the logic of Paley's original claim that adaptation proved design; but because the process of evolution was purposeful, one still might be able to argue that it had been initiated by a benevolent Creator. Although one would no longer be able to *prove* that He was responsible for every case of adaptation, one might still *believe* that He was indirectly the cause. This position was taken up by the leading figure in the effort to reconcile Darwinism with theology, the American botanist Asa Gray. Although he became one of Darwin's leading supporters in America, Gray was a deeply religious man who felt it necessary to show that his acceptance of the theory did not destroy his faith. He published a number of essays defending Darwinism from the charge that it promoted atheism, later collected in his *Darwiniana* of 1876 (reprinted 1963; see Dupree, 1959, and on the general theological response in America, Persons, 1956).

Gray argued that Darwin's theory was no more atheistic than Newton's physics: it merely showed how the universe worked and was neutral on the question of whether or not it had been set up in that way by design. A theist thus could retain his faith while accepting the theory. The scientist only describes the order of nature, leaving the religious thinker to seek an explanation in terms of divine purpose. In the case of natural selection, the laws of nature give a process that is purposeful in the sense that it keeps life adapted to changing circumstances and continually improves it toward a higher state. Surely this is compatible with the faith that the universe has a purpose given by its Creator. It might even be argued that an omnipotent Creator could foresee all the events that would result from His laws and hence would be personally responsible for their occurrence. Gray used the analogy of a billiards game, asking if the ball's course is determined by necessity or design. Obviously, by both: the player uses the determinate laws of physics to carry out the plan he had in mind when he made the shot. In the same way, the Creator may be using the laws of nature to carry out His ultimate purpose.

Gray seems to imply that the existence of any mechanism for maintaining adaptation is evidence for design. But this view is difficult to uphold in the case of natural selection, which makes suffering an integral part of nature. Paley had tried to minimize the extent of animal suffering, but many had come to believe that these evils could not be dismissed so easily and had begun to doubt the concept of a moral Creator. Gray tried to turn this to his advantage by arguing that natural selection at last explained the purpose of all this suffering. The elimination of the unfit is essential to ensure the evolution of life and therefore achieve God's purpose. Suffering was not the chief foundation of nature's activity but a regrettable by-product of the fact that a rational Creator must allow His designs to be worked out by a process of law.

Darwin himself was tempted by this line of thought (Gillespie, 1979).

He sometimes tried to minimize the amount of suffering caused by the struggle for existence, comparing it with the overall happiness of most individuals born into well-adapted species. But there were still awkward questions that could be asked. Why should there be a process of elimination at all—were the Creator's hands so tied that He could not design a more humane mechanism of evolution? And was the overall direction of evolution quite as purposeful as Gray implied? The appearance of parasites living on the suffering of other organisms suggested that self-interest was the sole guiding force of adaptation—no moral purpose was involved. Darwin doubted that any theologian would want to attribute such horrors to a process directly supervised by God. Even if one accepted a universe set up by a purposeful Creator, it would be preferable to believe that He had no way of predicting or controlling the details of the realization of His ultimate design.

Eventually, Gray seems to have conceded that it was impossible to argue for the existence of a benevolent God from a mechanism as selfish and wasteful as natural selection. In one of his papers, he seized on Darwin's failure to account for the origin of variation and suggested that the Creator might somehow direct the appearance of new characteristics along lines beneficial to the species. In a series of letters, and in the conclusion to his *Variation of Animals and Plants under Domestication* (1868), Darwin insisted that Gray now had gone too far. If variation were directed along beneficial lines, selection would become superfluous; the variations themselves would direct the course of evolution. In any case, we observe that the majority of variations are meaningless. Did God provide the pigeon breeders with the variations that led toward bizarre breeds such as the fantail? To suppose this was to dispose of design by a *reductio ad absurdum*. Given enough random variation, selection can effect evolution without any outside help. The suggestion of "directed" variation merely readmits the possibility of supernatural interference and leaves the naturalist once again without a truly scientific theory of evolution.

By suggesting that God somehow could influence the course of variation, Gray had in effect turned to a form of theistic or designed evolution. Other scientists who were concerned with retaining the connection between God and nature followed the same course more explicitly. Charles Lyell at last had been converted to evolutionism but could not approve of natural selection and suggested instead that there must be divine control over the process (Lyell, 1863). The anatomist Richard Owen had long been suspicious of miracles, and, although he wrote a vindictive review of the *Origin* (reprinted in Hull, 1973*b*), he proposed a theory of evolution unfolding by divine plan in his *Anatomy of the Vertebrates* (1866–68). Ellegård's survey of the popular press (1958) reveals the wide support for such ideas outside the scientific community.

Gray had begun by defending natural selection and therefore was committed to a utilitarian view of nature, stressing adaptation as the driving force of evolution. But those who saw theistic evolution as a positive alternative to selection were free to claim that some aspects of nature cannot be accounted for by any mechanism based solely on adaptation. In effect, they revitalized the idealist version of the argument from design, the belief that there is a pattern or harmony in the world that defies natural explanation and must stem from the will of the Creator (Bowler, 1977*a*, 1983). St. George Jackson Mivart claimed (chap. 7) that resemblances among various branches of evolution defy natural explanation (1871; Gruber, 1960). Mivart insisted that such parallel developments must result from predispositions to evolve in certain directions imposed on life by its Creator. The physiologist William Benjamin Carpenter tried to substantiate this belief on a smaller scale by demonstrating what appeared to be regular patterns of evolution in the shells of minute sea creatures, the Foraminifera (1888). One of the most popular writers in this field, the Duke of Argyll, argued in his *Reign of Law* (1867) that the beauty displayed by some groups of animals could not be explained naturally. Hummingbirds, for instance, exhibited a degree of coloration that was of no use to them even in the search for a mate. Argyll insisted that the Creator has a sense of beauty and has deliberately expressed this in His control of the evolutionary process.

Mivart suggested that evolution took place by a series of sudden mutations, each advancing the species a significant distance along a predetermined path. The same idea was the basis of Owen's theory of "derivation" (1866–68, vol. III). Appeal to such hypothetical mutations allowed one to dismiss the everyday variations studied by the Darwinists as irrelevant to the real process of evolution. Perhaps these variations are the result of accidental causes; but there may be another kind of variation, discontinuous and directed along a predetermined path, brought into play only when the pressure for change requires the species to move on to the next stage in its development. This also has the advantage of allowing the naturalist to preserve the traditional concept of species as real entities, each a distinct step along the path designed by the Creator.

But how did God impose His will on the course of evolution? Was it possible to set up a theory that would retain design and still be scientifically respectable? Early versions of theistic evolutionism such as Chambers's *Vestiges* had confused things by presenting "creation by law" as though it were only a series of disguised miracles, and the post-Darwinian debates revealed the impossibility of clarifying the issue. An obvious approach was to explore the idealist view of causation. Argyll suggested that the concepts of law and cause had to be understood at two levels. The scientist used the empiricist approach equating cause with the constant conjunction between one kind of event and its consequences. At the philosophical level, however, this had to be supplemented by the idealist view that an active power of will is needed to effect the change. Argyll seemed to imply that if the laws of nature depend on the will of God, then we should be able to see how He directs the evolutionary process. But even if we accept the laws as an expression of the will of God, they still must work in a purely determinate fashion if the scientist is to be able to study them. To embody design in the sense required by Argyll, the laws would have to anticipate future goals and direct nature toward them, something hardly compatible with the scientists' experience that causation depends on control of the present by the past.

The question of anticipating the future brought the problem to a head. Argyll believed that evolution could anticipate future requirements of a species and begin to direct variation appropriately. How do we know, he asked, that rudimentary organs are vestiges of once-useful structures? Perhaps they are actually incipient stages of organs being prepared for *future* usefulness. Here is the stumbling block designed evolution falters on in its efforts to gain scientific respectability. Experience shows that the laws of nature work by allowing the past to control the present, not the present to be drawn toward a future goal; and the scientist cannot accept so obvious a reintroduction of teleology as would be required by Argyll's view. When challenged by Wallace, Argyll insisted that he was not reintroducing supernatural control; the Creator worked only through law and not through miracles. Yet a concept of law that included future goals chosen by the Creator was scientifically meaningless, because the goals were explicable only in supernatural terms.

The only alternative was to assume that design was imposed not by continuous action of divine will but through predispositions built into life when it was first created. This was Gray's idea of God using the laws of nature to complete His plan, just as the billiards player allows the laws of motion to complete his shot. Could one believe that the first forms of life contained the information needed to bring about every change in the millions of years of evolutionary history? Any attempt, furthermore, to attribute the details of evolution to divine plan again would raise the question of the Creator's responsibility for parasites and other unpleasant forms of life. Theistic evolutionists tried to play down this aspect of nature, but this in itself alienated them from the scientific community.

In the end, consideration of these issues convinced scientists that theistic evolutionism was an unworkable compromise. By the end of the century, only a few conservatives such as Argyll still were advocating this approach. Professional scientists had at last recognized that it was not part of their job to seek evidence of design and that any attempt to reintroduce teleology would compromise the acceptability of their theories. The moral problems created by selectionism remained, however, and a number of scientifically viable alternatives were suggested in the hope that they still might allow a purpose to be seen in evolution. The most promising was the Lamarckian mechanism of the inheritance of acquired characters. This avoided the harsher implications of natural selection, because each individual was supposed to adapt itself to the new environment by its own efforts, requiring no elimination of the unfit. Lamarckism could be seen as the kind of mechanism that a benevolent God would have chosen to ensure a comfortable life within an ever-changing world. It also was thought that the mental functions would be constantly stimulated by the necessity of responding to the changes, ensuring an overall progress. American naturalists led the way in fusing evolution with religion by means of Lamarckism (Pfeifer, 1965; see also Moore, 1979, and chap. 9). Similar views were expressed in Britain by Samuel Butler and became widely popular toward the end of the century.

The American paleontologist Edward Drinker Cope stressed that Lamarckism allowed living things to be in charge of their own destiny. Instead of being a mechanical puppet at the mercy of the environment, the organism could respond to any challenge by choosing to adapt to a new way of life. The element of choice contained in the traditional notion of creative design thus was transferred from an external God to the life-force that He had built into nature. In Cope's view, consciousness itself was a driving force of evolution, constantly elevating itself to new heights by exercising its faculty of choice. Yet the actual mechanism of the inheritance of acquired characteristics by which the body was adapted to the new life-style seemed to satisfy the requirements of scientific naturalism, permitting a genuine reconciliation between science and theology. For naturalists such as Cope, a new form of Christian evolutionism could be founded on the universal tendency for life to advance under its own steam toward the goal foreseen by the Creator.

Lamarckians in general repudiated Spencer's laissez-faire approach to regulating human behavior. They too hoped for future progress in the human race but believed that this could best be promoted by a cooperative political system in which the best ideals of civilized man were implanted in each generation by state-controlled education. Man should take charge of his own future development, instead of following blindly where nature led. The Lamarckian view of progress differed from that of Spencer (who was himself a Lamarckian in biology) only in that it saw human consciousness capable of perceiving the true nature of the goal toward which evolution was working. Spencer opted for laissez-faire precisely because he thought that the goal could not be reached more rapidly through human interference—the process of social development was so complex that it was better to let nature take its own slow course. Lamarckians such as Cope, by contrast, believed that in man, consciousness had reached a new level of awareness, that man could actually perceive the goal toward which the Creator had directed evolution. Man thus could take charge of evolution and speed up the achievement of that goal by artificially directing the characteristics of future generations through education.

The Lamarckian emphasis on cooperation was not just a political program aimed at combating the harshness of laissez-faire capitalism. It also could be linked directly to the process of organic evolution, by eliminating natural selection's reliance on struggle so that cooperation could be presented as the true driving force of progress. The moral values we cherish are transferred to nature, which allows the intensification of these values in man to be seen as a necessary consequence of evolution. Those animals that cooperated among themselves have generally been more successful, and therefore the instinct to cooperate was intensified until it became enshrined in the human conscience. The American, John Fiske, led the way with his *Outline of Cosmic Philosophy* (1874). This was written under the influence of Spencer and presented a similar vision of universal progress. Yet Fiske introduced a significant twist into the theme by presenting altruism as the guiding feature of human evolution. For him, altruism, the willingness to sacrifice oneself for others, represented the most successful evolutionary policy for the species and had become the essential stimulus for the growth of human civilization. Henry Drummond's widely popular *Ascent of Man* (1894) drew an explicitly religious message from the argument that altruism, in the form of a willingness to cooperate with others for the good of all, was the driving force of all evolution (Moore, 1985*b*). The emergence of genuine altruism in man thus was not a violation of nature's laws but a direct consequence of the fact that our species has been shaped by laws intended to promote that very factor in all living things. Peter Kropotkin's series of articles later collected under the title *Mutual Aid* (1902) relied on the author's own observations of wildlife to confirm that there is little sign of a struggle for existence in nature. Animals generally improve their chances of survival by cooperating with one another, which makes it plausible to claim that evolution actually has progressed by enhancing the level of cooperation in each successive generation. Significantly, Kropotkin also wrote in favor of biological Lamarckism, arguing that characteristics developed successfully in a group of animals can be inherited by their offspring and intensified in later generations. Without this, his kind of theory would degenerate into a form of Darwinism, in which successful groups eliminated those with a lesser degree of cooperation in the struggle for existence.

EVOLUTION AND MAN

Darwin knew that application of his theory to man would be its most controversial aspect. He was convinced that the theory must include man along with other animals, thereby challenging the most fundamental assumptions of Christianity. The reaction to Chambers's *Vestiges* had confirmed the strength of popular feeling on this issue and ensured that everyone understood the materialist implications of human evolution. Darwin decided not to deal with man in the *Origin*, to avoid inflaming an obviously controversial situation, but he felt it would be dishonorable to conceal his beliefs. He compromised by inserting a single sentence: it hinted that evolution would throw light on the origins of man. This was enough to ensure that the whole question of man became an integral part of the debate from the beginning, even though Darwin did not speak out on the issue until his *Descent of Man* (1871).

The central problem concerned the nature of man's mental and moral attributes. Traditionally, these were assigned to the spiritual world as characteristics of the soul—which was only temporarily joined to the flesh of the body. Unlike some religions, Christianity always had insisted that the animals have no spiritual characteristics. If evolution were accepted, this distinction would break down and all of man's qualities would become parts of nature. To preserve our unique status, we would either have to reject evolution or assume that something very special had occurred in the branch of development leading toward man. Without some unique relationship to the Creator, the human situation would have to be reassessed and meaning for our existence sought in our position at the head of the evolutionary process. Essentially, the problem faced by Darwin and the evolutionists was that of creating a culturally acceptable explanation of how nature had generated the higher faculties of the human mind (R. J. Richards, 1987).

Conservative thinkers were at first appalled by the claim that we might be no more than superior apes. The strength of feeling on this issue can be judged from the fact that two of Darwin's closest supporters, Lyell and Wallace, refused to endorse a mechanistic account of human origins. Ever since his geology had first opened up the question of the origin of species, Lyell had been disturbed at the prospect of denigrating mankind through a link with the animals (Bartholomew, 1973). In his *Antiquity of Man* (1863), Lyell at last accepted a progressionist account of the fossil record and conceded some plausibility to Darwin's theory—but he insisted that the appearance of the human race required a sudden leap that would take life onto an entirely new plane. Darwin wrote that this part of Lyell's book made him groan. He saw no point in making an exception to the general rule of continuous evolution just to protect the traditional view of mankind's uniqueness. Lyell continued to be disturbed by this issue, however, and later endorsed Wallace's claim that the creation of mankind had required supernatural interference with the evolutionary process.

Wallace's early views on human evolution fitted the Darwinian picture, but in the course of the 1860s, he came to doubt that natural selection alone could account for the production of certain characteristics. Significantly, this occurred just as he began to develop an interest in spiritualism, which would have suggested to him that we do indeed possess a soul capable of existing independent of the body (Smith, 1972; Kottler, 1974; Turner, 1974). His doubts were expressed in a paper entitled "The Limits of Natural Selection as Applied to Man" (in Wallace, 1870). He believed that even some of our physical characteristics, such as lack of hair on the body, confer no biological advantage and cannot have been developed by selection. A number of mental qualities seem similarly useless, especially when we think of the primitive state of culture of early humans. What, for instance, was the purpose of the musical sense or the ability to perform abstract mathematical calculations? Unless the selectionist can show that such faculties would have been of use to a primitive savage, he will have to admit that their origin cannot be explained naturally. Unlike most of his contemporaries, Wallace believed that modern primitives are mentally the equals of the white race, but he insisted that they do not use their higher abilities. If their life-style is a relic of the earliest phase of human culture, then our ancestors too could have gained no benefit from acquiring the same abilities. Since natural selection cannot develop an ability that is not in constant use, Wallace concluded that some supernatural agent must have taken control at a crucial stage in human evolution.

Wallace's opinion shows that reluctance to confront the implications of a naturalistic account of human origins was not confined to conservative thinkers. In general, though, this reluctance was gradually overcome in the course of the late nineteenth century. As we shall see, this was made possible by stressing the progressive character of natural development, so the human race could be seen as the predestined end product of a morally purposeful system. As a prerequisite to general acceptance of human evolution, it was necessary to establish a scientific case for the link between mankind and the animal kingdom. It had been obvious since the eighteenth century that our closest physical relatives are the great apes (chap. 4), and the evolutionists now had a vested interest in emphasizing the closeness of this relationship. The most effective kind of evidence would have been fossils showing a direct transition from the ape to the human form. If the "missing link" was not available, the evolutionists would have to start from a comparison of the modern ape and human forms.

The first account of these issues from the evolutionist side was T. H. Huxley's *Man's Place in Nature* (1863; reprinted in Huxley, 1893–94, vol. VII). Huxley had already clashed with Owen on the question of man's resemblance to the apes, at another session of the famous 1860 British Association meeting. Owen defended Cuvier's classification of man and the apes into two distinct orders, the Bimana and Quadrumana. The apes were considered "four-handed" because their feet resembled the hands far more closely than do those of man. Owen also had claimed that parts of the brain found in humans were totally lacking in the chimpanzee brain. Huxley challenged Owen on these points and went on to expound his own interpretation in *Man's Place in Nature*. He showed that Cuvier and Owen had exaggerated the resemblance of the hand and foot in the apes. Owen's failure to find certain structures in the ape brain, furthermore, was the result of his dependence on artificially preserved specimens. Altogether, Huxley argued, man's physical characteristics place him clearly in the same order of animals as the great apes, the order today called the Primates.

Close physical resemblance suggested an evolutionary connection, but this could only be confirmed by the fossil record. So Huxley turned to ancient specimens of man in the hope of discovering what the public would call the "missing link." He knew, of course, that man could not have descended from the living great apes. Darwin's theory predicted only that the apes and man must have a common ancestor, which might have general apelike characteristics; but this ancestor would not resemble any of the modern apes, because the latter have undergone specialization for life in the forests. Was there any sign that early forms of man retained the characteristics of their animal ancestors? Until quite recently, even geologists as radical as Lyell had insisted that there were no really ancient human fossils. Now, Huxley had at least two fossils he could study in an effort to understand man's evolution, the Engis and Neanderthal skulls (Eiseley, 1958; Leakey and Goodall, 1969; Reader, 1981; for a collection of primary sources, see Leakey and Prost, 1971). The Engis skull had been known since 1833, but its true age was just coming to be recognized. Unfortunately for Huxley's purpose, the Engis skull appeared to be no different from the skull of a modern man. He realized that if the true story of human evolution ever were uncovered, it would have to trace our ancestry back an enormous distance in time.

At first sight, the Neanderthal skull seemed more promising. It had been discovered in Germany in 1856 and was clearly of great age. It had thick brow ridges, not unlike those of an ape. Some naturalists had dismissed it as a pathological specimen, a deformed creature, perhaps an idiot or a wild man. Others insisted that even as such it might throw light on the origin of man, because the deformities could be atavisms, or throwbacks to primitive stages of man's history. Huxley could not accept the specimen as pathological, a view amply confirmed by subsequent discovery of many similar remains. He acknowledged that the brow ridges made Neanderthal one of the most apelike human skulls he had ever seen, yet he was forced to admit that it *was* fully human. The cranial capacity was as large as any modern skull, indicating that Neanderthal man had a brain as big as our own. He could not be a true link between man and his presumably small-brained animal ancestors; indeed, there were living races differing just as much from the average character of modern man.

The fossil record thus offered no firm evidence for human evolution, and Darwin's followers were forced to discuss the process in a purely hypothetical manner. Only in the last decade of the century was the search for fossil evidence successful. Darwin had predicted that Africa would turn out to be the cradle of the human race, but the Dutch naturalist Eugene Dubois decided to try his luck in the East Indies, home of the orangutan. After a search of some time on the island of Java, Dubois in 1891 turned up the skull and thighbone of a manlike creature far more primitive than Neanderthal man. Convinced that he at last had discovered the link between ape and man, he named the creature *Pithecanthropus erectus,* although to the world-at-large it became known as "Java man." At a zoological congress at Leiden in 1895, however, the scientific community greeted the discovery with skepticism. Only Ernst Haeckel, whose name for the hypothetical link had been borrowed to christen the new form, proclaimed that it was final confirmation of human evolution (translation, 1898). Dubois became secretive, refusing to let others examine his finds. The significance of his discovery thus was obscured until other specimens were unearthed in the present century.

Lack of fossil evidence had to be dismissed as yet another example of the record's imperfection. There was, at least, a growing body of archaeological evidence for the antiquity of man and the primitive state of his technology in ancient time. The traditional belief in the recent creation of the human species had been challenged first in the 1840s with the discovery of stone tools and weapons alongside remains of extinct animals, such as the mammoth. Particularly important was the work of Boucher des Perthes among the gravel beds of the Somme river. At first the implications of these discoveries had been ignored by a scientific world still mesmerized by the reputation of Cuvier, but further discoveries in the 1850s made it clear that man indeed had inhabited the earth for a vast period of time. Charles Lyell, originally an adherent of the recent creation theory, now published a survey of the new evidence in his *Antiquity of Man* (1863). The work of John Lubbock (1870) and others established an outline of the modern view of prehistory. The archaeological evidence revealed a steady improvement in technical ability from the earliest crude stone tools to the discovery of bronze and iron. In the absence of fossil evidence for the *biological* improvement of man, evolutionists seized on the evidence for *cultural* progress as at least indirect support for their claims. The great development of prehistoric archaeology that took place in the late nineteenth century allowed the construction of a sequence of cultural periods that were supposed to have succeeded one another as the human race progressed. Little thought was given to the possibility that different cultures might exist side by side in the same epoch. Gabriel de Mortillet, in particular, linked the rigid sequence of cultural developments to the biological evolution of man, eventually including both Pithecanthropus and Neanderthal man as steps in a progression from the apes (on developments in archaeology, see Daniel, 1975; Grayson, 1983; Hammond, 1980). At the same time, anthropologists such as Edward B. Tylor and Lewis H. Morgan began to use modern savages to illustrate what they believed to be the early stages of human cultural development.

The anthropologists and prehistoric archaeologists thus constructed a linear model of cultural development that exerted considerable influence on Victorian evolutionary thought (Burrow, 1966; Stocking, 1968, 1987). Many accounts of the debate on "man's place in nature" fail to acknowledge the significance of these fields, perhaps because the linear model of progress is distinctly non-Darwinian in character. For Lubbock, Tyler, and Morgan, culture necessarily advanced through a predetermined hierarchy of stages. Each race progressed at its own rate, so the slow developers are now left stranded at levels so primitive that they illustrate how the white race's ancestors lived in prehistoric times. Here was a linear model of progressive evolution that would have major implications for anyone trying to understand how the lowest form of mankind had emerged from an animal ancestry (Bowler, 1986).

Darwin's own approach to the problem of human origins was thus conditioned by two contradictory influences. On the one hand, contemporary developments in cultural evolutionism seemed to imply that development should be treated as a linear progress along a predetermined scale. On the other, his own theory treated biological evolution as an open-ended process in which each branch's history is shaped by unique factors. Strictly speaking, the theory of natural selection did not imply that evolution advances inexorably toward a particular goal. If getting bigger brains or more social habits was so useful to our own ancestors, why did the apes not seize on the same opportunity? Darwin's *Descent of Man* (1871) was the first major attempt to tackle the problem of explaining human origins in evolutionary terms, but its approach to the topic tended to vacillate between the progressionist and the truly Darwinian alternatives. As we shall see, Darwin did provide what modern evolutionists would call an "adaptive scenario" to explain the unique character of human evolution. But at many points in his account, he too succumbed to the prevailing enthusiasm for progressionism, thus encouraging the tendency by which his theory was absorbed into a more optimistic view of natural development.

Darwin needed to convince his readers that the difference between human and animal mental powers is a difference in *degree* but not in *kind*. Those who wished to retain man's unique spiritual status insisted that with the appearance of man certain totally new characteristics had been introduced into the world. In Darwin's view, there could be no such discontinuity. Man may be superior to the animals; but, if he has evolved from them, his mental and moral faculties must be merely extensions of theirs. Darwin also had to show that improvement of these powers could be accounted for by a process of natural evolution, based on selection and a generous helping of Lamarckism. There should be no need to postulate supernatural guidance, only the blind operation of the laws of nature.

Darwin could draw on his own wide range of observations in his efforts to diminish the apparent differences between man and the animals. He tried to show that every one of our supposedly unique faculties is possessed in at least a slight degree by the higher animals. He argued that animals display true intelligence, as well as instinctive behavior; they experience the whole range of emotions including rage, boredom, awe, and so forth; they communicate with one another through a simple form of language; and they have moral instincts that allow them to work for the good of others in the species. Modern observers agree that Darwin greatly exaggerated the human qualities of animals; he fell into the all-too-obvious trap of anthropomorphism in his anxiety to make the case for evolution. It is difficult to find hard evidence that animals make truly constructive use of intelligence, for instance. Wild chimpanzees will sometimes use sticks as tools, but in the laboratory, they show little evidence of being able to think out solutions to simple problems and usually depend on trial and error. Evidence for emotional states based on facial expressions—to which Darwin devoted an additional book (1872)—easily can be confused with accidental similarities between man and the apes. In the case of language, Darwin and his followers tended to identify the instinctive warning cries of many animals with the far more sophisticated use of sounds to denote abstract concepts, which is the basis of true language. They also assumed that the languages of many primitive peoples amount to no more than collections of clicks and grunts like those made by animals.

Moral feelings were, of course, the most delicate point. Darwin emphasized the willingness of parents in many species to sacrifice themselves for their offspring. A dog defending its master at risk of its own life showed that such instincts can be generalized even among animals. But Darwin's opponents naturally felt that such purely instinctive behavior fell far short of the human capacity for altruism and the recognition of universally binding moral laws.

Whatever the lasting value of his comparisons, Darwin felt that he had demonstrated at least the plausibility of breaking down the gulf between man and the animals. He then had to tackle the second problem: to show how unaided nature could have raised man's faculties to a level so high that many people assumed them to be qualitatively different from those of the animals. In addition, man had certain purely physical characteristics that even Wallace thought were incapable of natural explanation. How, for instance, could one explain the loss of body hair, which seemed to confer no biological advantage? Darwin attributed this to sexual selection, the subject of the second part of his *Descent of Man*. Because the degree of hair loss is different in the two sexes, Darwin argued that such a characteristic had become associated with sexual attraction and thus had been enhanced over many generations.

On the question of human intelligence, Darwin could call on the plausible assumption that natural selection would develop this character because it was useful. But he also realized that he must explain why our ancestors were more forcefully affected by this trend than were the apes. He speculated that the greater development of human intelligence was the result of our ancestors adopting an upright posture, which freed their hands for toolmaking. The transition to upright walking on the ground had triggered the growth of the human mind by accidentally providing an additional stimulus to the use of intelligence. The apes had been trapped by their continued use of their hands as a means of locomotion among the trees. Haeckel also endorsed the view that the upright posture was the crucial breakthrough in human evolution, while Wallace (despite his rejection of a totally naturalistic theory) suggested that our ancestors had been driven out of the trees by a climatic change. Darwin and a few of his followers thus anticipated the modern view that the transition to bipedalism, not the acquisition of a large brain, was the definitive step in the emergence of the human family. Having made this point, however, Darwin then went on to discuss the enhancement of our mental powers as though it were an inevitable consequence of natural selection.

Could the development of the moral faculties also be explained naturally? At first sight, it would appear that natural selection would work against the production of any instinct that would lead an individual to sacrifice his own good for that of others. Darwin tried to solve this problem by appealing to an idea first suggested by Wallace which corresponded roughly to what is now called "group selection." In those animals where the parent must care for the young, the family group will become important and selection may well favor the instinct to defend that group and preserve the offspring. Darwin argued that in man such instincts have been extended to include the willingness of the individual to work for the good of his tribal group. There would be a struggle for existence among the groups themselves, in which those tribes with strong cooperative instincts would exterminate those less firmly united. The inherited effect of habit due to Lamarckism could also promote the same trend. Darwin believed that early man saw his obligations as strictly limited to members of his own tribe, as do some savage tribes today. Only when man developed intelligence, and the leisure to apply it to abstract questions, did philosophers begin to extend the social instincts into general moral laws imposed through religious sanctions.

Darwin thus made it possible for his followers to argue that moral values are an inevitable product of human social evolution. Since living in social groups is natural to many species, this allowed them to believe that the evolutionary process itself has been designed to achieve a morally significant result. The growth of intelligence had merely allowed the moral instincts to be articulated. But Darwin's suggestion that our mental powers are the result of a unique transition to bipedalism threatened the logic of this attempt to turn evolution into a purposeful process. In fact, much of Darwin's own discussion of cultural evolution followed the progressionist model created by anthropologists and archaeologists such as Lubbock. Small wonder, then, that the majority of his contemporaries ignored the suggestion of a unique turning point in human evolution and constructed generally progressionist schemes of mental and moral evolution (R. J. Richards, 1987). G. J. Romanes, who inherited the Darwinian mantle in the area of mental evolution, expounded a developmental system in which social activity—via the emergence of language—was the real cause of mental progress (1888). Most late-nineteenth-century writers on human origins ignored what modern scientists see as the key implication of Darwinism, preferring to concentrate instead on the inevitability of progress toward the human mind (Bowler, 1986).

The anthropologists' decision to treat modern savages as illustrations of how the white race's ancestors had lived in the remote past had obvious implications for Victorian attitudes toward other races (chap. 10). For Lubbock and the majority of evolutionists, this decision had been taken quite deliberately as a means of filling in the gap in the fossil record of human origins. Modern primitives became, in effect, the equivalents of the missing link sought by the paleontologist. The resulting cultural and racial hierarchy merely served to define the ladder of progress that the most active branch of human evolution (the white race) had ascended. Even when human fossils were discovered in the later part of the century, they were interpreted in a way that would allow them to be fitted into this progressionist scheme (chap. 11).

To make room for this image of the human past, Lubbock had set out to destroy the traditional belief that the history of man, as a fallen creature, could only be one of degeneration. It had been widely believed that man could not civilize himself: he must have been taught the various arts and sciences by supernatural means; afterward, many races had steadily degenerated. Inspired by this view, Archbishop Richard Whately of Dublin openly challenged Lubbock's archaeological evidence for human progress at the British Association meeting in 1868. Lubbock had little difficulty in disposing of the case for degeneration but soon was challenged by Argyll on a subtler point (1868; Gillespie, 1977). Argyll freely admitted that man had progressed in some ways but pointed out that increasing sophistication of technology—which was really all that the archaeologist could demonstrate—was not necessarily a sign of mental or moral progress. Early man may have been just as human as we are today but simply had not had time yet to make some of the more important inventions.

Although Argyll's real purpose was to claim that man had been created suddenly in his present form, he had put his finger on a genuine weakness in the evolutionists' view of society. Eventually, the claim that cultural progress is a direct continuation of biological evolution was challenged by the new social sciences of the twentieth century. Although prepared to accept the idea that man must have evolved in the remote past, modern anthropology was founded on the belief that his character was already fixed at the beginning of cultural history (chap. 10; Cravens, 1978). In a sense, Argyll's refusal to accept the basic idea of human evolution had led him to anticipate an important principle of modern thought. The early evolutionists indeed had overextended the principle of biological progress and had confused the quite distinct processes of biological and cultural evolution.

EVOLUTION AND PHILOSOPHY

Those who accepted a wholehearted evolutionism had to elaborate a new philosophy to replace the faith by which men had lived for centuries. The opportunity was seized gladly by radical thinkers who flourished in the late nineteenth century. If man could no longer see his life sanctified by God, he would have to find meaning in his position on the evolutionary scale. A new definition of morality could be introduced, in which evolutionary success would be the only criterion of good. But did this approach reform morality or merely abrogate all moral standards? Throwing the emphasis on individual success too easily encouraged the worst kind of behavior. The only hope of justification was in the claim that evolution itself was aimed at some ultimate goal. Thus, many evolutionary philosophers betrayed the real spirit of Darwinism and saw man as a key step in the universe's progress toward perfection. This was true of both the liberal theologians who saw progress as the fulfillment of God's plan and the materialists who ignored the element of design altogether. Only a very few thinkers succeeded in breaking this mold to grapple with the possibility that Darwinism's real implication was the destruction of any world view based on necessary progress along a hierarchy of complexity (Collins, 1959; Passmore, 1959; Randall, 1961; Mandelbaum, 1971).

Radical thought was not short of inspiration. Comte's positivism, with its evolutionary view of human knowledge, had been introduced into Britain by writers such as G. H. Lewes. Tyndall and the materialists proclaimed their intention of explaining everything in terms of matter and motion. Similar forces were at work in Germany (Gregory, 1977). But the most popular of the new philosophies, that of Herbert Spencer, arose from a combination of evolutionism with the British tradition of utilitarianism and laissez-faire. Bentham already had challenged Christian morality by arguing that happiness is the only good; he had pictured the mind as a conditioned-reflex machine operating by the association of ideas. By mid-century, utilitarianism was in a state of crisis: its leading exponent, John Stuart Mill, began to question the possibility of a moral theory based solely on pleasure and pain. Associationism also was being undermined, as Alexander Bain tried to incorporate mental faculties with which it was not really compatible (Greenaway, 1973). The possibility of a truly materialist psychology based on the location of mental faculties in the brain was just beginning to be recognized (Young, 1970*a*). Thus, Spencer's combination of utilitarianism and universal evolutionism turned out to be the most positive response to the crisis.

Although Spencer is little read today, in his own time he was one of the most highly regarded philosophers. Evolution was the key to his "Synthetic Philosophy," and indeed it was Spencer who popularized the word "evolution" and created the popular belief that it denoted an essentially progressive process. Although frequently associated with Darwinism, the phrase "survival of the fittest" was coined by Spencer, who cared little for the details of science. To him, biological evolution was just one aspect of a universal process, and he was convinced that Lamarckism played a much greater role than Darwin admitted. Although Spencer had been seen as a leading social Darwinist (Hofstadter, 1959), the link between his social philosophy and the selection mechanism is tenuous. It is certainly true, though, that his philosophy created a morality based on the individual's success in contributing toward the inevitable progress of evolution.

Spencer had little formal education and once worked as an engineer (Spencer, 1904; Duncan, 1911; Greene, 1959*b*; Peel, 1971; Kennedy, 1978). He turned to journalism while putting together the outlines of his new philosophy. *Social Statics* (1851) elaborated on the utilitarian model and gained him a reputation. As early as 1852, Spencer published an article favoring Lamarckian evolution, and at about the same time he came across von Baer's concept of embryological growth as a development from general to specialized structures. This he decided was the pattern for all natural processes, the kind of development he was to call "evolution." In 1855, he published a study of psychology based on an evolutionary model. The philosophy of progress was sketched out in essay form in 1857 (reprinted in Spencer, 1883). The first part of the Synthetic Philosophy, *First Principles*, appeared in 1862, followed by its application to biology in 1864.

Spencer dismissed questions about the ultimate purpose of the universe to the realm of the "unknowable" about which the philosopher could not speak. The material world would have to be understood exclusively in terms of natural laws; the most fundamental of these were the indestructibility of matter and the persistence of force. To answer whether these laws produce merely a random pattern of motion or whether there was some meaningful direction to natural change, Spencer introduced his idea of evolution as an increasing complexity of structure, a trend from homogeneity to heterogeneity. He argued that in the long run the laws of matter ensured that all changes must follow such a course. The first appearance of primitive life forms was an inevitable product of the tendency for matter to organize itself, as was the evolution of such forms toward higher levels of organization. Spencer was not a simple progressionist: he appreciated that life evolved through a branching process and that not every branch was progressive. Progress was thus slow and irregular and could not be seen as the unfolding of a plan of creation toward a single goal. Yet progress was inevitable in the long run: evolution gradually must push life toward higher states of organization and cause the appearance of new qualities. Intelligence would increase until eventually a form such as man would appear, capable of initiating a new phase of social evolution.

As the author of the phrase "survival of the fittest," it would be easy to imagine Spencer as a biological selectionist. Yet, in fact, he was convinced that Lamarckian use-inheritance was the main basis of animal evolution (Freeman, 1974). He also denied the role of individual competition in the earlier phases of social evolution, believing that the first complex societies were built up along rigidly militaristic lines. Modern industrial societies emerged only in later stages, based on freedom of individual enterprise. These societies had a higher level of organization that permitted more rapid progress through the combined initiatives of many individuals, each seeking his own profit. Even here the purpose of free enterprise was not so much to eliminate the weaker members of society as to encourage them to stronger efforts in the hope of alleviating their misfortunes. We should not fall into the trap of labeling Spencer a "social Darwinist" as though his whole philosophy were nothing more than an application of the selection mechanism to political economy.

What were the implications of Spencer's social evolutionism for traditional moral philosophy? Bentham's utilitarianism already had pointed the way to a redefinition of morality by judging behavior solely in terms of its value in producing happiness. A good action was one that promoted happiness, not one that conformed to some higher moral law sanctioned by God. With a certain amount of prompting from legislation, individuals would work naturally together for the good of all; each would help society as he tried to help himself. Spencer now adapted this individualism to an evolutionary situation. Human nature was not a fixed entity, and there was no single most perfect kind of social organization. The psychology of the individual has been shaped by the evolution of the race and must constantly be reshaped to conform with an ever-changing social situation. Happiness is achieved when the individual is a well-adjusted, productive member of his society, and the purpose of the moralist is to show how the greatest number of individuals can be brought into this condition. People must not be taught some absolute, fixed law of conduct based on the outdated form of society in which religion flourished but how to adapt themselves to present changes in their own society. The moralist must show them how to adapt, confident in the belief that in so doing he is encouraging the progress of the race toward an ever-higher state.

At one level, Spencer's philosophy eliminated morality as it had traditionally been understood. Instead of following transcendental ethic principles, the individual must simply conform to the current stage of social evolution. Since we cannot foresee the future course of progress, the individual will only know if he is "right" if nature rewards him with success, "wrong" if he is punished with misery. Those who suffer the penalty of failure will be encouraged to take more effective action in the future. In principle, this is a complete ethical naturalism: we follow our own self interest, secure in the belief that whatever nature does is right. Yet Spencer saw himself as a moral philosopher and his faith in progress meant that in practice his ethical system was compatible with the harsher aspects of traditional religious values. The claim that nature rewards success was, after all, only an extension of the Protestant work ethic. Liberal Christians could thus adapt themselves to Spencer's evolutionary ethics by arguing that God rewards useful behavior in this world as well as the next (Moore, 1985*b*). The gulf between evolutionary ethics and traditional morality was thus not as great as some historians have assumed. Spencer merely naturalized the moral values that the middle class had at first tried to justify by religion. Nature now became God's agent for rewarding the liberal virtues of thrift and enterprise.

Spencer's philosophy defined the natural counterpart to the Lamarckian position outlined earlier. Both took it for granted that evolution is progressive, that the development of human society represented a continuation of the biological hierarchy driven by essentially the same forces. They disagreed over how to encourage the future progress of man. Spencer held that it was essential for nature to take its course, while the Lamarckians thought that man already could see the goal of the process and could speed up development toward it. The religious faith of Spencer's opponents helped them transfer a more orthodox view of morality into evolution, yet they shared his faith in progress and the notion of a hierarchically structured universe. The concepts of hierarchy and progress, however, were those most threatened by the details of Darwin's theory. Branching evolution made it difficult to define how one form could be ranked higher or lower than another, especially when some apparently "low" forms have survived over vast periods of time. Natural selection worked toward adaptation, not progress, and was opportunistic in its exploitation of new avenues of development. If these points were accepted at face value, much of the late-nineteenth-century's evolutionary philosophizing would be worthless.

A few thinkers tried to grapple with the nondirected character of evolution. Perhaps the most imaginative use of Darwinism was made by those who seized on the lack of direction as a guarantee of human freedom. Normally, we think of selection as a deterministic theory in which the individual has no control over his success or failure. Yet theories of inevitable progress are themselves deterministic, because they hold that the path of future development is marked out by the hierarchy of organization. By encouraging a breakdown of the hierarchical relationship, Darwinism undermined the notion of a goal that all must advance toward or perish. John Dewey (1910) argued that this requires us to ask new kinds of questions in philosophy. Darwinism shows us that man has freedom to shape his own destiny, because there is no preordained pattern of development. The concept of freedom also was important to pragmatists such as Charles Peirce and William James (Wiener, 1949). Both saw the real lesson of Darwinism in its destruction of the whole idea of determinism. Nature is inherently creative, and the lack of constraints on evolution guarantees the freedom of the human individual. Peirce also saw evolution as the growth of "cosmic reasonableness," through which a rational order emerged from a primitive, unpersonalized chaos. Here the idea of progress was reintroduced, but in a subtler form that did not impose a preordained structure on the final goal. Progress is possible, but it need not proceed inevitably in a single direction, because life has the freedom to create its own future.

A typical example of the effort to grapple with the unstructured nature of biological progress can be seen in the writings of the French moral philosopher Henri Bergson. The result he called *Creative Evolution* (translation, 1911; Gallagher, 1970). Bergson realized that there is neither a harmonious plan of nature nor any hope of perceiving the hand of an intelligent Designer in the creation of every species. The history of life has been progressive but only in a very irregular way. This could be accounted for, however, if we assumed a constant state of tension between the original creative life-force, the élan-vital, and the resistance of the inert matter from which that force must construct living bodies. Creation is thus a constant striving upward of the life-force, divided and redivided into a host of separate branches by the practical necessity of coping with the material world. The creative impulse impels life to always try to reach higher levels, yet it always has to fall back, unable to overcome completely the resistance of matter. If there is a God, He would have to be understood not as a distinct entity controlling evolution but as the ongoing creative process itself. In the end, the life-force can be seen as consciousness, penetrating matter in the effort to manifest its potentialities on an ever-higher scale. Intelligence and moral feelings have naturally increased as long-term results of evolution, and although man has advanced far beyond the animals in these capacities, he remains bound to the whole system of life that created him. Although now we have lost the comfort of believing in a God who cares for us, we can take heart in the realization that our consciousness symbolizes the spiritual heart of the creative process that is itself both nature and God.

The term "emergent evolution" was coined by the psychologist C. Lloyd Morgan (1927). Originally sympathetic to Darwinism, Morgan always had suspected that some more positive force must aid selection in directing evolution. Emergent evolution stressed the creativity of the process but in a manner quite different from that advocated by Bergson. "Emergence" denoted the capacity of nature to generate totally new qualities at certain levels of organization, qualities that could not have been predicted from a study of lower levels. Life itself was such an emergent quality, suddenly appearing at a certain stage in the increasing complexity of material structures and possessing capabilities far beyond those of a purely material entity. Similarly, mind appeared as an emergent quality when life had advanced to a certain level of organization. It thus was possible to treat the human mind as unique to man, and, in fact, the bulk of Morgan's book was devoted to psychological studies with little reference to biological evolution. Other writers also took up the theme of emergence, including Samuel Alexander (1920) and Roy Wood Sellars (1922). The movement was an effort to retain a faith in the values of human nature while admitting that man had been framed by evolution. Instead of assuming that mind is the driving force of evolution, it was presented as an unexpected product of biological progression. Evolution was no longer the cosmic unifying principle it had been for earlier writers but was merely an accepted fact. Morgan himself still derived religious comfort, however, from the belief that nature contains higher qualities latent within herself and thus can be said to have a spiritual purpose.

The philosophy of organism developed by Alfred North Whitehead also had strong religious overtones. Originally, Whitehead had contributed to the new wave of logical analysis that had begun to replace the old fashion for cosmic philosophizing during the early twentieth century. In the 1920s, he began to broaden his interests in an effort to sketch out a new solution to the traditional problems. In *Process and Reality* (1929), he suggested that the world is best seen not as a collection of distinct objects but as a complex of ongoing processes in which nothing is ever isolated from the rest of nature. Atoms themselves were quasi-organic entities, capable of interacting at all times with their surroundings. Whitehead believed that the processes of nature are ultimately meaningful, orderly, and harmonious, with man as their highest product. Life and mind are not qualities standing above matter, as in emergent evolutionism, but are essential components of a universe in which nothing is completely "inorganic" or lacking in awareness. Thus, Whitehead went beyond Bergson to make spiritual qualities *part of* matter, rather than separate forces *acting on* matter. It should still be possible, therefore, to recognize a kind of Platonic order in the way processes unfold, evidence of the God who stands as an ideal toward which the whole aspires. Not that there is a single unified plan of creation with a fixed goal, but the world nevertheless progresses by creating its own orderliness in each epoch of its history.

The philosophies outlined above all managed to avoid the earlier preoccupation with fixed hierarchies of development. They nevertheless assumed that progress toward higher levels of organization was possible within evolution, in this way, they retained some element of the faith that nature is a system with a moral purpose. For some Darwinists, however, even this went too far. Natural selection turned evolution into a chapter of accidents, brought about by a combination of chance and life-or-death struggle. How then could it be *intended* to produce higher states or exhibit any moral purpose? Even Darwin could not adjust fully to the prospect of a totally meaningless universe, and he retained lingering hopes that evolution in the end worked for the good of living things. It was left for T. H. Huxley to explore for the first time the prospect that nature might indeed be without purpose or moral significance.

Although Huxley never missed an opportunity to oppose the claim that nature reveals evidence of supernatural design, he was not an atheist. He argued that science is neutral concerning the existence of a Creator: it could neither prove nor disprove it. He coined the term "agnosticism" to describe this state of active doubt on the religious question. But precisely because the existence of God could not be proved, it had to be accepted that there is no obvious purpose in evolution. Huxley opposed any attempt to show that the fossil record confirms a progress in the history of life. All efforts to demonstrate such a progression were based on an oversimplification of biological classification, and the great advantage of Darwinism was precisely that it allowed change without progress. By eliminating progress, however, Huxley was brought face to face with the prospect that nature lacks direction and hence purpose. Evolution kept life adapted to changing conditions, but it did so only by causing a vast amount of suffering and apparently without steering toward any meaningful goal.

Huxley's pessimism made him extremely suspicious of Spencer's efforts to found a new ethics based on evolution. His campaign against Spencer culminated in his Romanes lecture of 1893 entitled "Evolution and Ethics" (Huxley, 1894; Helfand, 1977; Paradis, 1978). In it, he developed the theme that nature is without purpose, a vast mechanical system grinding along inexorably without reference to human values. Attempts to impose a spiritual value on nature are illusions created out of vain hope and anthropomorphism. But because evolution is not even progressive in the purely biological sense, there is no hope for the future if we simply surrender to it and accept its harsh values as our own. Spencer advocated a policy of laissez-faire on the grounds that it would improve the race, but, Huxley argued, if the same principle gives no progress in nature, his claim that human progress will be inevitable is unjustified. Why should we violate our deepest sense of moral responsibility to others, to follow the dictates of a natural system that is without meaning?

Huxley believed that the highest qualities of the human mind are intrinsically valuable, even though they are neither produced by nature nor sanctioned by God. By a cosmic accident, man has been given faculties that allow him to recognize the meaningless character of the system that gave him birth. We must cherish our moral feelings precisely because they go beyond nature to establish a sphere of activity that has become an integral part of our humanity. The mind may be a cosmic accident, but its value to man is crucial because it is the only thing that makes him human, the only thing that can give meaning to an otherwise meaningless universe. Civilization is all the more valuable because it violates the basic principles of nature. To uphold his moral standards, man must break the very laws of evolution, protecting the weak instead of leaving them to be eliminated. We must struggle to retain our integrity in a hostile world, not in any hope of ensuring progress but because to do so is what it means to be human. In more modern times, a similar recognition of the meaninglessness of nature induced a sense of insecurity verging on moral paralysis in the philosophers of existentialism. But Huxley actively threw himself into the struggle for social reform, making the best use of his abilities to stave off for a while the encroachments of a blind and mechanical nature.

For totally different reasons, then, Huxley joined with the supporters of creative or emergent evolutionism in resisting Spencer's new morality. At one level, the debate concerned the very nature of moral philosophy. Spencer urged the rejection of traditional definitions of right and wrong and their replacement with strictly utilitarian standards: actions would be judged solely in terms of individual happiness and anticipated progress of society. His opponents admitted that man was linked to nature but wished to preserve by one means or another the significance of the traditional moral feelings. Whether derived from or in direct resistance to nature, moral law should be an autonomous guide to decent conduct by establishing standards independent of physical necessity. At a more practical level, however, the debate concerned actual social policies. For Spencer, morality was essentially equivalent to a successful social policy; he believed that success would be achieved only by eliminating all restrictions on personal initiative. His opponents rejected laissez-faire not only as an abrogation of society's moral duty but as a political system that would cause untold misery to those unable to compete with more ruthless individuals.

In effect, Huxley and the others were arguing against what has become known as social Darwinism, and the debate with Spencer points us toward the whole complex issue of the social implications of biological theories. This issue cannot be understood as a simple application of the Darwinism selection mechanism to society. Spencer, who has been seen as a leading social Darwinist, was a biological Lamarckian; Huxley, who was certainly a leading Darwinist, opposed all efforts to model social policy on a biological mechanism. These two facts in themselves demand caution in our approach to the issue. Discussions of social Darwinism all too often have ignored another fact—that the evolutionary model could have been of a non-Darwinian character. Darwinism was not the dominant evolutionary theory of the late-nineteenth century; the original opposition to it had intensified rather than died away. A number of alternative evolutionary mechanisms were available, and these must be outlined before exploring the social implications of the biological theories.

9

The Eclipse of Darwinism

The "eclipse of Darwinism" is a phrase used in Julian Huxley's survey of evolution theory (new ed., 1963) to describe the situation before genetics and selectionism were combined to give the "modern synthesis." Huxley was involved in the creation of this synthesis, and he knew just how precarious the state of Darwinism had been around 1900. J. B. S. Haldane made the same point in the motto chosen to head his book on the genetic theory of selection (1932): "Darwinism is dead—any sermon." But if the preachers were rejoicing, it was the scientists themselves who had brought Darwinism into eclipse. From the high point of the 1870s and 1880s, when "Darwinism" had become virtually synonymous with evolution itself, the selection theory had slipped in popularity to such an extent that by 1900 its opponents were convinced it would never recover. Evolution itself remained unquestioned, but an increasing number of biologists preferred mechanisms other than selection to explain *how* it occurred. Classic surveys of the situation, even those by writers sympathetic to Darwinism, admitted the strength of the opposition (Romanes, 1892–97; Plate, 1900, 1903, 1913; Kellogg, 1907; Delages and Goldsmith, 1912). Those unsympathetic to the selection theory rejoiced in its decline—one attack, translated from the German, had the hopeful title *At the Deathbed of Darwinism* (Dennert, 1904).

Modern historians have paid little attention to this outburst of anti-Darwinism. Eiseley (1958) notes the declining fortunes of selectionism but mentions only Hugo De Vries's mutation theory as an alternative. The best accounts are in Philip Fothergill's little-known book (1952) and in the classic surveys of the history of biology by Rádl (translation, 1930) and Nordenskiöld (translation, 1946). The latter, however, are as much primary as secondary sources, written under the impression that Darwinism was indeed dead. Later historians have paid considerable attention to those aspects of post-Darwinian biology which subsequently contributed to the rise of the modern synthetic theory. In general, though, they have turned a blind eye to theories that have been repudiated by modern biology. However important these theories may have been at the time, they have been dismissed as blind alleys that cannot possibly have influenced the growth of modern evolutionism. As we gradually emancipate ourselves from a historiography so obviously based on hindsight, it becomes increasingly apparent that non-Darwinian theories did indeed play a major role in shaping late-nineteenth- and even early-twentieth-century views of evolution (Bowler, 1983, 1988).

Analysis of evolution theories in the decades around 1900 is complicated by the fact that so many alternatives were under consideration. Research programs inspired by Darwin's biogeographical perspective were undertaken, while August Weismann's "neo-Darwinism" hailed natural selection as the only viable mechanism of evolution. Far from upholding the fortunes of Darwinism, though, Weismann's dogmatism merely alienated many biologists who were already inclined to invoke at least some non-Darwinian factors. The rise of neo-Lamarckism and the theory of orthogenesis marked the transition to a more explicitly anti-Darwinian form of evolutionism. These theories were clearly supported, in part, because they seemed to preserve an element of teleology that would counteract the apparent materialism of neo-Darwinism. Naturalists who were reluctant to concede that evolution is a haphazard, trial-and-error process argued that living development is constrained to advance in a purposeful or orderly manner by forces affecting the production of new variations. They suggested that variation is guided by the intelligent activity of individual organisms, or by forces inherent to the process of individual growth. Such theories represent a direct continuation of the developmental philosophy proclaimed in Chambers's *Vestiges* and Mivart's *Genesis of Species*. The continued use of the recapitulation theory by many non-Darwinian paleontologists illustrates the popularity of the view that evolution is somehow analogous to the growth of the organism toward maturity.

The collapse of this developmental view of nature came about in a somewhat paradoxical manner. It did not arise directly from a challenge by Darwinism—indeed, the selection theory was at its lowest ebb in the crucial years around 1900 when the plausibility of the recapitulation theory was undermined. In the early years of the new century, Mendelian genetics succeeded where Darwin had failed; it demonstrated that individual growth was not a suitable model for evolution, thus threatening the indirect evidence that had been used to support both Lamarckism and orthogenesis. Evolution now became a process that could only be controlled by the introduction of new genetic factors into the population. At first, the geneticists refused to admit that the spread of such "mutations" could be controlled by the selective effect of environmental fitness. They repudiated both Lamarckism and Darwinism as equally outdated products of the old, preexperimental tradition in natural history. Yet the hostility to Lamarckism was in fact far more profound, and it was on the revolution in heredity theory that a new basis for the selection theory would eventually be erected.

NEO-DARWINISM

Darwinism had derived its first support from areas such as the study of variation and geographic distribution. Investigation of these areas continued, especially in connection with the problem of speciation, that is, of how a single original population is divided into many. Eventually, the importance of geographic isolation for speciation came to be appreciated, despite Darwin's refusal to admit this as an essential factor. The field naturalists who continued these investigations often remained loyal to the old, flexible kind of Darwinism, but even some of them adopted a more dogmatic stance, insisting on selection as the only mechanism of evolution. In Britain, the application of statistical techniques to the study of variation led to the creation of the school of "biometry," whose supporters were among the staunchest defenders of Darwinism—and the bitterest critics of the new Mendelian genetics. But it was the German biologist August Weismann who became known as the most dogmatic neo-Darwinist. Weismann proclaimed the impossibility of Lamarckian use-inheritance because it was incompatible with his theory of heredity. This polarized opinions by alienating all those naturalists who suspected that mechanisms other than selection were involved.

Before Darwin, naturalists had tried to play down the amount of variation within each species in the hope of establishing a typical form. The selection theory allowed them to recognize that each species generally will exist as a group of more or less distinct varieties, each of which may be potentially a new species. But how did a single original population divide itself into several distinct new ones? What prevents two varieties from interbreeding and blending together as they separate to the point where crossing is no longer possible? What is the relationship between the increasing difference in physical appearance of the two forms and their ability to interbreed? These problems were very difficult to solve, largely because Darwin and his followers had not exploited fully the implications of what Ernst Mayr has called "population thinking." The inability of naturalists to solve these problems in the context of the selection theory certainly must have helped to lower the theory's prestige in the scientific community.

The Galápagos had shown Darwin that geographic isolation would encourage the production of distinct forms through the prevention of intercrossing; but in other cases, varieties of the same species coexist in the same area. Mayr (1959*b*) and Sulloway (1979) have shown that Darwin eventually came to neglect isolation, believing instead that ecological specialization within the same area will produce distinct varieties. Wallace also accepted sympatric speciation (speciation without geographic isolation) but disagreed with Darwin over how the separate groups acquired mutual infertility (Mayr, 1959*b*). George John Romanes argued that the selection mechanism could explain the origin of adaptations but not the origin of species. In other words, it was able to explain how a single population could change to meet an environmental challenge but not how a single population could divide itself into several distinct forms (Romanes, 1886; Lesch, 1975). He proposed a mechanism of "physiological selection" by which variants infertile with the parent form may be produced spontaneously. But such a concession exposed a fatal weakness: if mutations producing sterility had to be admitted in addition to selection, why not accept the idea that the variations themselves might produce *all* the new characteristics without any need for selection?

As early as 1868, Moritz Wagner had suggested that an initial period of geographic isolation was essential to allow the buildup of mutual infertility (translation, 1873). Isolation would prevent intercrossing while it was still possible, so that the separate populations could acquire distinct characteristics. If sufficient differences were built up, no intermingling would occur even if migration or changing geographic circumstances brought the forms back into contact. Unfortunately, isolation was seen as an alternative, not as an addition to selection, so that Darwin and most of his followers rejected Wagner's views. In fact, both sides in the debate had failed to appreciate the full implications of selectionism for the species concept. Species and varieties still were being defined in terms of morphological differences (i.e., differences in physical form) as though they were created by God. But the crucial difference was really whether or not two populations could interbreed, no matter how much difference in form there was between them. Species and varieties must exist as distinct breeding populations, or they will mingle and lose their character. In the case of varieties, interbreeding still is possible theoretically and must be prevented either by isolation or through the acquisition of what now are called "isolating mechanisms"—differences in behavior that will prevent mating. Isolation will be necessary, not until complete sterility has been built up but at least until isolating mechanisms have been formed which will prevent interbreeding in practice.

Only toward the end of the century did opinion begin to change in favor of Wagner's belief that an initial period of isolation is necessary. In the 1880s, the work of John Thomas Gulick on Hawaiian land snails established the correlation between varieties and distinct geographic locations (Gulick, 1888; Addison Gulick, 1932; Lesch, 1975). Karl Jordan developed a modern populational concept of species at the end of the century, coupled with an appreciation of the role of isolation (Mayr, 1955). Jordan refused to be sidetracked by the polarization of Darwinism and Lamarckism, realizing that it was far more important to clarify the basic nature of species, whatever the mechanism of change. By 1905, the American David S. Jordan conceded that the majority of field naturalists now accepted the importance of isolation. He also lamented the lack of interest shown by the new breed of experimental biologists in these developments.

Complaints about the experimentalists' lack of concern for the problems of the naturalist also occur in the writings of another important Darwinist, Edward B. Poulton (1890, 1908). As a student of animal coloration, Poulton was convinced that camouflage and mimicry had definite adaptive value. Because animals cannot control their colors, Lamarckism was impossible in this case, and natural selection seemed the only explanation for the development of protective colors. It was easy, Poulton claimed, for the laboratory biologist to dismiss such effects as mere coincidence—he has no idea of the living conditions of animals in the wild. The fact that the adaptive significance of coloration *was* widely challenged indicates just how far anti-Darwinian feeling had developed. Only field naturalists such as Poulton refused to give in, convinced that their observations showed the validity of selection, whatever the theoretical problems.

To some extent, the surge of anti-Darwinian feeling had been precipitated by the dogmatic selectionism of August Weismann. This had emerged from Weismann's new initiative in the study of variation and heredity, an initiative that, however, must be evaluated with care. Mayr (1985) regards Weismann as the most important nineteenth-century evolutionist after Darwin, and it is certainly true that his concept of the "germ plasm" as the material substance of heredity helped to clarify the modern concept of natural selection. Nevertheless, it should be noted that Weismann's assault on the problem of heredity was very much a continuation of Darwin's own program, in which the study of generation (individual reproduction and growth) was an integral part of evolution theory (Hodge, 1985). Weismann shared the fairly common distrust of Darwin's theory of heredity, pangenesis, and sought an alternative model of how parent and offspring are related. To clarify the issue, he began to exploit the new area of biology known as cytology, or cell theory, seeking to understand how the cells of the offspring are shaped by material inherited from the parents (Robinson, 1979; Farley, 1982).

Frederick Churchill (1986) has shown that Weismann's radical views on the nature of heredity flowed from a very traditional approach inspired by the recapitulation theory. By applying this theory to his early work on Hydrozoa, Weismann became convinced that all organisms contain a potentially immortal kernel of reproductive material that can be transmitted to future generations. Failing eyesight forced him to abandon the microscope, but his work had convinced him that the material responsible for transmitting characteristics in reproduction is located in the chromosomes, minute rod-like structures in the nucleus of cells revealed by staining. This important insight has been incorporated into the framework of modern genetics. Indeed, the whole concept of a material substance that can be encoded with the information necessary for reproduction is an important one. Weismann did not originate such a concept; in a rather different form known as the "idioplasm," it had been incorporated into Carl von Nägeli's theory of 1884 (Gillispie, 1960; Coleman, 1965), but in the course of the 1880s, Weismann refined his ideas on its basic nature to create his theory of the germ plasm (translations, 1891–92, 1893*a*; Romanes, 1899).

Weismann argued that the germ plasm is completely isolated from the body of the organism that carries it. The structure of the body, the "soma," is constructed according to information supplied through the germ plasm donated by the parents. The structure then carries the germ plasm within itself until it is ready to pass on to a new generation. Once the soma has been formed, it acts only as a "host" to preserve the germinal material. No changes affecting the body are communicated to the germ plasm, so that the organism can pass on to the next generation only what it received from its parents. Weismann's germ plasm enshrines the notion of "hard" heredity, the belief that heredity cannot incorporate responses that the body makes to its environment. By contrast, Darwin's pangenesis had allowed for "soft" heredity, which responded to the environment. Weismann suggested that the germ plasm is composed of units called "determinants," each responsible for producing a certain part of the body. In sexual reproduction, the determinants from both parents are mixed to provide the information used to construct the offspring's body.

Weismann's theory had a crucial implication for evolution: it made Lamarckism impossible. The parent's body does not *produce* its germ plasm; it merely transmits it. Changes in the body due to use or disuse are not reflected in the germ plasm and therefore cannot be inherited. Weismann dismissed the popular belief to the contrary as mere superstition. The normal variation in a population is due to the recombination of determinants as they are shuffled by sexual reproduction. Natural selection will favor those determinants responsible for producing useful organs and suppress any that are harmful. New variation can be introduced only by distortions in the determinants' structure caused by the occasional accident of duplication; but the body has no control over such accidents, and the variation so produced is totally random. Selection thus becomes the only conceivable mechanism of evolution.

In a famous experiment, Weismann cut off the tails of a group of mice and then continued to cut off the tails of their offspring over many genera-

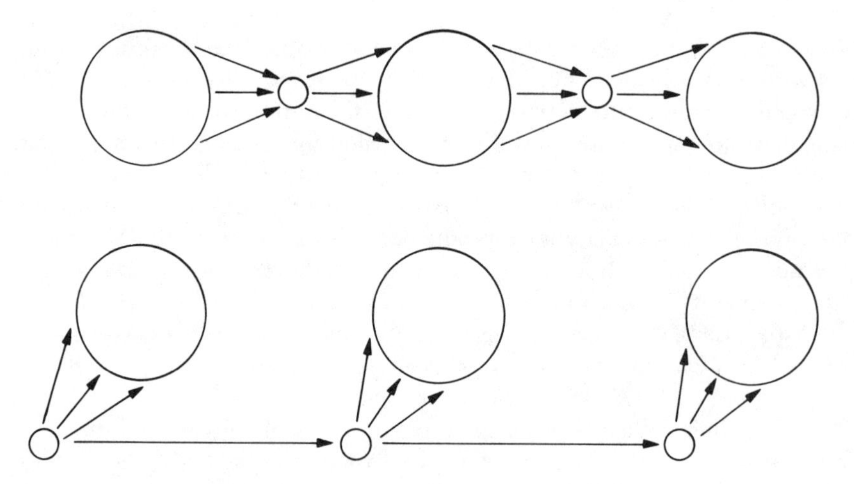

Fig. 21. Pangenesis and the Germ Plasm.

Two diagrams illustrate Darwin's and Weismann's concepts of the relationship between the body (large circle) and the hereditary substance responsible for transmitting characteristics to the next generation (small circle). For simplicity, the combinations required by sexual reproduction are not represented. In Darwin's theory of pangenesis (top), each part of the body produces a gemmule to transmit its characteristic. Gemmules are collected in the reproductive system and then passed on to form the basis for constructing the body of the offspring. In Weismann's theory (bottom), the body is developed from the germ plasm but then merely transmits it to the next generation without being able to affect it in any way. Because the parts of the body do not manufacture their own hereditary material, any changes caused by use and disuse cannot be transferred to the germ plasm and thus cannot be inherited.

tions. There was no indication that this orgy of mutilation produced mice with tails any shorter than normal. The Lamarckians argued that this was not a fair test. The acquired characteristics they supposed to be inherited were purposeful responses to the environment, not accidental mutilations. Yet Weismann's experiment had a valid point, because Lamarckism must rest on a theory of soft heredity. His test proved that heredity is hard: mice deprived of their tails still carried the complete germ plasm for this characteristic. Indeed, some of the very rare experiments that were supposed to have confirmed Lamarckism did depend on the inherited effect of mutilations, because these were so much easier to produce and check in the laboratory.

To those naturalists who had decided—generally for other than experimental reasons—that use-inheritance must have a role in evolution, Weismann's theory symbolized the increasing dogmatism of neo-Darwinism. If this was to be the new attitude, they would have nothing to do with selectionism. The germ plasm theory was based on insufficient evidence, they claimed, and hence Weismann's insistence on the "all-sufficiency of natural selection" was misplaced. Herbert Spencer, who always had insisted on a role for Lamarckism, now felt it necessary to challenge Weismann and proclaim his separation from the Darwinian camp (1893; Weismann, 1893*b*). Other Lamarckians rushed to attack the germ plasm concept, citing various lines of evidence that were supposed to demonstrate use-inheritance. Lamarckism had been gaining popularity for some time, but Weismann's rigid opposition now stimulated its supporters into making an open breach with Darwinism.

Weismann never admitted a role for Lamarckism, but even he was forced to make concessions on other fronts. To explain how useless rudimentary organs have been so drastically reduced in size, he postulated a mechanism called "germinal selection" (1896, 1904; Bowler, 1979). Selection occurred within the germ plasm itself, as the determinants for various characteristics competed for the limited supply of germinal nourishment. In the end, Weismann even admitted that germinal selection might produce variation trends capable of creating new organs. In most cases, he insisted, such trends would only become really important if the new organ turned out to be useful and could be seized on by natural selection. To his opponents, this concession exposed the weakness of his exclusive selectionism. Perhaps such internally directed trends played an autonomous role in evolution—the theory of orthogenesis. In any case, germinal selection seemed an arbitrary extension of the original theory, allowing the whole thing to be dismissed as a product of Weismann's overactive imagination.

Weismann had tried to solve the problems of variation and heredity by setting up a model for the physical process by which characteristics are transmitted from one generation to the next. The school of biometry had its origins in a different line of attack on the same problems: the application of statistical techniques to describe the range of variation within a population and the effect of selection on the range. The use of statistics in the study of the human population had been pioneered earlier in the century by Quetelet but only now became exploited on a wide scale. The theory of evolution stimulated the revival of interest in what seemed to be an eminently suitable way of trying to understand those phenomena most crucial for the selection mechanism.

The founder of the biometrical movement was Darwin's cousin, Francis Galton. Early in his career, Galton had performed an experiment that seemed to disprove the theory of pangenesis; but soon he began to apply his mathematical interests to obtaining an accurate picture of the spread of variation within a population. An important stimulus was Galton's concern with investigating the implications of heredity for the human race. He believed that the idea of selection should open the eyes of social thinkers to the prospect of racial improvement or degeneration resulting from the tendency of different classes to breed at different rates. To gain support for this "eugenics" policy of controlled human breeding, he pioneered the techniques that his followers would apply to animal populations in an effort to find experimental support for natural selection (Galton, 1889, 1892; Pearson, 1914–30; Wilkie, 1955; Swinburne, 1965; Froggatt and Nevin, 1971*a;* Provine, 1971; Cowan, 1972*a,* 1972*b*; De Marrais, 1974; Forrest, 1974; Mackenzie, 1982).

It has become commonplace to picture the range of variation by a frequency distribution curve depicting the proportion of the population occupying each part of the range. For most continuously variable characteristics, such as height in the human population, the curve follows the bell-shaped "normal" or "Gaussian" curve depicted in the diagram. But Galton and his followers, having developed a technique for representing the variation in a single generation, needed to find out what happened to the curve if selection was applied to the population over a number of generations. Galton proposed a "law of ancestral inheritance" to describe the proportion of an individual's characteristics derived from each of his ancestors: half from the parents, a quarter from the grandparents, and so on. This law is false by modern standards, but it was accepted by his followers as the basis for their efforts to measure the effects of selection.

Galton himself believed that his law would prevent selection from having a permanent effect on a species. To illustrate this, he introduced the concept of "regression." Imagine a sample of individuals from a particular part of the range of variation, such as a group of people with above average height. What will happen to the sample if we allow the individuals to breed only among themselves for a series of generations? Galton believed that the mean value of the characteristic for the sample would regress back toward that of the species as a whole. After a number of generations, descendants of our sample of tall people would have an average height equivalent to the normal for the human race. If this is the natural tendency, Galton argued, selection can have no permanent effect on a population. Regression provides a limit to variation, always pulling a characteristic back toward the old mean, despite the effects of selection. Selection can improve the health of a population but cannot permanently improve the breed. Galton believed that species normally remain static for long periods of time and evolve by sudden mutations only when confronted by a major environmental challenge.

Although Galton did not believe in the effectiveness of selection, his followers in the biometrical school argued that he had misinterpreted his own law of ancestral heredity. Two of them, in particular, became leading defenders of Darwinism, the mathematician Karl Pearson, best remembered for his discussion of the philosophy of science (1900), and the biologist W. F. R. Weldon. Pearson showed that on the basis of Galton's own

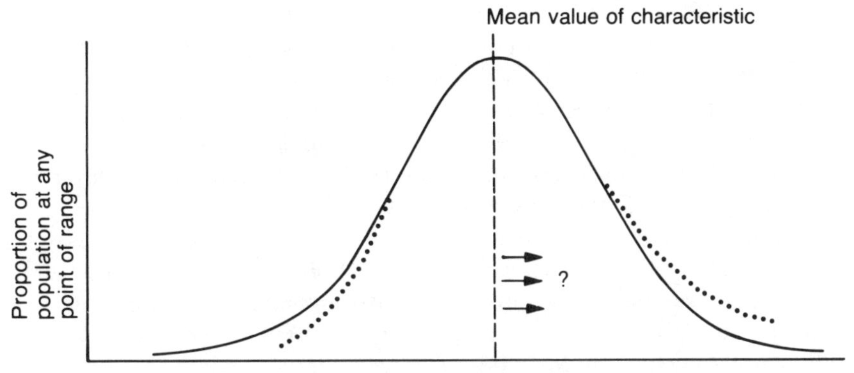

Fig. 22. Continuous Variation and Selection.

The solid curve represents the normal distribution for a continuously varying characteristic, such as height in the human population. This curve plots the proportion of the population occupying any point of the range of variation against the measurement of the varying characteristic. Thus the largest proportions are clustered around the mean value of the range (the average height of a human being), while at the extreme ends of the range on either side the proportions become vanishingly small (very tall or very short persons). The crucial question for the Darwinists was: what is the permanent effect of selection on such a distribution? If one extreme is favored in the struggle for existence (e.g., tallness), those individuals at that end of the range will breed more than normal; conversely, those at the opposite end will breed less. This will affect the distribution as shown by the dotted lines. Will this have any permanent effect in shifting the mean for the whole population in the direction favored by selection? Weldon's experiments were intended to show that such a permanent effect indeed would be produced.

law, selection would have a permanent effect (1896, 1898, 1900; Froggatt and Nevin, 1971*a*; Provine, 1971; Norton, 1973). The next step was to test whether this was true in nature as well as in theory, a task undertaken by Weldon. He set out to demonstrate a correlation between variation and death rates among crabs and, later, among snails (Weldon, 1894–95, 1898, 1901). The first efforts were unconvincing, but in later tests Weldon was able to show clear effects of selection. In waters containing a high level of sediment, larger crabs survived more readily than small ones; and over a number of generations, the cumulative effect of selection produced a lasting change in the population.

There was much opposition to the claims of the biometrical school of Darwinism. Weldon's observations showed the effect of selection only on a very small scale and did little to counter the belief that there is a limit beyond which selection cannot drive the species. Both Pearson and Weldon held views on the philosophy of science considerably in advance of their time, which alienated them from the majority of biologists (Norton, 1975*a*, 1975*b*). Because adaptation was not a function that could be defined objectively, Weldon simply correlated death rates with some easily measurable characteristic such as the width of the crab's shell. He argued that there was no need to give a causal explanation of why larger shells were favored. In response to criticism, he eventually provided an explanation based on the tendency of smaller crabs to have their gills clogged by muddy water, but his later work on snails again provided only a simple correlation between shell size and death rate, on the assumption that the utility of the characteristic would be far too complex for easy investigation. Weldon's unwillingness to deal with the nature of adaptation did not appeal to the field naturalists who also lent their support to Darwinism—or to those who opposed the selection theory altogether.

The most important factor limiting the success of the biometricians was their clash with the newly emerging Mendelian genetics. Even before the rediscovery of Mendel's work, Weldon had fallen out with William Bateson over the latter's claim that discontinuous variation was the real source of evolution. The techniques of biometry were adapted to the study of continuous variation, while Bateson stressed those characteristics inherited on an all-or-nothing basis and denied the significance of selection. When Mendel's work on heredity was rediscovered in 1900, it fitted neatly the category of discontinuous variation, allowing Bateson to become one of its leading champions. The biometricians were committed to rejecting discontinuous factors as atypical of the normal situation and thus opposed the new genetics. For this reason, Mendelism emerged not as the savior of Darwinism but as yet another alternative to it. The two sides battled back and forth in a debate that did credit to neither and which we can now see to have been futile, because each had access to part of the true picture (Provine, 1971; Cock, 1973; Darden, 1977).

The biometricians were misguided in their rejection of Mendelism, yet ultimately justified in their support for the selection of continuous variations as the mechanism of evolution. Indeed, their most important innovation was to recognize that variation and heredity are not mutually antagonistic forces. Even Darwin had seen heredity as a conservative force trying to maintain the original character of the species, while variation tried to disturb the process by introducing new characteristics. Galton and his followers at last had come to recognize the fallacy of this conflict analogy. Variation and heredity are in fact two different manifestations of the same process, the constant circulation of a large number of hereditary factors within a population as a result of sexual reproduction. Their problem was that without a suitable theory of how heredity worked, they could not exploit this insight in the defense of Darwinism.

NEO-LAMARCKISM

The first major challenge to the selection theory came not from Mendelism but from Lamarckism. In his own lifetime, Lamarck had gained little support for his mechanism of the inheritance of acquired characteristics. Only when Darwin made the basic idea of evolution acceptable did the possibility of rehabilitating Lamarck's idea emerge. Those who accepted evolution but were unhappy about natural selection began to cast around for alternative mechanisms. Because Darwin's own theory was in difficulties, perhaps the inheritance of acquired characteristics would make a more plausible alternative. At first, it was not even realized that Lamarck had pioneered the idea, and only toward the end of the century did the name "neo-Lamarckism" appear (Packard, 1901). Even then, few can have paid much attention to Lamarck's long-outdated writings. The complete Lamarckian system was totally incompatible with the post-Darwinian world view. Neo-Lamarckism consisted of a number of basic concepts, some of which had been suggested first by Lamarck, adapted to a system of evolutionism that would have been quite unthinkable at the beginning of the nineteenth century.

In its most obvious form, Lamarckism is based on the assumption that changes of structure produced by the activity of the adult organism can be reflected in the material of heredity and passed on to the next generation. Exercise, use, and disuse are known to affect the size of various organs: the weightlifter develops strong arms, the hypothetical giraffe stretches its neck reaching for leaves. Such changes occur in response to new habits and are generally adaptive, so that use-inheritance can serve as a simple replacement for natural selection. But it is the assumption that such acquired characters can be inherited that is the most controversial. Does the weightlifter's son inherit his father's muscles? Weismann's germ plasm made this theoretically impossible, and despite the scorn they poured on his system, the neo-Lamarckians were never able to devise a satisfactory model for their own kind of soft heredity. Most simply assumed that inheritance was possible because of all the other arguments they could bring forward to support their theory.

It must be emphasized, however, that Lamarckism was not a unified movement. Use-inheritance was not the only mechanism attributed to Lamarck by those who considered themselves his followers. Other naturalists argued that the growth of an animal or plant can be affected directly (i.e., without conscious reaction) by the environment and that such changes can be inherited. A plant grown in dry conditions, for instance, will develop characteristics that help it to retain water. Another group of Lamarckians were attracted to the movement from an entirely different direction. In America, the pupils of Louis Agassiz began with a strong interest in embryology and became convinced that evolution proceeded by a series of additions to the growth of the individual. They then saw that use-inheritance would explain why such additional stages were added on at certain points in a species' history. The American neo-Lamarckians always retained certain interests marking them off from other members of the movement, particularly the beliefs that evolution occurred in regular patterns and by isolated bursts of sudden change.

With almost no experimental evidence for the inheritance of acquired characteristics, why was Lamarckism so popular at the turn of the century? The answer must lie in its broader philosophical implications. It has been suggested that American neo-Lamarckism was a direct continuation of natural theology (Pfeifer, 1965). Certainly, some members of the movement adopted use-inheritance because it seemed more compatible with the Creator's benevolence than did natural selection. Yet one could no more *prove* design with Lamarckism than with Darwinism, because in both cases species are shaped by natural means rather than divine fiat. Some neo-Lamarckians had no interest in religious matters, and their motives must be sought elsewhere. The key point seems to be that Lamarckism allows life itself to be seen as purposeful and creative. Living things are in charge of their own evolution: they *choose* their response to each environmental challenge and thus direct evolution by their own efforts. With or without any religious implication, this is certainly a more hopeful vision than that derived from Darwinism. Life becomes an active force in nature, no longer merely responding in a passive manner to environmental pressures. Thus, there is a connection between the aspirations of many neo-Lamarckians and those expressed by Bergson's creative evolution.

An element of Lamarckism had been present in the original version of Darwinism. Toward the end of his career, Darwin himself conceded a significant role for use-inheritance. Some of his followers, particularly Haeckel in Germany, allowed Lamarckism an even more prominent role. Herbert Spencer also had supported a combination of selection and Lamarckism in works such as his *Principles of Biology* (1864). Only after the appearance of Weismann's exclusive selectionism did an open split develop between the neo-Darwinians and the neo-Lamarckians. In Germany, naturalists such as Theodor Eimer began to support use-inheritance at the expense of any role for selection in the production of adaptations (translation, 1890). Spencer now wrote attacking Weismann and proclaiming the importance of Lamarckism. By the end of the century, a host of naturalists had begun to support the movement, and all the more frequently they were advocating Lamarckism as a complete alternative, not as an addition, to selection (surveys in Packard, 1901; Kellogg, 1907; see Churchill, 1976; Limoges, 1976; Bowler, 1983).

Darwin himself had received a foretaste of what was to come in his dispute with the writer Samuel Butler (Willey, 1960; Pauly, 1982). Originally a convert to Darwinism, Butler was alerted to the possibility of a non-Darwinian mechanism for evolution by reading Mivart's *Genesis of Species*. Increasingly concerned with allowing a sense of purpose in the evolutionary process, Butler began to see in Lamarckism the prospect of retaining an indirect form of the design argument. Instead of creating from without, God might exist within the process of living development, represented by its innate creativity. Butler's main interest was the mental life of animals and of men, although unlike some other Lamarckians, he did not regard consciousness as the highest form of activity. He argued that instinct was the most fully developed kind of mental activity and incorporated this view into his Lamarckism. Conscious choice first directed an animal's response to a new situation, after which the appropriate actions were reduced first to a habit and then to an instinct requiring no conscious effort at all. Bodily structure would soon conform to the new instinct through use-inheritance. Butler suggested that man himself would be much happier if he could eliminate all conscious thought and reduce his life to instinctive activity. As he elaborated these views, he became convinced that Darwin had deliberately played down the significance of earlier evolutionists such as Lamarck in the historical sketch added to later editions of the *Origin*. In *Evolution Old and New* (1879), Butler criticized Darwin and emphasized the importance of Lamarckism, and he developed the same themes in a series of later books with titles such as *Luck or Cunning*?

Butler's writings gained little popularity until the emergence of a fully fledged neo-Lamarckism toward the end of the century. A representative sample of the arguments used at this later stage can be found in Spencer's attack on Weismann's claim for the all-sufficiency of natural selection (1893). In it, Spencer declared that "either there has been the inheritance of acquired characters or there has been no evolution." His arguments had two purposes: to undermine selection and the germ plasm theory and then to show the superiority of Lamarckism. Against selection itself Spencer used an argument that had considerable force when measured against the pregenetic selection theory (Ridley, 1982*a*). He pointed out that when a new structure evolved, all the rest of the body would have to accommodate the new development. Thus, a series of variations would be required to adjust the overall structure in a manner correlated to the new organ. What would be the chance of all these variations appearing together at the right time, if the species had to depend on random variation? Selection might explain the changes in a single organ but not an integrated transmutation of the whole body. In addition, there was the problem posed by the disappearance of organs that became useless because of changing habits. Selection might explain a reduction in size but hardly a complete elimination—a point later conceded by Weismann. Lamarckism encountered no difficulties in explaining such developments. In the case of complex changes, the whole body would be exercised in a new way so that the acquired characteristic itself had all the necessary correlations. The disappearance of useless organs was similarly a direct consequence of their actual disuse.

Spencer conceived Lamarckism as a utilitarian mechanism that adapted the structure of a species to its function in the environment by use and disuse. Indeed, by its very nature, use-inheritance is best suited to explaining adaptation in the active behavior of animals. But it was not impossible to accommodate Lamarckism to plant evolution, as shown by the writings of George Henslow (1888, 1895). Henslow believed that all natural variation is directed toward a goal, usually an adaptive response to the environment. A plant automatically acquires appropriate changes in structure when grown in unusual conditions. Henslow made no attempt to explain why the changes were adaptive, although he clearly regarded this as evidence of the Creator's wisdom. Although much of his work was devoted to proving that such responses do occur, he does not seem to have appreciated the need to show that they are inherited as true Lamarckism demands. Obviously, if the plants are kept under the new conditions, the effects will be produced in every generation. The crucial test would show whether they retained the abnormal characteristics for some generations after having been returned to their normal environment. Henslow's failure to appreciate the need to demonstrate this point illustrates a common weakness in the Lamarckian arguments: proving the existence of acquired characteristics and then *assuming* their inheritance.

Henslow also observed how the shapes of flowers are related to the ways insects enter them. He argued that this again demonstrates a direct response of the plant to an external stimulus. In each generation, the insects have exerted pressures on the flowers, and the accumulated effect of the distortions has shaped evolution. Again, his argument contained a logical fallacy committed frequently by the Lamarckians. The shaping of the flower to accommodate the insect does not prove a direct response by the plant to the stimulus. If the shape is useful in encouraging insects to fertilize the plant, selection of random variations may have produced the same effect, because in every generation those flowers that by chance were better adapted to the insects' requirements would have been fertilized more readily and would have produced more seeds.

In America, Alpheus Packard supported use-inheritance on the bases of entomology and a study of the cave-inhabiting blindfish, whose loss of eyesight he attributed to disuse (1889, 1894). The origins of the distinctive American form of neo-Lamarckism, however, go back to the dissatisfaction of a number of naturalists with the selection theory as early as the 1860s (Pfeifer, 1965, 1974; Dexter, 1979). The paleontologists Edward Drinker

Cope and Alpheus Hyatt, in particular, developed a characteristic approach to evolution based on the recapitulation theory.

Cope is best known for his dispute with O. C. Marsh over opening up the rich fossil beds of the American West (Osborn, 1931; Schuchert and Levene, 1940; Plate, 1964; Lanham, 1973; Shor, 1974). Although converted to evolutionism in the 1860s, both Cope and Hyatt found natural selection unacceptable. Significantly, both had contact with Agassiz early in their career, Hyatt as a student at Harvard. Neither began as Lamarckians; instead, they sought a way of reconciling Agassiz's idealist vision of development modeled on embryology to the new evolutionism (Bowler, 1977*b*, 1983; Gould, 1977*b*). The key was the recapitulation theory, the belief that the growth of the embryo repeats the evolutionary history of the species. For Agassiz himself, this parallelism of the two modes of development was a divinely planned harmony modeled on the hierarchy of classes leading up to man as the goal of creation. Cope and Hyatt were products of a later era and could not accept a single goal for creation. They realized that evolution must be a branching process but saw the possibility of using the recapitulation theory as a guide to the evolution of each major group. The result was a theory that ignored the concepts of Darwinian natural history to concentrate on displaying the development of life as a regular sequence of fossil forms paralleling the goal-directed process of embryological growth. While they abandoned the belief in an overall harmony of creation, Cope and Hyatt sought to retain an element of regularity in the evolution of each living group.

The law of "acceleration of growth" was first published in Cope's "On the Origin of Genera" of 1867 (reprinted in Cope, 1887) and in Hyatt (1866). According to this law, evolution progresses by a series of sudden additions to the growth of the individual. At certain points in time, every individual in a species begins to exhibit a new phase of growth that advances all to the form of a new species. To make room for this addition, the old adult form is compressed back to an earlier phase of growth, hence the "acceleration" of growth to accommodate an extra stage before maturity. Cope denied that evolution on a small scale is a branching process, claiming instead that each genus represents a group of species that have reached the same point in the historical development of their group. Their close relationship is not a sign of common descent but of identical position in the scheme of development. The evolution of a group thus consists of a number of lines advancing in parallel through the same pattern, the whole pattern revealed in the embryological growth of those that have reached farthest in modern times.

What decides the direction of the next stage in development? There could be no question of random variation and selection, because Cope and Hyatt supposed each step to be a collective move in a predetermined direction. Cope even denied any adaptive value for new characteristics, claiming they were just steps in a regular pattern designed by the Creator. Hyatt had no religious ax to grind, but he, too, in his first paper, implied that evolution is the unfolding of a predetermined sequence. Soon both men realized that by failing to give a clear indication of why new characteristics were added on at certain points in time, they were begging the whole question of a natural cause for evolution. Cope now abandoned his opposition to the principle of utility and conceded that most changes had an adaptive purpose. Both he and Hyatt began to see that Lamarckian use-inheritance would serve as the guiding force they were seeking. Lamarckism was well suited to the recapitulation concept, because it required that new stages are developed by the adult organism and then compressed back into embryological growth so that they can be inherited. Cope postulated a growth-force named "bathmism"; concentrated in those parts of the body most in use, it developed them at the expense of other areas. By the last decade of the century, this Lamarckian alternative had been developed to considerable depth (Cope, 1887, 1896; Hyatt, 1880, 1884, 1889).

Deeply religious, Cope went on to argue that Lamarckism left room for consciousness to be seen as the guiding force of evolution. Instead of being designed by an external Creator, the species design themselves as consciousness gradually extends its manifestations in the organic world. Evolution itself thus acquires a spiritual character through its ultimate purpose in developing the role of mind (Moore, 1979). Animals control their own evolution, because their conscious choice of how to respond to an environmental challenge determines the form of their body through use-inheritance. Mind itself, furthermore, will be promoted through evolution of ever-more-intelligent forms, and mankind is the ultimate expression of this trend. Hyatt, however, was suspicious of such metaphysical elaborations and thought it ridiculous to ascribe conscious motivations to the extinct cephalopods on which he worked.

Despite Hyatt's skepticism, Lamarckism's ability to generate a sense that life controls its own destiny was an important moral point in its favor. So vital was this issue that efforts were made to show that the selection theory could also accommodate a certain amount of conscious direction, avoiding the image of species driven helplessly by the power of the environment. The mechanism of "organic selection," later known as the "Baldwin effect," was suggested independently in the 1890s by the paleontologist H. F. Osborn and the psychologists James Mark Baldwin and C. Lloyd Morgan (Baldwin, 1902; R. J. Richards, 1987). According to this theory, individuals faced with an environmental challenge would choose the most appropriate response, and their bodies would acquire characteristics suited to the new behavior pattern through exercise. But it was not necessary to assume that these acquired characteristics were inherited, as long as they gave the species enough flexibility to adapt on a short-term basis. Eventually, random variation would produce a heritable version of the same characteristic, and selection would be able to move the species in the direction already marked out by the new behavior. As a pupil of Cope, Osborn hailed the mechanism as a compromise between selection and Lamarckism, but Baldwin and Morgan more perceptively realized that it gave the whole game to Darwin. Organic selection denied the *inheritance* of acquired characteristics, while using them to show how selection could be stimulated by positive response of living things to their environment.

The most distinctive feature of the American school was its insistence on the regularity of development within each branch of evolution. This was not a necessary requirement of Lamarckism—because for most Europeans, use-inheritance simply replaced natural selection as the driving force of an irregular, branching process on the Darwinian model. But Cope and Hyatt had begun by assuming that there are regularities in evolution and had only later applied Lamarckism to explain them. The cephalopods studied by Hyatt were well suited to this approach, because their evolution displays patterns that puzzled many later Darwinists. Cope extended the same approach to the vertebrates, arguing that the evolution of the modern horse, for instance, has proceeded far too regularly in a single direction to be explained by random variation. Almost certainly, this emphasis on regularity was a belated expression of the idealism originally introduced into America by Agassiz.

The same influence seems evident in Cope's desire to defend the distinct nature of species. Originally, he had insisted that evolution takes place by a series of sudden additions to growth, each marking the species off quite clearly from its ancestor. Even after he turned to Lamarckism, Cope argued that the species would resist environmental pressures until it reached an "expression point" when a new form would be produced quite rapidly. This is not quite evolution by sudden mutation, but the idea allowed the naturalist to treat species as distinct entities that did not blend into one another in the course of gradual evolution.

The most fascinating expression of the belief in the regularity of evolution is Hyatt's concept of "racial senility." He believed that a group, such as the Ammonoids, began with a relatively simple form that, during a period of favorable conditions, developed by regular stages into various more advanced forms. Then degeneration set in: conditions became unfavorable, and members of the group, unable to cope with the challenge, began to lose their advanced characteristics and descend back toward the previous level. The final stages before extinction bore a strong resemblance to the original, very simple starting point. This was a form of Lamarckism—the degeneration was a direct response to external factors—yet it presented life reacting blindly to challenges beyond its powers of adaptation. In its use of the analogy with growth, senility, and death, Hyatt's theory was the most complete expression of the view that the life cycle of the individual organism forms the best model for evolution. The possibility that all branches of evolution ultimately might run out of steam and face the prospect of degeneration and extinction was to be explored more fully by the advocates of orthogenesis.

If orthogenesis was the final expression of the spirit of American neo-Lamarckism, the alternative, less highly structured form of the theory continued to flourish into the twentieth century. The one point lacking was experimental proof of the inheritance of acquired characters, a deficiency that began to seem all the more important as the new trend toward experimental biology gained ground. The French physiologist C. E. Brown-Séquard reported positive results with the inheritance of epilepsy induced by brain mutilations in guinea pigs. This was widely discussed (e.g., Romanes, 1892–97), but it was not confirmed that the results were due to genetic inheritance. Possibly the mutilation produced a toxin transmitted from mother to offspring in the uterus. Too often, this kind of alternative explanation emerged to mar the plausibility of the few apparently successful efforts to prove Lamarckism. Gradually, in the early twentieth century, the majority of experimental biologists began to lose patience. Starting in 1900 a new factor had entered the scene with the rediscovery of Mendel's laws of heredity and the beginnings of modern genetics. As this movement grew, it inevitably undermined theoretical support for Lamarckism by endorsing Weismann's claim that the germ plasm is insulated from any direct influence by the body that carries it.

The twentieth-century decline in the popularity of Lamarckism was rapid in experimental biology, especially in the English-speaking world. The rigid hereditarianism of modern genetics allowed no room for Lamarckian effects, and biologists were actively discouraged from exploring more flexible systems of heredity. The claim that transmission could occur through the cytoplasm in addition to the cell nucleus was rigidly suppressed (Sapp, 1987), thus blocking the search for a mechanism that would allow Lamarckism to operate. Only in Germany, where genetics developed in a less dogmatic form, was there an interaction between experimental biologists and paleontologists searching for non-Darwinian mechanisms of evolution (Reif, 1983, 1986). In America and Britain, paleontologists and field naturalists continued to support Lamarckism and orthogenesis, but in so doing, they became alienated from the experimental branches of biology. The resulting split between the disciplines was not healed until the emergence of the modern synthesis in the 1930s and 1940s (Mayr and Provine, 1980). Positive evidence occasionally was reported and led to great controversies. In addition, there was still strong support for Lamarckism outside science, among writers who were more deeply affected by the theory's emotional appeal. The first decades of this century may have seen the greatest popularity of Lamarckism among those who wished to explore its broader implications.

The playwright George Bernard Shaw was a leading nonscientific advocate of Lamarckism (Smith, 1965). In *Man and Superman* of 1901 and more explicitly in *Back to Methuselah* of 1921, Shaw proclaimed Lamarckism the philosophical salvation of the evolutionary movement. Because the selection theory turned everything into a senseless accident produced by a nightmare of struggle, in Shaw's eyes it could not be true. Lamarckism offered hope that life would strive by its own efforts toward higher forms, hence it was the philosophy for all right-thinking people. We must learn from this and get on with the job of promoting our own development, or nature will give some other species the chance to dominate the earth. This concept of the inherent purposefulness of life Shaw called "creative evolution," appropriating the name already given to Bergson's similar but not quite identical philosophy. Shaw also praised Samuel Butler's earlier opposition to Darwin, although he disagreed with the idea of eliminating conscious thought in favor of instinct. Shaw believed that he was riding on a wave of anti-Darwinian feeling, apparently unaware that scientists now were turning their backs on Lamarckism; but it is not surprising that he was unaware of the latest developments in biology. Yet unless he was completely wrong in his assessment of feeling in the literary world, we must conclude that Lamarckism retained some degree of popular support long after it virtually had been eliminated from science.

Within the scientific community, a few desperate efforts still were being made to find positive proof of Lamarckism. The psychologist William McDougall performed an experiment in which rats trained to run a certain maze seemed able to pass on this knowledge to their offspring, as though it had become an inherited instinct (McDougall, 1927). It was shown later that he had unconsciously selected out rats who were better at running *any* maze. Perhaps the most controversial episode of twentieth-century Lamarckism is the one centered on the experiments of the Austrian biologist Paul Kammerer; interest in the episode has been revived by the publication of Arthur Koestler's *Case of the Midwife Toad* (1971). The modern scientific community generally has rejected Kammerer's results as a delusion, but Koestler—who shared the hope of many earlier writers that Lamarckism would give us a more hopeful philosophy of life—had urged that the case be reconsidered.

Kammerer's experiments were performed before the First World War, when Lamarckism still was not a totally lost cause, and they appear to have attracted wide attention. The famous "midwife toad" experiment took place at this time, although it was not necessarily the most important part of Kammerer's work. Most toad species return to the water to mate, and the male uses special pads on the forelimbs to grasp the female. The midwife toad, *Alytes obstetricans,* has adapted to mating on dry land and the males have lost the characteristic pad. Kammerer, who possessed unusual skill in breeding amphibians in captivity, reared the midwife toad in artificially moist surroundings and claimed to have produced males whose limbs were supplied with mating pads like any other species of toad. The effect also was supposed to be inherited. Kammerer's career was disturbed by the war, and in the 1920s, partly in an effort to raise funds, he tried to revive interest in his work in Britain and America. In 1923, he visited these countries, bringing with him a preserved specimen of the midwife toad which had supposedly inherited the mating pad. Translations of his works also were published (1923, 1924).

By this time, the scientific community had become much more suspicious of Lamarckism. Kammerer was unable to suggest any plausible mechanism of soft heredity and even used Mendelian language in his own writings. His claims were greeted with considerable skepticism, particularly by the geneticist William Bateson. Fully aware of the broader implications of Lamarckism, Kammerer painted a glowing picture of man in control of his own destiny. This led to irresponsible press headlines about breeding supermen, exactly the kind of exaggeration the scientific community distrusts. After Kammerer returned to Austria, Bateson demanded that the midwife toad specimen be subjected to closer examination. Kammerer refused to send it through the mail, which led Bateson to voice suspicions about the validity of the experiment. No one else could duplicate the effect, because the toads were extremely difficult to breed in captivity. Only a few scientists now defended Kammerer against Bateson's insinuations; they were led by E. W. MacBride (1924), one of the last supporters of the recapitulation theory. At last, when an independent test was made, the mating pads of Kammerer's specimen were shown to have been marked with india ink. Kammerer protested that he had not faked the original marks—an assistant must have injected the ink when the original marks began to fade because of artificial preservation. Shortly afterward he shot himself, however, leaving the bulk of the scientific community convinced that he was a charlatan.

Was the india ink added by someone wishing to preserve the original marks, or was it deliberate sabotage, perhaps a Nazi plot to discredit evidence hostile to their racial theories? Koestler certainly has suggested that Kammerer's experiments may have been genuinely successful, although others think he was simply dishonest (Aronson, 1975). Even those scientists who think that no fraud was involved believe that Kammerer was mistaken in his interpretation of the results (Waddington, 1975). As in other cases where Lamarckism at first seemed to have been proved, another explanation is possible. It is unlikely that we shall ever know the true story.

When Kammerer shot himself in 1926, he was about to take up a new position in Moscow. This was no coincidence, because in a few years Lamarckism would score its greatest success in Russia under the leadership of T. D. Lysenko. It has been suggested that Kammerer was meant to spearhead a Lamarckian takeover of Russian biology, a move delayed by his suicide until another figurehead could be found (Zirkle, 1949, 1959*b*). There was certainly an earlier, abortive attempt to impose Lamarckism on Russian biology (Gaissinovitch, 1980). In fact, Darwinism had never taken hold in Russia (Rogers, 1974), and the Marxist philosophy of the revolutionaries was positively hostile to what was considered the capitalist image of the selection theory. Lamarckism was the obvious alternative, and Lysenko succeeded in having it incorporated into official communist philosophy in the course of the 1930s. Was his rise to power merely a crude attempt to impose ideology on science or a consequence of the political leverage he gained by promising to end Russia's chronic wheat shortages (Medvedev, 1969; Joravsky, 1970)?

Lysenko's claim to fame was based on his discovery of "vernalization" of wheat, a process in which seeds are frozen so that they will germinate more rapidly in spring. The process had been known for some time in the West, but Lysenko claimed that the effect was inherited, just as Lamarckism would imply. Once the wheat had been processed, it would germinate earlier in all future generations. This would be of considerable value in areas with short growing seasons. Soon Lysenko gained enough political support to make open attacks on genetics and the selection theory. Genetics was blasted as idealist nonsense and the selection mechanism as reduction of nature to chance. One kind of rhetoric that Lysenko did not use, however, was that of the earlier Lamarckians who had foreseen an improvement of the human race itself through direct control of evolution. Marxism promised to give a perfect society without a genetic change in the nature of man. But the power granted Lysenko to try to relieve the food shortage allowed him to begin a purge of the established geneticists. All were forced to recant their allegiance to "bourgeois" Mendelism or were exiled to Siberia and in some cases never heard from again.

Only in the 1950s did Russian biology begin to recover from this nightmare. It became clear that, far from ending the food shortages, Lysenko had cut Soviet agriculture off from improvements the West was making through use of the despised Mendelian genetics. The whole bizarre episode leaves many questions unanswered. What is the role of ideology in science, and how big a role did it play in Lysenko's rise to power? Was he just an adventurer who exploited Stalin's credulity, or did he really appeal to the hope of creating a Marxist science? If there is to be such a thing as a Marxist science, it will obviously have to take a subtler approach than the deliberate enthronement of questionable experiments in the name of dialectical materialism. Marxists are now trying to see if the mistakes of the Lysenko affair can help them to work out a better perspective (Lewontin and Levins, 1976;

Lecourt, 1977). Certainly, the failure of Lysenko should not lull us into an automatic assumption that Western science is totally free from ideological connections.

ORTHOGENESIS

Closely associated with Lamarckism in the eyes of many naturalists at the turn of the century was another mechanism, known as orthogenesis. The term "orthogenesis" was popularized by Theodor Eimer, who was originally a Lamarckian, and it came to be used in the context of the regular evolutionary trends described by paleontologists of the American school. Literally, the term means evolution in a straight line, generally assumed to be evolution that is held to a regular course by forces internal to the organism. Orthogenesis assumes that variation is not random but is directed toward fixed goals. Selection is thus powerless, and the species is carried automatically in the direction marked out by internal forces controlling variation. The similarity with American Lamarckism is evident, in the postulation of both active forces in evolution and the regularity of development. The crucial difference is that the trends of orthogenesis are not adaptive. Far from being a positive response to the environment, they represent a nonutilitarian force that can in some cases drive the species to extinction. In this there is a similarity to Hyatt's concept of racial senility.

To assume that directed variation can force evolution into nonadaptive trends was to undermine the most fundamental principle of Darwinism. Use-inheritance and selection were compatible in the sense that both assumed adaptation to the environment as the major factor requiring species to change. This is why a Lamarckian paleontologist such as Cope could make contributions to our modern understanding of how evolution took place. The regularities of orthogenesis, however, occurred independent of the environment. The extinction of species unable to keep up with environmental change was an integral part of the Darwinian viewpoint; but a species that caused its own extinction was the antithesis of everything that Darwin symbolized. In their fascination with the regularity of development at the expense of utilitarian factors, the supporters of orthogenesis reveal the last vestige of the influence of idealism on modern biology.

Carl von Nägeli's theory of an "inner perfecting principle" driving evolution toward nonadaptive goals was an example of what would later be called orthogenesis (translation, 1898). But the idea was actually popularized by Eimer during the 1890s (Bowler, 1979, 1983). Eimer studied the color variation of animals, first among lizards and then among insects. At first a Lamarckian (translation, 1890), Eimer soon became convinced that there are patterns of evolution with no adaptive significance, and to these he applied the term "orthogenesis" (translation, 1898). He arranged the butterflies, for instance, into series of species that were supposed to represent the course of evolution, each series passing through the same sequence of color changes in the wings. This parallelism explained the similarities between unrelated forms attributed to mimicry by the Darwinians. Eimer claimed that similar sequences of color changes could be seen throughout the animal kingdom, apparently indicating that the pattern of development was inherent to life itself. Yet the patterns were all of *living* species—Eimer had no guarantee that the various modern forms corresponded to different stages in a temporal sequence.

Only the fossil record could provide a convincing demonstration of regular development through time (Rainger, 1981). Hyatt's concept of racial senility already had exploited the evidence that nonadaptive trends exist, and such examples came to form the main line of evidence for orthogenesis. The orthogenetic trend usually was supposed to be produced by a force internal to the species, driving its variation in a direction ultimately headed for extinction. A famous case was that of the recently extinct "Irish elk," thought to have died out because its antlers became too large as a result of an internal trend (Gould, 1974*b*). It seemed as though the trend that produced the antlers, perhaps originally for some useful purpose, had acquired a momentum of its own which had carried it far beyond the point of utility. This "overdevelopment" theory of extinction became widely popular among non-Darwinian paleontologists in the early twentieth century.

Strong support for orthogenesis came from the Russian biologist Leo S. Berg (translation, 1926), but perhaps its best-known exponent was the American paleontologist Henry Fairfield Osborn. As a pupil of Cope, Osborn had begun as a neo-Lamarckian but had soon become dissatisfied with this approach. Along with Baldwin and Morgan, he suggested the mechanism of organic selection to allow the conscious choice of animals to direct the course of natural selection. In the early twentieth century, he began to develop the theme of orthogenesis, to which he eventually applied his own term "aristogenesis" (Osborn, 1908, 1912, 1917, 1929, 1934). He knew, of course, that the basic evolution of any class must be a branching process in which lines of specialization branch out from a common ancestry. Osborn coined the term "adaptive radiation" to describe this diversification of a class into many different forms, but he believed that once the various orders within the class had become established, their subsequent evolution was a stable, linear process without the continuous small-scale branching postulated by Darwinism. His evidence came from apparently regular lines of development revealed by the fossil record, particularly in the horns and teeth of certain mammalian groups. Although such organs eventually become useful, Osborn insisted that when small they had no adaptive value and must have been developed by a force other than selection. Eventually, this orthogenetic force would cause the structures to grow so large that they became a positive hindrance, perhaps enough to cause extinction.

Most supporters of orthogenesis made no effort to explain why the trends occurred and left themselves open to the charge of postulating mysterious vital forces. This was perhaps unfair—their theory was the very antithesis of the more hopeful philosophy of creative evolution. But Osborn was disturbed by the lack of any natural explanation and at first attempted to supply one. He suggested that the interaction of energies within a living body might allow a variation-trend to become fixed in the germ plasm. Many early geneticists in fact believed that mutations had a tendency to occur in certain preferred directions, but they could envisage no mechanism for implementing a trend aimed at the future production of a useful characteristic. Osborn himself soon conceded that his own explanation was unsatisfactory, hinting that evolutionists might have to give up the search for material explanations and rest content with the abstract trends revealed by paleontologists. At this point, he was accused of toying with mysticism by the geneticist T. H. Morgan (see Allen, 1969*a*). Osborn's difficulties reveal the final dilemma of the idealist way of thought in an age of experimental science.

Most of the fossil evidence used to support orthogenesis has now been reinterpreted in a manner compatible with modern selectionism. Gould (1974*b*) has pointed out that even very large structures, such as the antlers of the Irish elk, may have had a positive purpose. During the emergence of the modern synthesis, it was also suggested that the selection theory might accommodate some apparently nonadaptive trends through the phenomenon of relative growth rates or "allometry" (Huxley, 1932). But in most cases, further discoveries have revealed that the supposedly linear trends were merely the paleontologists' oversimplified interpretations of insufficient evidence. It is easy to draw a straight line linking a few specimens, but further information usually shows that the evolution was in fact branching and irregular. Orthogenetic trends existed more in the minds of their exponents than in nature herself (Simpson, 1944, 1953*b*).

MENDELISM AND THE MUTATION THEORY

The origins of the modern genetic theory of natural selection are centered on a number of curious incidents. Mendel's work on heredity lay unnoticed for thirty-five years and brought its author only posthumous fame. When the new emphasis on the study of heredity led to its rediscovery in 1900, the potentialities of Mendel's approach for the selection mechanism were not appreciated. Far from being seen as the salvation of Darwinism, Mendelism was presented as yet another alternative by those who felt that its emphasis on discontinuous variation contradicted the ideas of biometry.

When the concept of "mutation" was introduced to allow for the appearance of new genetic characteristics, it was assumed that this would render selection superfluous. Only in the 1920s was it realized that the new approach to heredity could be interpreted to solve the problems inherent in the original form of the selection theory. (For general histories of genetics, see Dunn, 1965; Sturtevant, 1965; Carlson, 1966; Stern and Sherwood, 1966; and Stubbe, 1972; on twentieth-century biology, see Allen, 1975*a*.)

Interpretation of Mendel's own contribution is complicated by the fact that his followers seem to have read their own ideas into his writings. It has recently been argued that far from being a pioneer, he actually was trying to extend a tradition that dates back to Linnaeus (Olby, 1979). This was the belief that a new species might be produced by hybridizing two existing ones, which Mendel apparently saw as a possible alternative to Darwinism. His famous laws of heredity were of interest to him only insofar as they allowed him to postulate the constancy of hybrids in founding new species. Nor is it clear that he thought in terms of paired hereditary particles—the alleles of modern genetics. The mathematical rigor of Mendel's work was new, and that was exactly what his rediscoverers were seeking. Their ideas, however, were shaped by the post-Darwinian controversies over continuous versus discontinuous heredity, and they automatically interpreted Mendel's papers in this light. The following account of Mendel's work, for the sake of convenience, presents it as it was understood in 1900.

Gregor Johann Mendel was an unsuccessful student who received only a limited scientific education (Iltis, 1932; Stern and Sherwood, 1966; Orel, 1984). Eventually he entered the monastery at Brünn in Austrian Moravia (now Brno, Czechoslovakia) where he eventually rose to become abbot. His work on the hybridization of plants was performed in the monastery garden and published in the journal of the local natural history society in 1865 (translation, 1965, and in Bateson, 1902; Stern and Sherwood, 1966). He received no support from the scientific community and was positively discouraged by the church. He died in 1884, worn out by efforts to prevent his monastery from being taxed by the Austrian government.

Mendel concentrated on discontinuous variation, characteristics inherited on an all-or-nothing basis, which he felt would simplify the problem of identification after hybridization. A considerable literature on hybridization existed, and the phenomenon of discrete inheritance had been noted from time to time (Roberts, 1929; Zirkle, 1951; Olby, 1966), but no one had focused on the point to see if regular mathematical laws might apply. It is clear that Mendel knew exactly what he was looking for and had picked out a suitable experimental subject—the garden pea. The suggestion has even been made that his results are a little too good, perhaps "improved" by an assistant who knew what Mendel was looking for (Fisher, 1936; Wright, 1966; Waerden, 1968).

Mendel picked out seven characteristics in the garden pea, the maximum number he could have chosen to obtain the results he wanted. These included the size of the plant, the color of the flower, whether the seeds are wrinkled or not, and so forth. In the case of height, our example, the pea exhibits two distinct forms, tall and short. The experiment began with pure strains for each characteristic, that is, strains known to have bred true over many generations. The first step was to cross-fertilize plants with the opposite form of each characteristic, pure tall crossed with pure short, and so on. When the resulting seeds were grown, Mendel could see how the characteristics had been inherited by the hybrid. In every case there was *no* blending; the hybrid showed only one form of the characteristic that had been crossed. The hybrids from the tall-short cross were all tall—there was no blending to give a medium stature, and all trace of the short characteristic seemed to have disappeared. The hybrids were then self-fertilized and the seeds grown to maturity. In the second hybrid generation, Mendel found his famous 3:1 ratio. There was still no blending, but the characteristic that had apparently vanished now reappeared in one quarter of the plants in this generation. On the average, there were three tall plants for every short one.

To Mendel's later followers, it seemed obvious that in such cases there must be a particulate mechanism for heredity; a single particle or unit in the germ plasm had to be responsible for transmitting the characteristic from one generation to the next. It was equally obvious that this unit must exist in two distinct forms, one for each state of the characteristic (tall or short). The unit is what we now call the "gene" and the different forms its "alleles." Two crucial assumptions were required to complete the theory. First, each organism must carry two units for each characteristic. This is required by the nature of sexual reproduction: if the offspring is to derive a part of its germ plasm from each parent, it must get one unit for each characteristic from the male and one from the female parent. When it, in turn, reproduces, it will pass on just one of its units; that unit will combine with one from the other parent to give a pair for their offspring. The second critical assumption is that when the two opposite forms of the unit for a certain characteristic are combined in the same individual, they do not blend their effects. Instead, one is "dominant," the other "recessive"—one takes complete control of producing that characteristic in the new organism, while the other lies dormant in the germ plasm. Only when both units correspond to the recessive characteristic will it actually be produced in the mature organism.

Now we are in a position to understand Mendel's experimental results. In the pure strains from which he began (tall and short), every individual must have *both* pairs of units corresponding to the appropriate characteristics. If we represent the tall allele by the letter T and the short by S, the genetic form of the two pure strains must be TT and SS. When these two are crossed, the hybrid must derive one unit from each parent and can only have the structure TS. In this case, the tall form is dominant and the short has no effect in the hybrid.

When the first hybrid generation is self-fertilized, there is an independent redistribution of the units which gives approximately equal numbers of all four possible combinations of T and S. Allowing again for the dominant-recessive relationship, one quarter of the second hybrid generation will show the recessive character.

TS x TS (Tall) (Tall) — First hybrid generation

TT TS ST SS (Tall) (Tall) (Tall) (Short) — Second hybrid generation

Thus, a twin-unit theory of heredity, coupled with the dominant-recessive relationship, will explain the results of the experiment on discontinuous variation. In addition, Mendel was able to show that each of his seven varying characteristics was inherited independent of all the others. This is not always the case, however, because Mendel failed to discover the effect now called "linkage" (Blixt, 1975).

Surprise is sometimes expressed that Mendel's work went unnoticed for thirty-five years. Although the journal of the Brünn natural history society was not the most prestigious, copies were circulated even in England, and there were occasional references to Mendel's paper in later literature (Olby and Gautry, 1968; Weinstein, 1977). Mendel was in contact with one leading biologist, Carl von Nägeli, who told him his results were of little significance and put him onto other subjects whose complexities defied analysis by any technique available at the time. The fact is that the scientific community was not able to make use of the work in the 1860s (Gasking, 1959; Posner and Skutil, 1968). The experiment was probably seen as a contribution to the now outdated debate on the production of new species by hybridization. Discontinuous variation within a species was considered a rare exception to the normal rule of blending. Most cases of heredity are, in fact, far more complex than the simple example Mendel so carefully picked out. The same rules apply, but they cannot be seen to apply unless the case is examined with techniques far more sophisticated than those available to Mendel. What was needed was a situation in which scientists would become convinced that something was wrong with the traditional view of heredity, and hence readier to seek out a new approach.

By the end of the century, developments in the study of heredity had created a new environment in which the modern interpretation of Mendel's results could be promoted. Weismann had defined the concept of hard heredity through his assumption that characters are somehow predetermined by the constitution of the germ plasm. He maintained that the germ plasm of the cell nucleus exerted absolute control over the growing organism; any modification of the growth process by environmental factors was trivial and of no evolutionary significance. This theory of "nuclear predetermination" was at first highly controversial, since it violated the prevailing assumption that the growth of an organism could be influenced by the environment (Gilbert, 1978; Maienschein, 1978, 1984). Even Weismann had started out from the traditional view that heredity and growth should be studied together as part of an integrated biological process. Some biologists now tried to apply the experimental method to the study of growth, in an attempt to create a "developmental mechanics" that would explain how the fertilized ovum generated the structure of the growing embryo. But this proved far beyond the capacities of nineteenth-century science, and emphasis gradually switched to the problem of how characters are transmitted from one generation to the next. Soon a number of biologists were actively looking for cases in which the inheritance of distinct characters could be traced over a number of generations.

The emergence of what would soon be called "Mendelism," or genetics, marked a decisive break with the past. The traditional view that growth and heredity were aspects of an integrated process now began to break down (Horder et al., 1986). Heredity could be studied without worrying about the problem of how the characters were actually produced in the growing organism. By the standards of this new science, growth was irrelevant to evolution: only the introduction of new genetic characteristics could change a population, and this was a process that could not be explained by analogy with growth. Lamarckism, the recapitulation theory, and the use of growth as a model for evolution could all now be dismissed as relics of an outdated conceptual system. Genetics eventually succeeded where Darwin had failed: it eliminated the teleological approach to evolution implicit in the analogy with growth.

The new science was not, however, hailed immediately as the salvation of Darwinism. The early Mendelians were inclined to dismiss the selection theory as yet another outdated product of old-style natural history. In principle, genetics solved some of Darwin's problems, especially Jenkin's claim that blending heredity would swamp out new characteristics, whatever their selective advantage (De Beer, 1964; Vorzimmer, 1968). If characteristics were preformed in the cell nucleus and transmitted unchanged from parent to offspring, selection would always be able to increase the proportion of individuals inheriting a favored characteristic within the population. But the new science was founded by biologists committed to the belief that discontinuous variation alone has genetic, and hence evolutionary, significance. They argued that evolution would have to work through the appearance of discrete new characters, the process soon to be called "mutation." The continuous range of variation observed in some characters by the supporters of biometry was of no real importance. The geneticists were also suspicious of the claim that the external environment might determine which new characters might be reproduced successfully.

By 1900, the search for a discontinuous model of heredity had led to a situation in which Mendel's work was being duplicated, now in a thoroughly modern context. Two biologists, Carl Correns and Hugo De Vries, "rediscovered" Mendel's laws and suggested that they held the key to a new theory of heredity (Wilkie, 1962; Zirkle, 1964). The claim that E. V. Tschermak was also a rediscoverer is no longer accepted, and even De Vries's role has been questioned (Zirkle, 1968; Kottler, 1979). Whatever the true circumstances, Mendel was now posthumously hailed as the hero of the new science and his experiments presented as a classic illustration of nonblending, discontinuous heredity.

The leading British champion of Mendelism was William Bateson (Beatrice Bateson, 1928). Originally a morphologist in the tradition of Haeckel, Bateson turned in frustration to the experimental study of heredity, hoping that this new direction would lead evolution theory out of the morass into which speculative Darwinism had drawn it. The work of Francis Galton and the American biologist W. K. Brooks convinced him that evolution took place by sudden steps and that discontinuous variation was far more significant that continuous. In 1894, his *Materials for the Study of Variation* emphasized these points by showing that discontinuously varying characteristics were far more numerous than the Darwinians admitted. This inevitably led to a debate with the supporters of biometry and culminated in a personal clash between Bateson and Weldon. Interested in how discontinuous factors were inherited, Bateson began his own breeding experiments and soon read of the rediscovery of Mendel's work. He published the first English translation of Mendel's papers (in Bateson, 1902) and went on to champion Mendelism as the key to a complete reorganization of the whole science of heredity (Coleman, 1970; Provine, 1971; Cock, 1973; Darden, 1977).

Because biometricians were already in dispute with Bateson over the significance of variation in evolution, they were almost inevitably led to reject Mendelism; and Bateson himself was now finding it difficult to reconcile the new genetics with his original concern for the mechanism of evolution. Mendel's laws deal with the transmission of existing characteristics, not the production of new ones, and Bateson had become suspicious of the claim that new genetic factors could be created by mutation. He was convinced that mutations were always degenerative, involving the breakdown of an existing characteristic. How, then, could positive evolution occur? Bateson noted that genetic factors sometimes are masked by inhibiting genes that prevent their characteristics from being manifested. If the inhibiting agent were removed by degenerative mutation, the factor it had once masked would show itself, apparently as a new characteristic. Bateson almost seemed to suggest this as a general theory of evolution (1914), despite its implication that all the characteristics developed in the whole of evolutionary history once were contained in the genes of the first living things, masked by a series of inhibitors that have since disappeared (Bowler, 1983).

Whatever Bateson's suspicions, it became clear to most biologists concerned with the mechanism of evolution that some process must introduce new genetic factors from time to time. Something like the modern approach eventually emerged from the widely popular "mutation theory" introduced by the Dutch botanist Hugo De Vries. In 1889, De Vries had published a theory of "intracellular pangenesis" based on the idea of discrete units for each characteristic (translation, 1910*a*; Darden, 1976). In following up this theory, he came across Mendel's work and thus became one of the "rediscoverers." Yet De Vries himself soon lost interest in Mendel's laws as he developed the concept of mutation. The theory was published in *The Mutation Theory* of 1901–03 (translation, 1910*b*) and in a series of lectures delivered at the University of California (1904; Allen, 1969*b*; Bowler 1978, 1983). To many of the naturalists who had become weary of the debate between the Darwinians and Lamarckians, the mutation theory seemed to have all the answers. It explained the origin of new characteristics *and* the separation of new varieties and species without the complicated isolating mechanisms required by Darwinism. It seemed, furthermore, to be based on hard experimental evidence, firmly in the tradition of the new biology.

De Vries did not use the term "mutation" in quite the modern sense. For him, mutation did not introduce a new factor into the existing population; rather, it created in one step a new form that would continue as a distinct breeding population. The mutated form was a new variety in the Darwinian sense, but instead of being formed gradually from the parent by the accumulation of small variations, it appeared suddenly when a number of individuals mutated to the new characteristic. Mutationism thus provided the ideal explanation of speciation without isolation, because the new form was distinct enough to breed only among itself from the beginning. De Vries insisted that there were positive mutations that would account for introduction of new characteristics in the course of evolution. He believed that species undergo occasional bouts of rapid mutation, which would account for the gaps in the fossil record and would also speed the whole process up to fit in with Kelvin's limit to the age of the earth.

De Vries's evidence came from his study of the evening primrose, *Oenothera lamarckiana*. He had found this species growing wild in Holland, apparently producing significantly mutated forms before his very eyes. Each new form, he maintained, bred true from the moment of its first appearance and thus formed a distinct variety. He actually gave the most important mutations their own specific name, implying an even greater degree of separation. Not until 1910 was it first suggested that there might be something wrong with the evidence De Vries had based his theory on, and only in the 1920s was it finally shown that *Oenothera* has an unusually complex genetic structure incompatible with the original interpretation. It was, in fact, a hybrid, so the "mutations" were caused by recombination of existing factors rather than production of new ones.

Meanwhile De Vries's theory gained wide popularity as an alternative to Darwinism. Yet De Vries himself insisted that he had no intention of challenging the whole structure of Darwin's theory, only of reworking it into a new form. Certainly, he had struck a blow at the biometricians' version of the selection theory, because he claimed that only mutations could produce significant hereditary change. Natural selection of individual variations was powerless. This meant that there was no need to assume that all characteristics of a species have adaptive value, because mutated characteristics are produced by a random change in the germ plasm. But De Vries's claim to be a good Darwinian was based on his willingness to admit that natural selection does operate at a higher level, among the mutated forms themselves. During a period of mutation, a species will throw off a large number of new varieties, most of which will be nonadaptive. These mutated varieties will have to compete among themselves for the limited amount of food or space available, and the weaker ones may be eliminated. Sooner or later, mutation will create a variety better adapted to prevailing conditions than even the parent form, and this will soon eliminate all its rivals. In the long run, adaptation does shape the course of evolution, a belief that De Vries held to be an essential safeguard against the resurgence of mystical or vitalist theories.

Many of De Vries's followers felt that his efforts to retain selection were unnecessary. In the case of Thomas Hunt Morgan, the mutation theory became the basis of an attack on the whole philosophy of Darwinism (Allen, 1968, 1978; Bowler, 1978, 1983). Although he later supported a combination of genetics and selection, at the beginning of his career, Morgan was strongly opposed to the Darwinian vision of nature based on struggle, perhaps in part for moral reasons. His *Evolution and Adaptation* (1903) attacked not only the selection mechanism but also the whole utilitarian view of nature. Morgan picked up De Vries's point that mutations are not created for any adaptive purpose and went on to argue that there was no need to imagine selection acting at any level. He believed that any mutated form not grossly incompatible with the environment would be able to survive and reproduce. Environment exerts no significant control over evolution: its course will be determined solely by the kinds of mutations that appear.

Also at this time, the Danish biologist Wilhelm Johannsen was publishing his experimental work on the breeding of "pure lines" in beans (translation, 1955). His work was widely interpreted as further evidence against selection and for the extreme form of the mutation theory. Johannsen, however, was not concerned with the wider problems of evolution and studied only selection acting within a population. The species he chose reproduced by self-fertilization, and by a "pure line" Johannsen meant all of the descendants produced by this means from a single individual. In his beans, variation showed a continuous range between two extremes; yet when this range was analyzed into its component pure lines, Johannsen was able to show that each bred true. The continuous range of variation was in fact made up of a number of overlapping but distinct elements, which did not blend and which maintained their position in the range over many generations. On this basis, Johannsen asserted that selection is effective only to the extent that it can pick out those pure lines lying at one extreme of the range of variation. Once an extreme pure line has been isolated, it cannot be affected by selection because any remaining variation is purely somatic, that is, it results from trivial, nongenetic differences in growth caused by external factors. Thus, selection must reach a limit beyond which it cannot pass—just as Jenkin and the opponents of Darwin had claimed long before. Johannsen believed that the only way in which genuinely new factors could be introduced into the genetic makeup of the species was by sudden mutation alerting the characteristics of an existing pure line.

The biometricians showed that Johannsen's results were not as clear as he maintained, but their opposition was lost in the enthusiastic reception of this new "proof" of discontinuous evolution. De Vries's original concept of mutation was forgotten as the Mendelians appropriated the term to designate the spontaneous appearance of new genetic factors *within* a breeding population. Johannsen himself went on to argue that his principles would still apply in normal sexual reproduction. The segregation of factors according to Mendel's laws would complicate matters, but selection still would be able to do no more than isolate the genes responsible for producing extremes of the range of variation. Positive evolution would occur only when a new factor was added by mutation, and thus it was mutation rather than selection that was the cause of evolution. In order to clarify the situation in the case of sexual reproduction, Johannsen introduced the modern distinction between an individual's "genotype" and "phenotype" (Churchill, 1974). The genotype is the genetic constitution of the organism, the phenotype its physical appearance. Because of dominance these may not be the same. Thus, in Mendel's original experiments, the tall parent form and the first hybrid generation had the same phenotype (both were tall) but different genotypes (TT and TS).

The best evidence that mutations feed new genetic characters into the existing population came from experiments on the fruit fly, *Drosophila melanogaster*, by T. H. Morgan and his colleagues (Shine and Wrobel, 1976; Allen, 1978). They identified a large number of different mutations appearing in their experimental populations, thereby demonstrating that mutation contributed to the species' range of variability. Many of the mutated characters were clearly not viable in the wild, although the mutated genes were transmitted according to Mendel's laws. Morgan had at first been suspicious of Mendelism, but his study of the chromosomes in *Drosophila* now convinced him that here was a mechanism that would explain why inheritance worked in accordance with these laws. Weismann's claim that the germ plasm resided in the chromosomes of the cell nucleus was now confirmed, Bateson alone refusing to accept this mechanistic interpretation of Mendelism (Coleman, 1965, 1970). It was shown that in the production of sperm or ovum by meiosis, the reproductive cell or gamete is given only one of the homologous pairs of chromosomes present in normal cells. The fertilized ovum derives one member of each pair of its chromosomes from the sperm and one from the ovum, allowing both parents to contribute an equal amount of genetic material to the offspring. It was even possible to work out relative positions of the genes corresponding to various characteristics in the chromosomes. The classic *Mechanism of Mendelian Inheritance* (Morgan et al., 1915) established the new theory of heredity with an entirely new line of evidence.

Morgan's work completed the elimination of Lamarckian ideas from experimental biology, at least in America and Britain. Although there were at first some dissenting voices, the new science of genetics now began to purge itself of all dissension from the claim that the chromosomes were the sole vehicle of heredity (Sapp, 1987). The possibility that the cytoplasm, the extranuclear material of the cell, might also affect the transmission of characters was suppressed. This, in turn, made Lamarckism seem impossible, since the chromosomes were thought to transmit genetic characteristics as fixed units that could not be modified by external factors. Curiously, this insistence on nuclear preformation was not typical of German genetics. Here the new discipline did not become so rigidly institutionalized, allowing cytoplasmic inheritance to remain plausible along with non-Darwinian factors such as Lamarckism and orthogenesis (Harwood, 1984, 1985; Reif, 1983, 1986). The complete elimination of the analogy between growth and evolution was thus primarily a feature of biology in the English-speaking world.

Within this rigid approach to genetics, mutation now seemed the only source of new characteristics. It was believed that all significant variation within a species was the result of a relatively small number of distinct genetic factors that combined according to Mendel's laws. These factors determined the limit of variation and hence the limit of selection. True evolution depended not on selection of the existing factors but on the introduction of new ones by mutation. The cause of the mutations was unknown, but they were believed to be spontaneous changes taking place within the gene. They were not caused by any purposeful influence exerted by the body, and hence Lamarckism was false. Gradually, in the 1910s, it began to be recognized that a complex system of Mendelian factors would account for the effects studied by the biometricians. Experimental work also showed that genetic recombination could itself lead to the appearance of new characteristics in the course of selection, breaking down the limit to variation set up by Johannsen. In the long run, mutation indeed adds new characteristics to the population, but this is not absolutely necessary for short-term evolution, nor are the effects of the mutations always as drastic as the early Mendelians had assumed.

Mendelism may have helped to destroy Lamarckism, but the conflict between Bateson and the biometricians prevented its incorporation into the selection theory. Because of this conflict, the Mendelians exaggerated the relevance of discontinuous variation and were prevented from seeing that their concepts could be treated in a more flexible way that would allow them to be synthesized with the principles of biometry. In addition, the exclusive Mendelian concern for laboratory work isolated them from the problems of traditional natural history, allowing them to oversimplify the complex questions of adaptation and speciation. Evolution was visualized in a highly artificial manner, without reference to the complexities of the real-life situation. Many geneticists thought that evolution would be driven not by selection but by the mutations themselves tending to produce new characteristics systematically in a particular direction. This notion of a "mutation pressure" driving evolution along nonadaptive lines produced, in effect, another explanation of orthogenesis. T. H. Morgan himself had raised this possibility during his early, anti-Darwinian phase, and a number of Mendelians continued to take it seriously. At least some of the Mendelian opposition to Darwinism thus shares a common ground with the more traditional line of attack initiated by the supporters of Lamarckism and orthogenesis. Mutation was seen as an internal process governed by purely biological laws and was expected to produce a more orderly kind of evolution that did not have to respond to haphazard changes in the environment.

The polarization of Mendelism and Darwinism continued until about 1920, when tempers at last began to cool down. Laboratory biologists and field naturalists became aware of the extent of their mutual isolation and began to gain a wider appreciation of the problems involved in the creation of a truly comprehensive evolutionary theory. It now began to seem that the most promising source of such a theory would be a synthesis of selection and a more sophisticated version of Mendelism. The concept of evolution by discontinuous mutations moving in a particular direction was still supported in the 1930s by a few unorthodox geneticists (e.g., Goldschmidt, 1940; see Allen, 1974). By now, though, most geneticists accepted laboratory evidence that mutations occur in many different directions; the scientists were put in a position that required them to adopt some form of selection in order to give any consistency at all to evolution. It was also realized that the vast majority of mutations produce only a small effect on the phenotype, while the same characteristic often may be affected by a number of different genes. The almost inevitable conclusion was that genetic variation provides the range of continuous variation studied by the biometricians and that the environment selects out those genetic combinations conferring adaptive fitness on the organism. The stage was now set for a complete reconciliation between genetics and Darwinism.

10

The Social Implications of Evolutionism

The phrase "social Darwinism" conjures up an image of mankind in the grip of a ferocious struggle for existence and of the social Darwinist applauding the elimination of the unfit as a necessary step toward progress. Thanks to Richard Hofstadter's classic *Social Darwinism in American Thought* (new ed., 1959), this image is frequently linked with the cutthroat competition that arises within laissez-faire capitalism. The image is best exemplified by the "robber barons" of late-nineteenth-century industry who competed to gain the greatest share of profits that could be made by exploiting the public. Use of a Darwinian metaphor to justify the gospel of struggle preached by these industrialists seems almost inevitable. If nature achieves progress through individual competition, then the survival of the fittest must be the key to economic and social progress for mankind. Yet Hofstadter's book refers to other situations in which the Darwinian analogy also could be applied. The most important area of struggle may not be among individuals at all but among nations or races, providing a justification for imperialism, colonialism, and slavery. Even within a single society, a slight twist of the argument allows the struggle for existence to be replaced by an artificial form of selection known as eugenics. The state would control breeding of human beings to discourage multiplication of the unfit.

Given this wide range of applications, it is hardly surprising that the term "social Darwinism" proved difficult to define precisely (Halliday, 1971; Rogers, 1972). A recent study of social evolutionism in Britain (Jones, 1980) points out that even liberalism and socialism were able to adapt the idea of natural selection to their own purposes. It is impossible to see social Darwinism as a simple and obvious application of the Darwinian theory to man. Links between biology and social thought could be established in many different ways, and in each case one aspect of the scientific theory was chosen to be stressed. Those who praised the evolutionary stimulus of warfare were convinced that individual competition would undermine the nation's ability to combat an external foe. Far from inspiring new social attitudes, biological ideas were exploited as a means of justifying ideological positions that already existed. The link between science and ideology is also a two-way process; scientific theories to some extent are modeled on the social environment in which they are created. If cutthroat capitalism looks like a direct application of natural selection to the economy, it must be remembered that Darwin's theory itself is often supposed to have an ideological foundation based on the influence of Malthus (chap. 6; Young, 1969, 1971*b*, 1973).

The historian must be aware of the many different ways of building social analogies on biological foundations, but little will be gained if the term "social Darwinism" is extended to cover so many different ideas that it becomes virtually meaningless. Darwin's name has often been applied to any system based on social progress, with the result that his theory has picked up all the bad publicity that ought to have been shared by other ideas. Bannister (1970, 1979) has challenged Hofstadter's claim that the capitalist form of social Darwinism was prominent in late-nineteenth-century American thought. The attack is two-pronged, aimed at both the supposed popularity and the internal logic of the Darwinian analogy. He suggests that liberal historians have deliberately exaggerated the reliance of the earlier generation on the struggle metaphor, in order to emphasize how far their own viewpoints rise above such deliberate ruthlessness. "Social Darwinism" always has been a term of abuse used by opponents of laissez-faire. The frequency with which Darwinian language was used by industrialists can be overestimated too easily, because many comments that seem to reflect the logic of natural selection actually can be ascribed to a rather different source of inspiration. This leads to debate on the true purpose of Herbert Spencer's philosophy of individualism. For Hofstadter, Spencer was the man who transmitted social Darwinism from Britain to America. Yet Spencer's acceptance of natural selection in biology was limited by his preference for Lamarckism, and it is possible to argue that he saw laissez-faire more as a stimulus to activity than as a means of eliminating the unfit.

At the opposite end of the political spectrum, it is important to clarify the relationship between Darwinism and Marxism (Heyer, 1982). Marx welcomed the idea of evolution because of its inherent materialism, but from the start, he suspected that Darwin's mechanism reflected the competitive atmosphere of Victorian capitalism. The story that Marx offered to dedicate a volume of *Capital* to Darwin is now known to have been based on a misunderstanding of the relevant correspondence (Colp, 1974; Feuer, 1975; Fay, 1978). Only the loosest of analogies can be drawn between Darwin's concept of struggle and Marx's belief that class struggle is the driving force of history, since the latter has its origins in the philosophy of the dialectic. In postrevolutionary Russia, the state has more often than not been hostile to Darwinism, as in the Lysenko affair (chap. 9).

The Marxists' rejection of the selection theory is understandable, but it should not mislead us into thinking that the link between capitalism and Darwinism is unambiguous. The fact that Spencer was a Lamarckian suggests that the model offered by free enterprise society has more than one biological analog. Spencer's position also confirms that non-Darwinian theories of evolution may have played a significant role in the social debates. The majority of Lamarckians, however, chose a social policy based on state intervention as a means of achieving better conditions for all. They hoped that inherited effects of resulting improvements would lead to genuine biological progress in the race itself. The eugenics movement also called for state control, but for the very different purpose of restricting the ability of the unfit to breed. Extensive modern literature on the origins of this movement reveals another source of scientific inspiration independent of Darwinism. Although founded in Britain under the influence of Darwin's theory, eugenics flourished most actively in America after it was able to exploit the new science of Mendelian genetics. Yet the early geneticists repudiated natural selection as a mechanism of evolution. Their claim that the "bad" genes in the human race must be eliminated by artificial selection was not based on the assumption that natural evolution worked through the more rapid duplication of the fitter genes.

The complexity of interaction between biological and social thought arises because scientific theories often were absorbed into existing debates, already generated by far more basic differences in attitudes. Many of the political differences mentioned above reflect opposing positions on the question of whether "nature or nurture" is the most important factor determining a person's character (Pastore, 1949; Cravens, 1978). Those who believe that nature, that is, heredity, is the most basic influence will reject the claim that improved conditions and education have a beneficial effect on character. If a person inherits a bad character, then nothing can help him and it is a waste of money even to try. This attitude leads to conservative political thinking based on the need to limit numbers and influence of the lower classes, who presumably represent the chief source of bad characters. Elimination of the unfit could take place by either natural or artificial selection, a choice of social Darwinism or eugenics. Darwinism and Mendelism both could offer a scientific basis for eugenics because both stressed the power of heredity to determine the biological character of the individual. On the contrary, the Lamarckian emphasis on soft heredity can be considered an extreme form of the belief that nurture can triumph over the limitations of nature. It believes that human beings *can* be improved by exposure to better conditions and goes beyond this to assume that improvements can be inherited over a series of generations. In fact, there is no need to assume any inherited effect for the supporters of "nurture" to be able to assert that better conditions will make better human beings. For this reason, liberal thought has continued to flourish in the twentieth century despite the collapse of biological Lamarckism.

Another attitude that went far deeper than any particular biological theory was the tendency to think of human differences in hierarchical terms. Eugenics and social Darwinism were based on a ranking of individuals within a single society. Some were assumed to be naturally abler than others, a fact reflected in the division of society into higher and lower classes. Even more pervasive was the tendency to rank the different races of mankind. Europeans almost invariably assumed that other races were inferior to themselves, and the degree of inferiority was measured by the level of technological and social development. It was very easy to assume that the "lower" races corresponded to earlier stages in the process by which the highest form of man has evolved from the apes. In theory, Darwinism should have undermined the concept of a linear hierarchy of progress, because it stressed the branching nature of evolution. Yet virtually all evolutionists accepted the linear image of human origins and used it to justify prevailing racial prejudice. Lamarckism could be adapted far more easily to the hierarchical approach, because it had close links with the belief that evolution is modeled on progressive development of the embryo toward its final goal. Those biologists who made the strongest efforts to establish a clearly defined racial hierarchy almost always incorporated a substantial element of Lamarckism into their biological thinking. The translation of a social prejudice into a "scientific" theory was achieved more easily by a non-Darwinian route, although in this case the underlying attitude was so widespread that no theory could escape it completely.

SOCIAL DARWINISM

The most familiar version of social Darwinism is surely the one that attempts to justify the competitive ethos of Victorian capitalism in terms of the struggle for existence. Yet the extent to which this line of argument was used has been questioned by some modern historians, and this disagreement is reflected in similar differences of opinion over Darwin's own beliefs. Greene (1977) has shown that interpretations range the whole gamut from open accusations that Darwin promoted aggressive individualism (Harris, 1968) to denials that he had any sympathy for such views (Freeman, 1974). Darwin's writings contain passages that can be interpreted either way. Greene argues that he was influenced by the current of thought favoring economic competition and was aware of that influence. Darwin saw a role for both individual and tribal struggle in the evolution of man and feared that the relaxation of selection in civilized communities could be harmful to the race. At the same time, he was unwilling to accept the more extreme implications of a laissez-faire policy. He was surprised to read in a newspaper the claim that his theory justified Napoleon and every cheating tradesman. Apparently, he did not want to see competition based on force or cunning, although he felt that something was needed to prevent the spread of inferior characteristics within the race.

The seeming paradox of Darwin's own thought can be resolved if we link him with the movement that Jones (1980) sees as the first significant effort to apply his theory in the political arena. This is not the aggressive form of social Darwinism but a more traditional liberalism; its principal goal was to restrict the power of the landed aristocracy. These liberals did not object to the existence of a ruling class, merely to the degenerate nature of existing rulers. They saw a new and *natural* aristocracy, which had emerged in the form of the professional class and which should be allowed to take its place as the dominant influence on society. The debate between Darwinism and religion can be interpreted at least in part as a result of this new class seizing on evolution theory as a means of attacking the church, which symbolized the traditional source of authority (Turner, 1978; Desmond, 1982). Darwin himself came to occupy the position of a country squire, superintending the affairs of the village community at Down. Men such as Huxley came to play an increasingly large role in public affairs on a national scale. Moore (1982) argues that Darwin's burial in Westminster Abbey was used by his followers to establish their position as the new ruling class, defending most of the old social hierarchy "under new management." The logical continuation of this movement was not the aggressive form of social Darwinism, because it accepted as already proven the superiority of the new ruling class. If the number of unfit individuals at the bottom of the social hierarchy was becoming too much of a burden on the community, the answer was not starvation but government control to restrict their breeding. Although Darwin himself was suspicious of artificial selection in man, the professional class turned increasingly to the eugenics movement as the century progressed toward its close.

The more extreme form of social Darwinism argued for a policy of complete laissez-faire in order to give free rein to economic competition. The state must withdraw from all efforts to limit the freedom of individual action, leaving everyone to rise or fall according to their ability. Progress would only occur if the fittest were allowed to fight their way to a dominant position in the economy, while those unfit to work would have to take the consequences. This is the image of social Darwinism portrayed by Hofstadter as an integral part of the capitalist ethic. While Darwinian rhetoric was used by some industrialists to defend the free enterprise system, we must take note of Bannister's objection that the popularity of the "struggle for existence" metaphor has been overestimated (see also Heyer, 1982). Debate inevitably centers on the true nature of the extreme form of laissez-faire individualism promoted by Herbert Spencer and his followers.

Spencer's support for laissez-faire was evident from his earliest writings, such as *Social Statics* (1851), to late works, such as *The Man versus the State* (reprinted 1969; Peel, 1971; Kennedy, 1978). He saw the development from primitive, authoritarian societies to modern capitalism as a key step in human evolution. Freedom was essential to allow progress through the cumulative effort of many individuals. It was also the necessary means by which all individuals would be brought into harmony with developments in society. Spencer insisted that the state should concern itself solely with external affairs; internally, it had no business trying to regulate the lives and activities of the people. There should be no state control of health care, education, or relief for the poor—many of the activities taken for granted today. Anyone who wished to set himself up as a doctor or teacher, for instance, should be free to do so and would be successful as long as he could persuade people to pay for his services. If he was no good at his profession , he would fail, after a few incautious customers burnt their fingers in trying him out. It was no more the state's responsibility to protect the public in such cases than it was to shield the failures from the misery brought on by their own inefficiency.

Similar views were advocated by Spencer's followers, particularly William Graham Sumner, the economist of Yale University. Sumner's philosophy was summed up in the phrase "root, hog, or die." He openly challenged anyone to displace him from his position by displaying teaching ability superior to his own. No sympathy was spared for those who failed in their endeavors: idleness and inefficiency were punished by nature herself, and any attempt to alleviate the suffering would allow the weakness to spread. Industrialists also absorbed Spencer's views, or at least claimed to operate according to his philosophy. Andrew Carnegie became his avowed disciple, while the phrase "survival of the fittest" was used to justify unrestrained competition by railroad magnate James J. Hill and by John D. Rockefeller.

Hofstadter's claim for the wide popularity of the selection analogy has now been disputed. Even Sumner did not share Spencer's optimistic view that real progress would result from free enterprise. Despite the examples cited, it has been argued that even among the business community there was only a limited acceptance of the survival of the fittest as a way of life (Wyllie, 1959; Bannister, 1970, 1979). Those who were successful in eliminating their rivals might have wished to justify their ruthlessness in this way, but the majority of ordinary businessmen were too conscious of their own vulnerability to feel the same. Outside the business community, explicitly non-Darwinian forms of social evolutionism also were gaining popularity.

Even accepting the fact that Spencer's philosophy did gain some support, we still must ask to what extent it presented laissez-faire as a counterpart of natural selection. In fact, the connection at best can be only a loose analogy. The ultimate goal of economic competition is to set up a monopoly, and at that point all benefits of natural selection would cease. The successful man dominates his fellows and thus directs society, but there is no corresponding biological effect if he does not have more children who will inherit his characteristics. In many cases, it is clear also that the inheritance of wealth does not correspond to the inheritance of ability, yet the wealthy continue to exert political power. This last point actually was used by A. R. Wallace as an argument for socialism (Durant, 1979). If all differentials of wealth are removed, he claimed, husbands and wives will choose one another for their biological qualities, just as nature intended.

Rather than a form of natural selection, Spencer's social philosophy can be viewed as an expression of the widely popular Victorian belief in the virtues of "self-help," which were exemplified also in Samuel Smiles's book (1859). This in turn was a naturalization of the Protestant work ethic, and Moore (1985*a*) points out that many liberal Protestants saw Spencer's philosophy as an extension of their traditional morality. The most important purpose of laissez-faire was not to eliminate the unfit but to encourage all men to improve themselves by their own efforts. Laziness was considered a greater obstacle to progress than stupidity, and the solution was not to eliminate the lazy but to force them to greater effort. This would be best achieved by preventing any alleviation of the suffering inflicted by nature as the penalty for failure. Spencer admitted that one consequence of laissez-faire would be the "shouldering aside" of the weak by the strong, and there can be no doubt that by the time he wrote *The Man versus the State* in 1884, he had become convinced that creeping socialism was allowing the worst elements in society to proliferate. In the absence of state support, these unfit individuals would never have been able to multiply to a level at which they became a burden. Yet the true source of progress, in Spencer's philosophy, remained the stimulus to individual initiative provided by the threat of failure and poverty. Socialism not only permitted the unfit to survive but also destroyed the quality of self-reliance so essential for social development. The fear of misery teaches everyone how to adjust to a new economic situation, and the next generation learns the lesson directly from its parents. Such a policy seems more a form of social Lamarckism than social Darwinism and suggests an explanation for Spencer's Lamarckian preferences in biology. It can be argued that much of what has passed for social Darwinism was in fact Spencerianism of this kind, owing little or nothing to the analogy with natural selection.

By the end of the century, Spencer's popularity was already waning. He denounced the wave of nationalist fervor sweeping many countries as a return to the less efficient form of militaristic society; but nationalism proved stronger than the old laissez-faire individualism. The thrill of pride felt by many at the prospect of conquest or colonial expansion illustrates the second level of competition that also can be called social Darwinism. The motto of the "survival of the fittest" inevitably was applied to the prospect of national struggle, each nation secure in the belief that it would prove to be the fittest. Expectation of future wars led to appeals for national unity that were inconsistent with the old emphasis on free enterprise. Thus, despite the Darwinian rhetoric, the concept of national struggle turned its back on the principal feature of the biological selection mechanism—individual competition.

The importance of national cohesion was emphasized in Walter Bagehot's *Physics and Politics* (1872). Setting out deliberately to apply the principal of natural selection to society, Bagehot argued that throughout history the strongest nations have dominated their weaker neighbors and that the strongest always have been fittest in the sense that they have contributed toward the development of civilization. Inferior nations, if not eliminated, have been subjugated and taught the advantages developed by their conquerors. But Bagehot's alienation from the true spirit of the selection theory is evident from his insistence that the chief agent of progress has been the growing power of governments. Far from hailing individual competition, Bagehot praised anything that has helped to bend the individual to the will of society, including religion. Church and state should be united for the strength of the nation, and freedom of thought suppressed. Throughout history those nations that have advanced toward higher levels of organization have been conquerors; the implication was that the same factor still must govern national competition.

The belief that national or racial competition was vital for the progress of humanity remained popular into the early twentieth century. The eugenics movement (see below) drew much of its appeal from the claim that the white race must prevent biological degeneration if it is to maintain its superiority over other races. Some paleoanthropologists now tried to argue that the fossil record of human evolution revealed the constant extermination of inferior types by those more highly developed (Bowler, 1986). The disappearance of the Neanderthal race from Europe was seen as an early example of the process by which the white race was now exterminating the native peoples of America and Australia. As late as 1949, Arthur Keith constructed a theory of human evolution in which tribal and racial struggle was the chief driving force. Significantly, Keith and many other advocates of racial struggle had no interest in natural selection as an explanation of how the higher races were produced. Selection was merely a negative process to eliminate those races that had lagged behind in the march of progress. The concept of struggle had taken on a life of its own, functioning independent of the Darwinian theory in biology.

Darwinism also exerted significant—but sometimes contradictory—influences outside the English-speaking world. In France, biological Darwinism was not very strong, and social Darwinism seldom stressed the concept of individual struggle (Clark, 1984). In Germany, thinkers for many different ideological backgrounds took up evolutionism and were often willing to call themselves Darwinists (Kelly, 1982). Historians have focused most attention on the militaristic form of social Darwinism, in which war was hailed as the means by which the strongest nation exerts its authority (Zmarzlik, 1972). Nor was this notion of the dominance of the fittest a mere perversion of Darwinism by the military elite. On the contrary, it was encouraged by one of the leading German Darwinists, Ernst Haeckel (Gasman, 1971). Haeckel had made evolution a foundation of his quasi-religious philosophy of "monism," a philosophy based on the unity of spirit and matter (Holt, 1971). An organization called the Monist League spread monism throughout Germany. Although disappointed that the nation was united by the military rather than by the middle class, Haeckel and his followers soon threw their weight behind the strong central government. Evolution taught them that throughout history progress has been achieved by superior races dominating their neighbors. The struggle must continue to ensure further progress, and the next stage would be the emergence of Germany as a world power. Haeckel was an ardent nationalist throughout World War I and was bitterly disappointed by Germany's eventual defeat. Members of the Monist League subsequently played a role in the emergence of Nazi ideology, emphasizing the innate superiority of the German people and the inevitability of their ultimate triumph.

Haeckel insisted that to be strong, the nation had to be united and centrally organized. This was the antithesis of Spencer's individualism, which often was dismissed in Germany as a reflection of petty British commercialism. A strong nation could not allow itself to be torn apart by such internal bickering; it must stand united behind the leaders who point the way to destiny. Curiously, Haeckel, like Spencer, was a Lamarckian as far as the internal development of both species and societies was concerned. But Spencer hailed the transition from militarism to capitalism, while Haeckel held that national competition was still the more important force. Thus, the state had to improve its citizens by molding them into a purposeful group. Spencer believed that nature must be left to take its course, because man cannot control the direction of progress; Haeckel believed that man can seize the initiative and forge powerful new nations in the light of his own vision of the future.

Apart from the bare concept of progress through struggle, there was little in German ideology that depended on a real analogy with the selection mechanism. Admittedly, Darwin had acknowledged the role of national conquest in human progress. But the biological origins of both Nazism and Spencerianism lie in two quite different combinations of Darwinism and Lamarckism. The Nazis drew on other aspects of German thought, including the racist sentiment of a *Volk* superior to the surrounding peoples, perversions of Nietzsche's superman concept, and Hegel's idealist philosophy of the state. Subordination of the citizen to the state and belief that the goals of the state can be symbolized by the aspirations of a great leader are certainly values more characteristic of idealism than of Darwinism.

EUGENICS

The claim that the state has a duty to limit the multiplication of its least fit citizens became the basis of the eugenics movement. Darwin himself had pointed out the problem created by the relaxation of natural selection in civilized societies, where the rigid weeding out of the unfit no longer takes place. The weakness of laissez-faire social Darwinism is that the least fit members of society are seldom left to die off completely; indeed, the slums of the great cities now were coming to be regarded as breeding grounds for the worst kinds of human characteristics. It naturally was assumed that the poorest class in society contains the greatest number of inferior characteristics, because poverty was considered a direct consequence of lesser ability. Darwin's cousin, Francis Galton, first began to argue that this problem was getting out of hand and that the state would have to play an active role in controlling the relative proportions of fit and unfit individuals. Proliferation of the inferior type of humanity in slums threatens degeneration for the race and is also a drain on the community's resources. Conversely, professional people with the highest level of ability tend to have relatively few children. The only solution was government intervention with the aim of reversing these natural tendencies. For Galton, the drive to improve the character of the race by means of what was, in effect, artificial selection took on an almost religious significance. His work would lay the foundations of the whole eugenics movement. (For general surveys of eugenics, see Blacker, 1952; Allen, 1975*b*, 1976; Bajema, 1977; Farrall, 1979; Kevles, 1985.)

Galton did not coin the term "eugenics" until 1883, but long before this, he had begun his attempt to establish the basic hereditarian position that ability is inherited and cannot be enhanced by education. His *Hereditary*

Genius of 1869 presented evidence drawn from a survey of prominent families to show that the abilities of the fathers were inherited by their sons. The lesson to be learned from this, he claimed, was the need to keep up the biological standards of the race by ensuring that individuals with high abilities have large numbers of children. This was what he later would call "positive eugenics," to be achieved by measures such as giving the professional classes income tax relief for each child. At the same time, Galton also sketched out what was to become the most influential aspect of the movement: "negative eugenics" aimed at restricting the number of children born to parents of below average ability. The worst elements of the poorer classes, those presumed to have subnormal mentalities, would have to be physically prevented from passing on their infirmities. The basic outline of the eugenics program had already taken shape in Galton's mind (Forrest, 1974; Buss, 1976; Cowan, 1977; Fancher, 1983).

The initial reaction to Galton's proposals was not very positive. Many were shocked at the idea of an artificial breeding program for man. There was also considerable skepticism over Galton's hereditarian claims. After all, the children of prominent parents usually receive a better than average education, so how can we be sure that their abilities are truly innate? Galton's subsequent efforts to develop statistical techniques in biology were a direct response to this initial skepticism (Cowan, 1972*b*). He set out to show in even more detail that heredity controls the nature of a population. His application of statistics to the study of variation laid the foundations of the biometrical approach that his student, Karl Pearson, would use in defense of Darwinism. It can be argued that the actual structure of Pearson's statistical techniques reflects his desire to provide apparently scientific evidence to support a eugenics policy (Mackenzie, 1982). A similar point can be made in the case of Pearson's student, R. A. Fisher (Bennett, 1983; Norton, 1983).

At the turn of the century, Galton and Pearson at last saw their movement begin to gather momentum in Britain (Mackenzie, 1976, 1982; Searle, 1976, 1979; Jones, 1986). Pearson pointed to the poor quality of many working-class recruits into the British army during the Boer War as evidence of racial decay. Galton founded the National Eugenics Laboratory in 1904 and soon afterward the Eugenics Education Society and the *Eugenics Review*. Throughout the early decades of the new century, eugenics formed a significant, although never overwhelmingly powerful, political lobby. Parallel developments occurred elsewhere. In Germany, the attempt to restrict the breeding of the unfit had become popular long before the Nazis came to power (Weiss, 1986). In America, too, eugenics flourished through the early twentieth century (M. Haller, 1963; Pickens, 1968; Ludmerer, 1972). The American Breeders Association, a Mendelian group, set up the Eugenics Committee in 1906, and in 1910, the Eugenics Records Office was formed (Allen, 1986). The first International Eugenics Congress was held in

1912. The factor uniting all these diverse national movements was the belief that control of heredity offered a way of applying scientific management to the human race. For this reason, eugenics—although normally associated with the political right—was occasionally supported by left-wing thinkers (Paul, 1984).

The purpose of all this activity was to persuade governments that they should enact legislation to improve, or at least protect, the biological standards of the race. A few suggestions were made for encouraging the professional classes to have more children, but generally the movement concentrated on trying to limit the birthrate among the lower orders of society. Galton's call for racial improvement was replaced by a more pessimistic attitude based only on the hope of preventing further degeneration. The insane and the feebleminded became the most common targets, and criminal tendencies frequently were seen as by-products of mental deficiency. To provide the necessary means of "accurate" discrimination, the science of mental testing was encouraged, and the first efforts made to establish a scale for measuring intelligence. This was done by comparing a person's mental and physical age to arrive at the intelligence quotient, or I.Q. It was assumed that innate intelligence could be distinguished easily, although the early tests failed completely to eliminate the effects of education and thus emphasized the apparently poor intelligence of the lower classes (Gould, 1981; Evans and Waites, 1981). The eugenics movement insisted that those at the bottom of the mental scale should be institutionalized and segregated by sex, or even sterilized to prevent breeding. Some American states indeed passed legislation requiring the sterilization of those below a certain standard of intelligence. Fears also began to concentrate on the multiplication of "inferior" racial groups coming in as immigrants from eastern Europe. The most comprehensive sterilization program was set up in Nazi Germany during the 1930s, as part of the effort to purify the "Aryan" race (Harmsen, 1955).

The sudden rise to popularity of eugenics during the early twentieth century was aided by both social and scientific factors. Social workers gradually were converted to the hereditarian belief that dishonesty and shiftlessness were inherited over many generations. The sheer extent of the phenomenon was assumed to show that a biological rather than a merely social level of causation was involved. Those in charge of institutions for the insane and feebleminded became convinced that these traits were transmitted by a hereditary factor. At the same time, the middle and professional classes became increasingly disturbed at what was portrayed as a massive increase in the number of mentally subnormal individuals. The middle and professional classes were convinced that they had gained their own social position through an innate ability that they hoped to transmit to their children. They did not want to see their financial gains dissipated by taxation to support an ever-increasing population of incapable and insane; and their reluctance to take on the burden of larger families themselves accounts for the decline in popularity of positive eugenics.

At the scientific level, eugenics profited from rapid development of Mendelian genetics after 1900. Galton had formulated his original proposals to replace natural selection within the human population, and his British followers—particularly Karl Pearson—followed a similar approach. The statistical techniques of biometry were used not only to confirm the power of natural selection but also to study the inheritance of characteristics in the human population as a means of confirming the need for a eugenics policy. Darwinism was not the only source of a hereditarian viewpoint, however; indeed, Galton himself did not believe that the selection of individual differences was the cause of natural evolution. Rediscovery of Mendel's laws provided an entirely new scientific foundation for hereditarian attitudes. The laws offered what appeared to be an experimental demonstration of the fact that biological characteristics are inherited as complete units over many generations. In America, where Galton's form of biometry had not established itself, Mendelism was taken up with great enthusiasm by the newly emerging eugenics groups. It was thought that an inferior characteristic, such as feeblemindedness, was the product of a single gene circulating in the population. This, in turn, seemed to offer the prospect of a rapid elimination of the characteristic, if those possessing it were prevented from breeding. Although most early geneticists did not believe that selection was the cause of biological progress, their theory nevertheless provided an ideal foundation for the less ambitious hope that the spread of harmful characteristics could be reversed.

The circumstances in which eugenics flourished suggest that it was only an indirect extension of the Darwinian approach. Galton may have been alerted to the problem by the suggestion that the relaxation of natural selection might lead to racial degeneration, and British eugenicists, such as Pearson, continued to see an analogy between natural evolution and the artificial form of selection they recommended for man. Yet Galton himself was not a true Darwinist, because he held that natural evolution proceeds by sudden mutations rather than the selection of individual variations. This was exactly the position adopted by the early geneticists. The fact that Mendelism was seized on by American eugenicists reveals that their underlying concern was the hereditarian view of human character rather than a fascination with progress through selection. In the debate over nature and nurture, eugenicists were definitely on the side of nature, that is, heredity (Cravens, 1978). Reluctant to pay additional taxes to benefit the poor, they argued that the poor could not benefit from improved conditions because their inferiority was genetically determined. Sterilization offered a cheaper way out, particularly as it was thought that the level of harmful characteristics in the population would soon be reduced by this means.

The eugenics movement went into decline during the 1930s, although it continued to operate at a reduced level of popularity. The reason for the decline at first sight might appear to be exposure of the scientific weakness of the claims made on behalf of negative eugenics. During the 1920s, it was realized that the simple image of a few "bad" genes causing all the unfit characteristics was a travesty of the true situation. Many characteristics are affected by a number of different genes, and the environment also can affect the growth of an organism whatever its genetic potential. Population genetics also revealed how immensely difficult it would be to produce a major change in the constitution of the race as a whole. The synthesis of Darwinism and genetics worked out during this period hindered rather than helped the cause of eugenics. Yet it seems unlikely that these scientific developments were the principal factor underlying the movement's loss of prestige. Most biologists had recognized the true state of affairs long before popular decline began. Many of them turned their backs on eugenics, although with a few exceptions they refrained from criticizing it, leaving the demagogues free to continue expounding their oversimplified panaceas. Serious research into human genetics went into decline, as though scientists were reluctant to get involved with a topic that would have controversial implications. Was this merely a distaste for public debate, or fear of confrontation with a movement that still had wide public support (Ludmerer, 1972; Provine, 1973; Gould, 1974*a*)? By 1930, most eugenics had degenerated into a pseudoscientific charade, yet it was only after this date that scientists, such as J. B. S. Haldane (1938), began openly to point out its inadequacies.

The real cause of the demise of eugenics seems not to have been scientific developments but growing public awareness of the potential dangers, brought about especially by the excesses of the Nazis in Germany (Searle, 1979). The Nazis demonstrated the frightening capacity of a totalitarian state to control its population in accordance with a racist ideology. This pointed out the moral dangers inherent in less-rigorous forms of eugenics. It had seemed obvious at first that the state had the duty to limit the breeding of the unfit, when this was confined to a few obvious defectives; but the Nazis showed how the definition of "unfit" could be widened to include whole sections of the population. Moderates everywhere now began to have second thoughts, and as scientists at last began to inform the public of the weakness in the genetic foundations, eugenics fell into disfavor.

Both eugenics and social Darwinism were based on the claim that certain segments of the human population, defined by race or class, are congenitally inferior to the rest. Their low status was the result of their inability to function adequately in the modern world. Hence no reform of the social structure would allow them to attain any significant gains. Such a view has a clear ideological content: it would naturally be favored by those in privileged positions and who were looking for excuses to prevent dissipation of their advantages among the population as a whole. Those who called for reform, conversely, would gravitate toward the belief that social background and education determine a person's character; the poor are not held back by heredity, and if we improve their conditions they will become better people. Because it has proved remarkably difficult to find objective ways to separate the effects of nature from those of nurture, the controversy is still with us today. Reforms of the last few decades seem to have had little effect—but does that result from the superficiality of the reforms or from the fact that those offered help were incapable of benefiting from it (Gould, 1974*a*)? From the perspective of some socialists, however, both the hereditarian and environmentalist policies represent merely different ways of achieving the same end. Whether the unfit are eliminated or absorbed into society, the aim still is to preserve a social structure in which the professional class constitutes a ruling elite (Werskey, 1978).

NEO-LAMARCKISM AND SOCIETY

Until its collapse in the early decades of this century, neo-Larmarckism offered the most obvious alternative to the hereditarian viewpoint in biology. Historians may have played down its role in social thought, feeling perhaps that the movement lacked a sound basis in biology. The Lamarckian alternative is noted by Hofstadter (1959) in the case of Lester Ward, while Stocking (1962) has discussed wider acceptance of such views in America; but there has been no extensive study of social Lamarckism comparable to those devoted to social Darwinism and eugenics. We now realize the influence of Lamarckism within scientific biology around 1900 and can see the need to take its social implications more seriously. Apart from Spencer's use of Lamarckism to defend laissez-faire, there was a widespread belief that inheritance of acquired characteristics would allow man to take charge of his own evolution in a far more positive way. Lamarckism became the philosophy of hope for the future. At the same time, most Lamarckians refused to extend their optimism to the nonwhite races; indeed, their theory was the chief foundation of the belief that the races can be ranked into an evolutionary hierarchy.

Although he has been labeled a social Darwinist, we have seen that Herbert Spencer's philosophy of complete individual freedom can be explained at least as well in Lamarckian terms. Spencer supported individualism on the grounds that experience is the best teacher and nature the only guide to progress. But for most neo-Lamarckians, social reform and state education would lead the race toward a goal chosen by man himself. Both positions, however, contain the seeds of a fundamental confusion. If people learn more effective behavior (by whatever means), we can assume that this will be passed on to future generations; but does such *cultural* inheritance via education have any logical connection with the process of *biological* heredity? Do our children continue a better way of doing things because we teach them, or because the new pattern has become ingrained in their physical constitution, so that they are born with a predisposition toward the new behavior? In the former case, social Lamarckism amounts to no more than the obvious antihereditarian position that people can be improved by exposure to better conditions; but if the improvement can be inherited biologically, we open the way to a new social policy based on the hope that the effect may become cumulative over many generations. By shaping human behavior through education, we actually are controlling the evolution of the species. In the nature-nurture debate, Lamarckians were on the side of nurture—but they went beyond a simple belief that human beings are capable of improvement, to argue that the improvements can become cumulative through heredity. In effect, this blurred the distinction between nature and nurture, with the consequence that Lamarckians tended to confuse the quite different processes of biological and cultural evolution.

There were moral objections that could be raised against Spencer's reliance on nature as sole guide to progress. Those who favored reform saw laissez-faire as merely a new way of defending the position of an elite class in society. Spencer held out the hope that eventually a perfect society would emerge, but in the meantime, the failures in every generation would have to learn their lesson through misery. Rapid progress was impossible, because Spencer believed that evolution was a slow and complex process beyond the power of man to control. His opponents wanted to speed up the process and argued that this could be done only if man took control of the system of formal education. We can choose the goals of our evolution and shape its course through the education of our children. The inheritance of acquired characteristics will ensure cumulative effects in each generation, leading to a far-reaching change in human nature.

This optimistic view of Lamarckism was especially popular in America and was first expounded by the sociologist Lester Ward in opposition to Sumner's extreme individualism (Hofstadter, 1959; Scott, 1976). The distinctive American school of neo-Lamarckism had a clear influence on the work of Joseph LeConte (1899; Stephens, 1976, 1978) and on the psychology of G. Stanley Hall (1904; Gould, 1977*b*). All of these figures gained prominence at the turn of the century, coinciding with the period of neo-Lamarckism's greatest success in biology. Their ideals were a direct extension of the views expounded by leading biologists, such as the paleontologist Edward Drinker Cope. To save the argument from design, Cope insisted that consciousness is an integral feature of the animal kingdom, and it has directed evolution of even the lower organisms (1887). Use-inheritance has gradually developed the power of consciousness up to the level of the human mind, which can now perceive the nature of the process and take control of it. Hall's emphasis on the positive role of education in shaping the child's mind was an extension of this view and was also based on the expectation that a positive improvement in human nature should be detectable over just a few generations. The appeal of such a view is obvious and accounts for its survival into the twentieth century long after most biologists had rejected use-inheritance. Man can be seen as no longer the slave of blind mechanical forces but an active participant in his own rapid evolution. This point was exploited by Paul Kammerer (1924) in his trip to America, and it still was capable of generating newspaper headlines about breeding a new race of supermen.

Those who insisted that Lamarckism was the only mechanism of evolution could present the future development of mankind as an inevitable extension of the natural process. But if it were conceded that selection played at least some role, one would have to be more careful. After all, if both selection and use-inheritance were natural processes, social Darwinism could be justified as easily as social Lamarckism. Both Ward and LeConte realized this and insisted that the use of education to shape the future of man would represent a break with nature. They admitted that man had evolved largely by selection, so their call for reform was a deliberate effort to go beyond nature's harsh and inefficient methods. Yet they also believed that there was a residual Lamarckian element in evolution which could be exploited for purposeful human development. Although the skills and attitudes acquired by an individual in the course of his education cannot be inherited, the *capacity* to learn might be improved by use-inheritance to the benefit of the race as a whole.

Even more than the social Darwinists, the Lamarckians left themselves open to the charge that they were allowing their concept of nature to be influenced by their desire for human progress. Biologists certainly fell into this trap, as did Alpheus Packard when he included in a highly technical defense of Lamarckism (1894) the claim that his theory *had* to work or the cumulative growth of civilization would have been impossible. He failed to realize that new ideas are transmitted to future generations by education, not by biological inheritance. Even Lester Ward declared that our efforts to ensure progress will be wasted if Lamarckism is not valid—although he should have been in a position to recognize that there is no logical connection. Perhaps the most extreme case of this wishful thinking is Bernard Shaw's remark in the preface to *Back to Methuselah*: if the theory of natural selection were true, "only fools and rascals could bear to be alive." The arrogant assumption that the universe must work in accordance with the

Shavian preference for Lamarckism highlights the kind of sloppy thinking that eventually brought the whole movement into disrepute.

Shaw defended Lamarckism as an essential bulwark against materialism in philosophy, a view that still has its adherents today (Koestler, 1971). His attempt to rally support for the theory in the 1920s was very much a rearguard action, conducted in defiance (or perhaps in ignorance) of its collapse as a credible movement in biology. Yet those who were concerned with the practical problems of social reform kept up their campaign knowing fully that they no longer could expect any inherited improvement in the race. Indeed, liberal attitudes flourished in the mid-twentieth century based on the simple assumption that people can be helped by improving their education and the conditions in which they live. It could be argued that the Lamarckian hope for eventual genetic progress in the race was an unnecessary encumbrance that obscured the logic of the claim that helping the poor is the right thing to do now—whatever its future benefits. Indeed, by assuming that the effects of nurture could be converted into nature (heredity), the Lamarckians blurred a distinction that should have been kept as clear as possible when stressing the moral basis of reform.

Social scientists, in any case, had begun to turn their backs on Lamarckism even before its final elimination from biology. About the time of World War I, academic sociologists and anthropologists became increasingly disenchanted with the whole evolutionary approach. A famous paper by A. L. Kroeber (1917) criticized Ward and many others for their blind assumption that nature must measure up to man's preconceptions. Kroeber argued that human cultures belong to a distinct field of activity he called the "superorganic." Our unique mental character ensures that the processes of the mind are unrelated to the process of biological evolution, Darwinian or Lamarckian, that first created mankind. Because ideas and behavior patterns are transmitted by learning, not by biological inheritance, cultural developments do not depend on an increase in the brainpower of the race. The implication of this is that the character of individual human beings is controlled as much by their environment as by their genes. This provided the necessary justification for policies of social reform. An equally important characteristic of the new social sciences, however, was their rejection of the nineteenth-century assumption that cultures and races can be ranked into a hierarchy corresponding to the evolutionary progress of mankind.

EVOLUTION AND RACE

The concept of a hierarchy of races with the white man at the top had emerged long before Darwin popularized the theory of evolution (chap. 4). Europeans almost invariably had assumed that they were biologically superior to the races they were subjugating with their military technology. Darwinism seemed to have implications for the question of how closely the races are related, but in practice it could be adapted to any preconceived opinion. The mechanism of the survival of the fittest could be used to justify a more ruthless approach toward conquered peoples, in which extinction was both a symbol and a consequence of inferiority. Little effort was made to exploit Darwinism's tendency to undermine the foundations of the linear chain of being. All too often, evolutionists simply accepted the traditional view that inferior races were intermediates between the apes and the highest form of man. The lower races were dismissed as frozen relics of an earlier stage in the upward progress from the apes. The link between evolution and embryology was pressed into service to uphold this image of linear development, although this fitted more naturally into the Lamarckian than the Darwinian scheme of development. (For general works on the growth of racial attitudes, see Stanton, 1960; Montagu, 1963, 1974; Mead, 1968; Stocking, 1968; Bolt, 1971; Frederickson, 1971; J. S. Haller, 1975; Stepan, 1982; Banton, 1987.)

European expansion around the globe had brought its people into contact with societies existing at levels of technology far below their own. Some of the more primitive savages were dismissed at first as little more than animals, although by the early nineteenth century this view had become untenable. If man was a single species, the races became merely local varieties, some of which had perhaps degenerated through exposure to unsuitable conditions. This was "monogenism": the belief that despite racial differences, man was a single species with a unique origin. Some naturalists, however, were so conscious of racial differences that they adopted a "polygenist" view, holding that the races had separate origins. Louis Agassiz came to believe that the black and white races were created separately as distinct forms of man, a view that could all too easily be exploited by those Americans who argued that slavery was a natural condition for the inferior race.

The theory of evolution allowed races to be considered strongly marked varieties, descended from a common stock but with widely different characteristics acquired in the course of their later separation. Grouping the races into a single species seemed necessary, because they still could interbreed. It was, nevertheless, widely held that hybrids from such a union possessed weak constitutions, confirming just how far the races have diverged. In Germany, Ernst Haeckel claimed that in all respects except the ability to interbreed, the differences were enough for the races to count as separate species. The crucial question was how far in the past the various branches of human evolution had begun to diverge. If the common ancestor of all the races lay very far in the past, one could argue that divergence had proceeded to such an extent that the varieties were almost distinct enough to be considered true species.

Darwinism injected a more deliberate emphasis on struggle into the existing view of the relationship among races. The supposed inferiority of blacks had been used to justify their enslavement, but in the case of the American Indian, the white man appeared to be eliminating the other race rather than merely exploiting it. Darwinism seemed to imply that this was an inevitable consequence of the struggle for existence, an implication of the subtitle of the *Origin of Species*: "the preservation of favoured races in the struggle for life." Throughout history, superior races have tended to eliminate inferior ones in any territory where they mix. The expansion of Europeans around the globe has allowed this struggle to begin on a worldwide scale. In the tropics, conditions might restrict the white man to a supervisory role over the natives, but wherever an inferior race occupied lands suitable for occupation by whites, it would be displaced. For a while, it even was believed that the Negro would die out in the United States, because he had survived only while protected in a state of slavery and would join the red man in extinction when faced with open competition. The "survival of the fittest" justified all this as the necessary price to be paid for progress. In the twentieth century, the Nazis carried this racial form of Darwinism to its ultimate conclusion. The Aryan was seen as the highest form of man, desstined to rule the world, while other races were dismissed as subhuman, fit only to be enslaved or exterminated (Tennenbaum, 1956; Poliakov, 1970; Mosse, 1978).

Although both Darwin and Spencer believed that racial struggle played a role in human evolution, those who supported imperialist policies tended to frown on individual competition *within* modern society. Karl Pearson openly predicted the extinction of lower races, yet he also proclaimed himself to be a socialist (1894). By this he meant that a centralized government was essential if the nation were to succeed in the worldwide struggle for supremacy (Semmel, 1960). This implied that one of the state's chief duties would be to maintain the biological standard of its people through eugenics. In particular, Pearson feared the consequences of the racial mixture that would result from immigration of inferior foreigners into civilized nations. The strength of the Anglo-Saxons would be undermined if they mixed with races they were meant to conquer. Pearson studied children in London slums, trying to show that immigrants were congenitally weaker and less intelligent than the native British population. In America, where immigration contributed far more obviously to the total population, similar fears generated what became the most popular goal of the eugenics movement. The white establishment viewed the prospect of rapid multiplication among "inferior" Oriental and Eastern European immigrants with alarm. In San Francisco, the Asiatic Exclusion League was formed (*Proceedings*, 1907–13), while a host of individual writers harped on the same theme (Schultz, 1908; Grant, 1918; Burr, 1922; Fairchild, 1926; Ross, 1927). Many regarded the need to exclude these inferior breeding stocks as more crucial than elimination of defective characteristics in the white population. Their campaign ultimately was crowned with success by the passing of the Immigration Restriction Act of 1924.

Calls for immigration restriction reflect a more pessimistic image of the white man's position in the world. Far from seeing the European as conqueror of the world, this new breed of racists saw him as a delicate flower that could soon be trampled underfoot by biologically more active races. The fear of the "yellow peril"—the fast-breeding Chinese—indicates an appreciation of the fact that the white man's much vaunted civilization might not count very much in the racial struggle for existence. The goal was no longer to expand but to consolidate, to ensure that where the European had gained a foothold he did not allow his racial character to be swamped. In a perverted way, this attitude at last recognized the true implications of Darwinism. In the struggle for existence, it is impossible to define a certain type as the "highest" that must inevitably dominate all others in all circumstances. Success goes to those who can best adapt to and dominate a new environment, and intelligence may not be the only deciding factor. Although still convinced of the white man's mental and moral superiority, the advocates of racial eugenics had come face to face with the harsher realities of struggle. A similar sense of insecurity may underlie the Nazi fear of a "stab in the back" by the race they dismissed so contemptuously as subhuman, with the horrific consequence that the process of extermination was taken out of the hands of an unreliable nature and entrusted to the state's own, more efficient machinery of death.

This more pessimistic attitude did not undermine the traditional belief that in all morally significant senses, the white race was the highest product of human evolution. Despite Darwin's emphasis that development is a branching rather than a linear process, the vast majority of his followers continued to believe that evolution of man consisted of a single line of ascent from the apes. The idea that the "lower" races might represent intermediate stages of the chain of being linking the apes to man had been developed in the eighteenth century and was still popular in the early nineteenth century (chap. 4; Priest, 1843). The new science of anthropology ranked societies in a natural hierarchy leading from the most primitive to the most civilized. Burrow (1966) shows that an evolutionary view in which societies progress gradually along this hierarchy became popular even before the idea of biological evolution. The theory of social evolution left one question unanswered: why have some races preserved an earlier form of society to the present? It was certainly not necessary to connect social progress with any biological improvement in the nature of man, as Wallace and Argyll had shown (chap. 8). One solution was to argue that a poor environment prevented some races from expressing their full human potential, but the prevailing belief in biological differences between the races offered a far more tempting way out of the problem.

The system of universal evolution worked out in the late nineteenth century treated biological and social progress as integral aspects of the same phenomenon. The technological backwardness of non-European societies indicated that they were culturally inferior, and this, in turn, was taken as a sign of the biological inferiority of the other races. Because both societies and races could be ranked in a linear scale of perfection, it seemed only natural to connect the two scales by assuming that cultural development was a manifestation of biological progress; non-European races were stuck at a more primitive level of society because they had not developed the intelligence needed to organize themselves more effectively. These lower races of the modern world were regarded as equivalent to earlier stages in the evolutionary process that had formed the white race. The white man had risen farther because the challenging conditions of northern Europe had stimulated his evolutionary development. Conversely, the enervating conditions of the tropics had retarded the development of those races exposed to them. These unfortunate people were links to the white man's evolutionary past, living fossils preserving both the mental and physical signs of their closer relationship to the apes.

This image of the various races occupying different positions on a linear scale of development was most consistent with the form of evolutionism promoted by the recapitulation theory. Lamarckians, in particular, were attracted to the view that evolution is a linear process mirrored in the goal-directed development of the embryo (chap. 9; Gould, 1977*b*). While Darwinism promoted a concept of ever-branching evolution, many neo-Lamarckians saw each group evolving through a linear pattern toward a predetermined goal; successive stages were added by an extension of the growth process in the embryo. It was believed that various modern forms could occupy different positions on the same scale. Applied to the origin of man, this allowed one to think in terms of a hierarchy of forms leading from the ape to the highest race of man; inferior races were trapped at a lower stage of development. Instead of progressing all the way up to the highest form, the lower races had somehow failed to add on the last stages of growth, retaining childlike (= apelike) characteristics to the present. G. Stanley Hall regarded the lower races as permanently childlike (Muschinske, 1977), and many Lamarckians looked for apelike characteristics in their mental and physical features.

The implications of this approach were not limited to the race question. The Italian C. Lombroso founded a science of "criminal anthropology" in which the criminal type was seen as a throwback to an earlier stage of evolution caused by a failure of growth (Nye, 1976). It was even suggested that women represented a stage of growth lower than that of men. But it was in

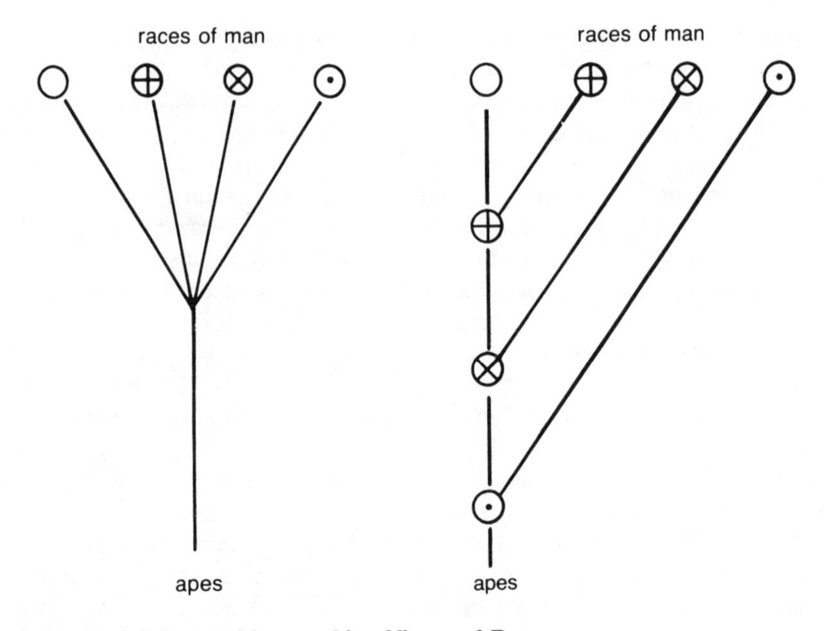

Fig. 23. Darwinian and Lamarckian Views of Race.

The diagrams illustrate two possible kinds of relationships between the different races of man, the left hand more consistent with the Darwinian view, the right hand with Lamarckism. The Darwinian concept is a branching one: the races are simply different and cannot be ranked as a hierarchy with one more closely related to the apes than another. In the other system, evolution of the highest form of man is treated as a linear process, recapitulated in the growth of the modern embryo because of the inheritance of acquired characteristics. Like embryological growth, the process is goal directed. Any one of the lower stages of evolution may have been preserved through to the present, if for some reason a variety of man failed to add on subsequent stages of growth to keep up with the forefront of evolution. Thus, there is no true branching, merely a series of parallel lines representing continuations of ancestral forms through to the present. Modern races correspond to the hierarchy of forms in man's evolution from apes.

the area of race that the Lamarckian concept of linear evolution had its widest application. American neo-Lamarckians such as E. D. Cope (1887) listed numerous features that were supposed to indicate the black race's "retardation of growth" (J. S. Haller, 1975). The same approach was employed in Germany by Haeckel (Gasman, 1971). Haeckel was a leading proponent of the recapitulation theory and a more linear concept of evolution. Although he exploited Darwinism to expound the notion of interracial struggle, Haeckel attributed the origin of new racial characteristics to Lamarckism. Through his writings, the hierarchical interpretation of human origins insinuated itself into the general viewpoint of late-nineteenth-century evolutionism. If the Nazi's subsequent denigration of other races to the subhuman level had any grounding in biological theory, it was in Haeckel's idealist version of Lamarckism as well as his Darwinism.

Lamarckism had the potential to alleviate some of the worst aspects of race theory. While it did not disturb the European sense of superiority, it could have offered the hope that other races might eventually be raised up to the same level through the cumulative effect of education and better conditions. This potential can be seen in F. J. Turner's "frontier hypothesis" of American development (Coleman, 1966). According to Turner, the stimulating environment of the frontier worked directly on the physical constitution of the pioneers, whatever their origin, to create a stronger race of mankind. Had the same principle been applied to races elsewhere in the world, it could have been argued that improved conditions would produce a similar biological progress. Paul Kammerer, whose involvement in the midwife toad affair brought Lamarckism into disrepute, indeed offered the hope of progress to all races of mankind. Yet the majority of Lamarckians refused to acknowledge this possibility, assuming that the lower races had been exposed to debilitating conditions for so long that there was no hope of significant improvement in the near future. The British embryologist E. W. MacBride defended Kammerer's Lamarckism, yet advocated an extreme form of eugenics aimed at sterilizing members of the "inferior" Irish race (1924; Bowler, 1984). Whatever their optimistic predictions for the development of the white race, Lamarckians were bound by the prejudices of their time and stressed another interpretation of their theory to deny this prospect to other races.

The collapse of evolutionary race theory came not because it was scientifically disproved but because the social sciences of the early twentieth century turned their backs on the whole evolutionary viewpoint (Harris, 1968; Hatch, 1973; Cravens, 1978). In Europe, scholars such as Max Weber and Emile Durkheim pioneered the technique of treating each society or culture as a functioning whole that cannot be evaluated by the standards of any other. The assumption that all behavior must have a rational basis (or at least must have had such a basis at a more primitive social level) was abandoned, and with it the need to rank all societies into a hierarchy with Europe or America at the top. Franz Boas and his students introduced a similar technique of cultural relativism into American anthropology. If societies cannot be ranked on a scale of perfection, there is no longer any reason to judge other races biologically inferior for failing to develop along European lines. Cultural forces alone account for the differences; and, as A. L. Kroeber (1917) proclaimed in his paper on the "superorganic," these forces have nothing to do with biological differences. The modern social sciences have repudiated not only the hierarchy of races and cultures but also the whole idea that concepts of biological evolution can be applied to cultural development (Greenwood, 1984). If the concept of evolution is still used, it implies a process of cultural development that need not be directed by biological processes (Ingold, 1987). Indeed, the social sciences' repudiation of biology has gone so far that immense controversy has erupted when modern biologists have tried to reclaim a role in assessing human behavior, as in the debate over sociobiology (chap. 11).

11

The Evolutionary Synthesis

During the early years of the twentieth century, Darwin's theory of natural selection had lost much of its popularity. Field naturalists preserved the original Darwinian emphasis on the role of geographic factors in evolution but were strongly tempted by alternative mechanisms of adaptation, such as Lamarckism. Paleontologists were convinced that evolution is directed along linear paths by either Lamarckism or orthogenesis. The new breed of experimental biologists had gone to the opposite extreme, using genetics to undermine Lamarckism but refusing to admit any role for adaptation and selection in controlling the flow of new characteristics produced by mutation. In the 1920s, the first moves were made toward a reconciliation between the divided branches of biology, and Darwinism emerged from its eclipse to provide the key to a new approach that would solve many outstanding problems. It was realized that the more sophisticated understanding of heredity provided by Mendelism would have to be applied to populations containing a wide range of individual variation and that selection might affect the relative frequencies of the genes. By 1940, many naturalists had come to recognize that their own work could be reconciled with this new form of selectionism, eliminating the need for unsubstantiated alternatives, such as Lamarckism. The resulting "evolutionary synthesis," or "modern synthesis," allowed Darwinism to reemerge as a driving force in biology.

No one doubts the significance of the evolutionary synthesis, but there had been some debate over how it was put together. At one time it was assumed that the decisive breakthrough came with the creation of population genetics as a new foundation for natural selection. This revived form of Darwinism was then exploited by field naturalists and paleontologists who had been looking for a way out of the confused state of their thinking on the evolutionary mechanism. William B. Provine's *Origins of Theoretical Population Genetics* (1971) provides a detailed account of this side of the synthesis. Ernst Mayr (1959*c*), however, has argued that the true story was more complex. In its original form, population genetics was a highly abstract science that often failed to include the geographic factors that field naturalists knew to be important. The evolutionary synthesis only came into being when the two approaches were integrated, each supplying something of importance to the end result. Provine (1978) has defended the role of population genetics, and a more balanced approach now has emerged (Mayr and Provine, 1980; Grene, 1983; for a more critical assessment, see Eldredge, 1985). Population genetics was important not so much because it supplied totally new concepts but because it destroyed the legacy of anti-Darwinian feeling and focused attention on new research opportunities. The study of microevolution (evolution within the species) was revolutionized by showing that genetic mechanisms were responsible for the geographic effects once attributed to Lamarckism. No proof could be offered that natural selection produced large-scale macroevolution, however, and in paleontology the synthesis was upheld mainly by showing that it was compatible with existing knowledge.

POPULATION GENETICS

Although many early forms of population genetics ignored vital Darwinian factors, such as geographic isolation, they enabled natural selection to reemerge as a viable mechanism of adaptive evolution. In the period before 1920, geneticists had been convinced that mutation was the only source of new characteristics in evolution; and they did not believe that differences in the adaptive value of new genes would control the extent to which they would spread in a wild population. Laboratory biologists were only too willing to ignore the pressures that might affect an organism living in the wild state. They played down the role of adaptation in evolution at the expense of processes, such as mutation, that could be studied in an artificial environment. The split with field naturalists was exaggerated by antagonisms that sprang up between early Mendelians, such as William Bateson, and the biometrical school, which had tried to preserve the Darwinian approach to the variation of wild populations. Recognizing that the genetic structure of a population is far more complex than had been imagined at first was a crucial step in the resurgence of Darwinism. Only then could it be supposed that selection for adaptive advantage might increase the frequency of some genes within the population at the expense of others. This development arose chiefly from the work of three men: R. A. Fisher and J. B. S.

Haldane in Britain and Sewall Wright in America. For all its mathematical sophistication, the work of Fisher and Haldane arose out of what Mayr (1959*c*) has termed the "beanbag" approach to genetics: selection was supposed to act on individual genes, each of which could be assigned a fixed adaptive value. Evolution would consist of no more than the addition and subtraction of genes from the "gene pool" of the population. Wright's approach stressed the interactions between genes as a source of additional variability within small, inbreeding populations. This subtler view of the role played by genetics blended far more easily with field naturalists' belief that geographic isolation was crucial for speciation.

The split between Bateson and the biometricians hardened into personal hostility that soon eliminated all hope of reconciliation. Yet there was a potential for combining the two approaches which eventually would be recognized as younger scientists came onto the scene (Provine, 1971). As early as 1902, G. Udny Yule pointed out that Mendel's laws were not necessarily incompatible with the biometrical approach to the measurement of variation. The Mendelians emphasized discontinuous variation because this was more accessible to their experimental techniques, dismissing continuous variation as of no genetic or evolutionary significance. Yule showed, however, that continuously varying characteristics might be explained by the inheritance of factors obeying Mendel's laws, provided it was assumed that more than a single pair of genetic alternatives could influence the same characteristic. If a series of genetic factors generates small degrees of difference in the same characteristic, the simple Mendelian ratios will be lost in the constant recombination of the factors. The interaction of the many factors will produce, in effect, a continuous range of variation.

Yule's suggestion had immense potential. If the biometricians could accept Mendelian inheritance, Fleeming Jenkin's classic "swamping argument" against the inclusion of mutations in selection would be overthrown. The effect of the new factor would not be halved in each generation by blending; instead, it would spread unchanged through the population, especially if it conferred an adaptive benefit. By adopting Yule's point, the biometricians would have gained a valuable advantage and still retained their own techniques for the study of continuous variation. Unfortunately, hostility between the two sides had become so bitter that neither could appreciate Yule's argument. It would be some time before a reconciliation along these lines would be possible.

In 1909, the Swedish biologist H. Nilsson-Ehle undertook a series of breeding experiments with grain that confirmed Yule's idea. He showed that some characteristics must be accounted for in terms of three of four genetic factors, each segregating independently according to Mendel's laws. Calculating what would happen if there were ten such factors, he showed that there would be nearly 60,000 different phenotypes. Because the differences between them would be slight, they would appear as a continuous range of variation. The same point was noted in America by Edward East (1910). Over the next decade, more and more geneticists came to accept the multiple-factor explanation of continuous variation, and some, including Nilsson-Ehle himself, now argued that it would be possible for selection to act on this wide range of genetic variation.

Although selection would act to increase the frequency of those genes that enhanced a useful characteristic, most biologists still accepted Johannsen's point that a limit to this process would be defined by the existing range of variation in the population. In this case, the production of new genetic factors by mutation would be the only source of new variation for long-term evolution. This is the essence of Mayr's beanbag genetics: mutation puts new genes into the population, and selection either increases their frequency or removes them if they are harmful. In fact, this is an extremely limited view that ignores the complex effects of interactions between genes, but it was the framework for the earliest forms of the genetic selection theory. Laboratory experiments showed that mutation produces a constant stream of small changes that are introduced into the population without disturbance. Selection could affect the rate at which a new gene spreads into a population, and evolution could be seen as the cumulative and virtually continuous effect of such small changes over a long period of time.

An early move in this direction can be seen in the thought of Thomas Hunt Morgan (Allen, 1968, 1978; Bowler, 1978, 1983). Originally, Morgan had followed De Vries in treating mutations as the instant basis of new subspecies; but he soon began to realize that they must feed new characteristics into the existing breeding population. At first, he still refused a role for selection, holding that any new characteristic would be able to spread through the population whether or not it conferred an adaptive benefit. In 1910, Morgan's experiments with the fruitfly *Drosophila* converted him to Mendelism and began to reveal just how small many of the naturally occurring mutations can be. He gradually began to admit that a new characteristic formed by mutation would not be able to spread unless aided by selection. By 1916, he was arguing for a primitive form of the genetic theory of selection: harmful or even neutral mutations would be prevented from spreading, while beneficial mutations would gradually take over the population, because they reproduced faster than the original gene.

Morgan was never willing to admit that harmful genes could exist in any quantity within a population, perhaps because he found the notion of a wholesale elimination of unfit individuals morally distasteful. His theory was an extreme form of beanbag genetics in which harmful new characteristics are removed from circulation almost as soon as they are formed by mutation. In a more sophisticated form, this view survived into the mid-twentieth century through the so-called "classical" hypothesis of population genetics defended by Morgan's student, Hermann J. Muller (1949). Here, the amount of genetic variability in a population was minimized: most individuals were assumed to have only a few genes that did not represent the normal, or "wild-type" of characteristic for their species. Such abnormal genes were produced from time to time by mutation but were steadily eliminated by selection because of their harmful effects. Only the very rare favorable mutations would spread to become the new wild-type for the species. Such an interpretation was supported by laboratory geneticists, who studied the more obvious—and hence more harmful—mutations and who believed that mutation would almost always destroy a gene's function. Most field naturalists were suspicious of the classical hypothesis, believing that natural selection played a far more active role by actually maintaining a high level of genetic diversity in all populations. Only with the development of molecular genetics did it become possible to prove that the classical approach had seriously underestimated genetic variability. Lewontin (1974) points out, however, that this did not automatically confirm the view that selection plays a more active role in evolution. The "neutralist" school of population genetics emerged, claiming that most of the observed genetic variability was adaptively neutral and had thus accumulated without reference to natural selection (Kimura and Ohta, 1971; Kimura, 1983).

The modern synthesis was based on the alternative, or "balance" hypothesis of population structure. Morgan and Muller repudiated what was to become one of the most important insights of the new population genetics: a wide range of genes affecting each characteristic already exists in any natural population for selection to act upon. Genetic characteristics produced by mutation continue to circulate within the population at a low frequency even if they confer no advantage, and the species thus builds up a fund of variability that can be exploited by selection when conditions change. Even in a stable environment, "balancing selection" actively helps to maintain the range of genetic variation that will later be exploited in adaptive evolution.

The existence of a wide degree of inheritable variation had been a central tenet of biometrical Darwinism. It was now necessary to interpret this in Mendelian terms through a multiple-factor explanation of continuous variation and to show that selection had the power to change the frequency of a beneficial gene. This last point already had been demonstrated in 1915, when R. C. Punnett published a study of mimicry in butterflies, with a table prepared by the mathematician H. T. J. Norton showing how selection would allow a favorable gene to spread within a population. Punnett himself believed in discontinuous evolution, but Norton's calculations showed that even a slight advantage soon would increase the gene's frequency. The way was open now for the creation of a new theory of gradual evolution based on the natural selection of a genetically variable population.

Ronald Aylmer Fisher was trained in mathematics at Cambridge and became interested in Pearson's biometrical techniques (Norton, 1975*b*; Box, 1978; Bennett, 1983). He soon broke with Pearson on the question of Mendelism, although Mackenzie (1982) argues that he remained loyal to the biometrical program. Fisher realized that Mendel's laws could solve many of the problems created by Pearson's reliance on blending heredity. Unit characteristics would preserve themselves without blending and thus would keep up the variability of the population. His first paper on this topic was rejected by the Royal Society of London, so sensitive were the feelings raised by the biometry-Mendelism dispute (Norton and Pearson, 1976), and eventually appeared in the transactions of the Royal Society of Edinburgh (1918). Over the next decade, Fisher applied his techniques to study the effect of selection on a genetically variable population, culminating with the appearance of his classic *Genetical Theory of Natural Selection* (1930).

Fisher made a number of assumptions to arrive at a workable mathematical model. Selection was supposed to act uniformly on a large population in which genetic recombination maximized variability. If a particular gene conferred an advantage resulting in a faster breeding rate, it was possible to calculate how rapidly its frequency would be increased. In Fisher's scheme, selection was a deterministic process, grinding away slowly but surely to increase the frequency of individual genes. Because the model worked with single genes, it was still based on the beanbag approach, although it showed that selection could only reduce the frequency of an unfavored gene, not eliminate it altogether if it were protected by being recessive. Fisher demonstrated that selection could act to maintain the balance between two alleles when the heterozygote is fitter than either homozygote. Most mutations were acknowledged to be deleterious, but because they tended to occur at a fixed rate, this balanced the efforts of selection to reduce their frequency. In a small population, a rare gene could be eliminated by chance. For this reason, Fisher believed that large populations benefited evolution by keeping up variability. A gene that at some point became favorable to the species would immediately begin to increase its frequency. Fisher assumed that new factors fed into the gene pool by mutation were generally very small, so that they became part of the regular variability of the species. Thus, although selection made use of discrete mutations, evolution was still a relatively continuous process with no sudden leaps.

J. B. S. Haldane published his first paper on population genetics in 1924 and an important survey of the field in 1932 (Clark, 1969). Like Fisher, he made certain assumptions in order to simplify the mathematics: an infinite population with random mating, perfect Mendelian dominance, and segregation. Again, the emphasis was on selection applied to single genes, but Haldane used practical examples to show that the process could work much more rapidly than Fisher supposed. The best known was the case of industrial melanism in the peppered moth '*Amphidasys*' *betularia* (now renamed

Biston betularia). A dark-colored or melanic form of this moth first had been noted in 1848 and had begun to spread in industrial areas of Britain where its color offered protection from predators by concealing it against the soot-covered background. By 1900, the melanic form almost completely replaced the normal gray version of the moth in these areas. Haldane showed that so rapid a spread must indicate a 50 percent greater production of offspring by the melanic form, a far more intense selective effect than anything imagined by Fisher.

Because Fisher and Haldane both assumed that selection was most effective when it acted on the wide variability of a large population, their theories dealt only with the question of evolution in a continuous, unbranching line. They ignored the field naturalists' interest in speciation, splitting of a population into a number of distinct branches, and were unwilling to concede that geographically isolated populations might play a significant role in evolution. In addition, their use of the beanbag approach treated each gene as an individual unit with a particular adaptive value. This did not take into account the possibility that genes interact with one another, enabling new interactions to extend the variability of a population without the creation of new genes by mutation and sidestepping Johannsen's claim that the amount of genetic variation is rigidly limited. The first step toward breaking down this limit was taken in America by William E. Castle (1911), who used breeding experiments with hooded rats to show that extra variation was available in some circumstances. Continued selection in a small population encouraged the formation of unusual genetic combinations, conjuring up variation beyond the normal bounds observed within a large, freely interbreeding population. It was also shown that certain "modifier genes" could influence the effect of those genes responsible for producing a particular characteristic. The breeding population now had to be seen as a complex system, capable of producing an enormous amount of variation when stimulated by selection and inbreeding.

This point was developed into a new form of mathematical population genetics by Castle's student, Sewall Wright (Provine, 1986). Experiments with coat color in guinea pigs had convinced Wright that interaction systems between genes were important, and his participation in the hooded rat experiments showed him how inbreeding within small populations could stimulate variation. By 1920, he had developed a powerful mathematical technique to analyze the effects of inbreeding, enabling him to show that gene interaction systems could be fixed by this means and then acted on by selection. He began to apply these insights to natural evolution, which he assumed would take place most readily in small populations whose inbreeding was intense enough to create new interaction systems naturally through random effects known as "genetic drift." The most effective situation for evolution would be a large population broken imperfectly into isolated local strains. Natural selection would then begin to act on the new interaction systems to give very rapid evolution. The random effects of inbreeding would allow a small population to move away from the "adaptive peak" for its species, so that it could cross a relatively nonadaptive intermediate zone until a new adaptive peak could be reached. Wright specifically criticized Fisher's views (1930) and then expounded his own theory of evolution at greater length (1931). The beanbag approach was abandoned now in favor of a more sophisticated view in which the creation of new interaction systems within the gene pool acts as an intermediary between mutation and selection.

THE MODERN SYNTHESIS

Few field naturalists could follow the complex mathematics employed by the founders of population genetics. They could only translate the mathematical conclusions into commonsense language and then see if the results offered any insights into their own level of research. A form of population thinking had arisen spontaneously among naturalists who were dealing with geographic variation within and between species. They had realized that the intricate patterns of variation could not be reconciled with a typological view of species in which a basic underlying form was only superficially modified by local conditions. Each local variety or subspecies had to be treated as a distinct breeding population with its own character, reproductively isolated from neighboring groups even where geographic barriers no longer separated them. These naturalists were sure that geographic isolation was crucial for the first splitting up of a once homogeneous population and that the conditions each subspecies was exposed to had shaped its unique character. Only when this kind of population thinking was fused with mathematical population genetics did the true outlines of the modern synthesis emerge (Mayr and Provine, 1980). The techniques used by Fisher and Haldane were less easy to adapt to the naturalists' geographic insights, but even so their work was sufficiently convincing that the majority of biologists now recognized that selection was a powerful mechanism of adaptive evolution. Field naturalists began to abandon their earlier reliance on unsubstantiated alternatives, such as Lamarckism, as news of the mathematicians' conclusions filtered through to them (Rensch, 1983). Not surprisingly, it was Wright's approach that proved easiest to apply to fieldwork, because his emphasis on the role of small, inbreeding populations exactly matched the geographic studies. Theodosius Dobzhansky exploited Wright's conclusions in his highly influential *Genetics and the Origin of Species* (1937), triggering an explosion of activity that initiated the modern synthesis during the 1940s.

Of particular importance in paving the way for the synthesis was the work done in Russia under the direction of Sergei S. Chetverikov (Adams, 1968, 1970; and in Mayr and Provine, 1980). Russian naturalists were not influenced by the anti-Darwinism so prevalent in the West around 1900, so that Chetverikov was in a strong position to appreciate the potential for a synthesis with genetics. The Russian school started from a conviction that natural populations contain much unseen variation in the form of recessive genes, a view that Chetverikov was able to confirm by crossbreeding wild populations of *Drosophila* with genetically pure strains brought from America (1926; paper translated 1961). His students, D. D. Romashov and N. P. Dubinin, developed a technique for studying statistical effects in various population sizes, confirming Chetverikov's belief that the underlying variability would be shown most readily in small populations. They introduced the concept of the gene pool as a reservoir of potential genetic combinations; those actually realized were dependent on the laws of probability (Adams, 1979). Chetverikov's school was eliminated as a result of the Lysenko affair (see chap. 9); however, his work influenced N. W. Timoféeff-Ressovsky, who moved to Germany in 1925 and was responsible for promoting more sophisticated ideas on the role of mutations in building genetic variability. Dobzhansky also acknowledged a debt to Chetverikov, although he had not belonged to the school before moving to America in 1927.

In Britain, the results of Fisher's work were explored in E. B. Ford's *Mendelism and Evolution* (1931). Ford was interested in ecological questions, and his subsequent work helped to show that selection could act much more rapidly than Fisher supposed. This confirmed the fact that Haldane's analysis of industrial melanism in the peppered moth did not constitute an exceptional case. Gavin de Beer's *Embryology and Evolution* (1930) undermined the evidence for the recapitulation theory the Lamarckians had relied on so heavily. Because genetic mutation does not work by addition to the existing growth pattern, there is no reason why the growth of the individual organism should pass through the adult stages of its evolutionary ancestors. Most influential of all was Julian Huxley, a grandson of T. H. Huxley, who began teaching at Oxford in 1920 (Huxley, 1970). Naturally enough, Huxley absorbed his Darwinism from an earlier generation of biologists and began to work for a revival of the theory. He collaborated with H. G. Wells to produce *The Science of Life* (1930), a comprehensive and popular account of the Darwinian view of evolution. His own work on animal behavior fell within the natural history tradition, but he also was interested in embryological growth and kept up with the latest developments in genetics. In 1940, he edited *The New Systematics,* which brought together contributions from all sides of biology, and his own comprehensive survey, *Evolution: The Modern Synthesis,* appeared in 1942.

In America, Francis B. Sumner did pioneering work that anticipated many aspects of the later synthesis (Provine, 1979). His studies of geographically isolated varieties of deer mice were begun with a Lamarckian bias, but Sumner soon began to argue (1924) that there was a genetic basis to the morphological differences among populations. In 1930, Ernst Mayr came to America from Germany, after several years' fieldwork with birds of New Guinea and the Solomon islands. Mayr was influenced by Bernhard Rensch, who had revived the belief that there was a strong correlation between geographic variation and climate. Rensch and Mayr began with a Lamarckian explanation of this phenomenon, but in the 1930s, they both realized that a Darwinian explanation was now possible. Mayr had not read the work of the mathematical geneticists, so that his own experiences form the basis of his claim that a populational view of species was developed independently on the basis of field research. For Mayr, as for many other naturalists, it was Dobzhansky's *Genetics and the Origin of Species* that pointed the way toward a complete reconciliation by presenting the mathematicians' conclusions in an intelligible form. Mayr's own *Systematics and the Origin of Species* (1942) was one of the founding works of the modern synthesis, emphasizing the role of geographic factors in speciation.

When Theodosius Dobzhansky joined T. H. Morgan's team at Columbia University in 1927, he brought with him the experience of the Russian school's populational approach. He thus was able to appreciate what the field naturalists required from the geneticists, with the result that *Genetics and the Origin of Species* (1937) played a vital role in bridging the gap between the practical experience of the naturalists and the abstract formulations of the experimental and mathematical biologists. In it Dobzhansky summarized the experimental evidence showing the true nature of mutations, emphasizing how small their effects could be and how they built up the natural variability of populations. He also summarized the mathematicians' conclusions, paying particular attention to Wright's work. His own studies of the genetic basis of geographic variation in insects were discussed, along with other demonstrations of the same effect. From 1938 onward, Dobzhansky collaborated with Wright in a series of investigations of the genetics of natural populations (collected in Dobzhansky, 1981). One aim of this research was to show that selection is not just a mechanism of change but can also act to maintain stability through balancing mechanisms, such as that based on superior heterozygote fitness. The widespread existence of such dynamic equilibria is important for modern Darwinism, because it demonstrates that populations do indeed contain substantial reservoirs of genetic variation that can be tapped under new conditions.

Dobzhansky coined the term "isolating mechanism" to denote nongenetic characteristics, such as behavioral differences, that can prevent interbreeding when two related populations come to occupy the same area. Founders of the modern synthesis were convinced that speciation occurred only through an initial phase of geographic isolation, for which Mayr introduced the term "allopatric speciation." With no interbreeding, the various populations could develop their own distinctive characteristics. Isolating mechanisms prevent them from interbreeding even if geographic barriers are subsequently removed; this allows the subspecies to remain distinct and to be driven apart further by selection. The new Darwinism was based on the assumption that speciation did not require special genetic mechanisms; given geographic isolation, natural selection alone can separate one species into many.

The evolutionary synthesis was extended to paleontology chiefly by George Gaylord Simpson, in his *Tempo and Mode in Evolution* (1944). Simpson's task was to demonstrate the plausibility of the claim that macroevolution as revealed by the fossil record took place through the accumulated effect of microevolutionary processes that were now being studied in modern populations. No proof could be offered; but it was possible to show that the available evidence from paleontology was at least consistent with the new theory, despite the anti-Darwinian claims of an earlier generation of biologists (Gould, 1980*a*). Simpson used a quantitative analysis to show that major evolutionary developments took place in the irregular and undirected manner predicted by Darwinism. He exposed the weakness of the evidence that had been used to argue for a linearity of development more appropriate to Lamarckism or orthogenesis. The evolution of the horse, for instance, was not a single advance toward the modern, specialized form; rather, it was an irregular tree with many side branches leading off to extinction. On the vexing question of the discontinuity of the fossil sequences, Simpson realized that something more than the imperfection of the record was involved. Substantial changes could take place through nonadaptive "quantum evolution," employing Wright's mechanism of genetic drift. Because these transitions would occur comparatively rapidly and in very small populations, they would be most unlikely to leave fossil evidence. In later works (e.g., 1953*b*), Simpson adopted a more rigorous Darwinian approach, placing more emphasis on the adaptive character of all evolution.

The modern synthesis ushered in an era of collaborative research of the kind typified by Dobzhansky's work on the genetics of natural populations. Different mathematical models were tested against one another by direct application to nature (Provine, 1978). Fisher was forced to concede that selection acted much more rapidly than he had supposed, but Wright's notion of nonadaptive evolution through genetic drift in small populations also came under fire. The synthesis now "hardened" to eliminate even the limited role for non-Darwinian mechanisms conceded by some of its founders (Gould, 1983). While it was clear that speciation took place most readily when small populations became isolated, many naturalists felt that differences between populations were built up by natural selection for adaptive characteristics. The importance of adaptation was emphasized by further studies of industrial melanism by H. B. D. Kettlewell (1955), showing the effectiveness of camouflage as a means of protection against predators. In more recent times, molecular biology has been used to confirm the fact that there is a wide range of genetic variability in populations (e.g., Lewontin, 1974). Efforts have been made to argue that much of this variation is not translated into phenotypic differences that selection in turn can act on (Kimura and Ohta, 1971), but this view has not been able to gain wide acceptance. Extensive research continues within the framework established by the evolutionary synthesis, despite the controversies that have broken out in recent years (see chap. 12; for surveys of modern Darwinism, see Simpson, 1953*b*; Mayr, 1963; J. Maynard Smith, 1976; Dobzhansky et al., 1977; Ayala and Valentine, 1979; Futuyma, 1979; Ruse, 1982).

THE ORIGIN OF LIFE

A parallel development that helped to boost the revival of Darwinism was the appearance of the first modern theory of the origin of life on earth. In the late nineteenth century, the ancient belief in the "spontaneous generation" of life from nonliving matter was finally discredited, leaving evolutionists with no plausible explanation of how the process first began. In 1936, the Russian biologist Alexander Oparin introduced a new approach to the question in which the notion of a single act of spontaneous generation was replaced by a process of chemical evolution through levels of increasingly complex organization. This not only extended the range of evolutionary thinking but also gave renewed support to Darwinism, because Oparin postulated a form of natural selection as the mechanism of improvement among preliving structures. When Stanley Miller and others provided experimental confirmation of some steps in the process during the 1950s, Oparin's view of the origin of life became accepted as an integral component of modern evolutionism (Farley, 1977).

In the *Origin*, Darwin had written as though the first forms of life were divinely created, although privately he admitted that this was a way of evading the issue. The concept of spontaneous generation had not been popular in the early nineteenth century, and Darwin did not want to see his theory tied to any wild speculations about totally unknown causes. His followers realized, nevertheless, that to be consistent, evolutionary theory ought to explain the origin as well as the development of life in terms of natural causes. Haeckel postulated a primitive form of life, the "monera," in which unstructured protoplasm already had acquired the basic properties of life. This formed the key link between inorganic matter and more advanced creatures composed of cells. In 1868, Huxley found a thin layer of jelly on the surface of mud dredged up from the ocean floor. He took it to be a living example of the first step in evolution, naming it *Bathybius haeckelii* (Rehbock, 1975). The seabed was thought to be still covered with the primordial slime, the *Urschleim,* from which all higher forms of life had evolved. It soon was discovered, however, that the jelly was precipitated when alcohol was added to the seawater as a preservative, and Huxley was forced to abandon the belief that the first step in the appearance of life still could be observed today.

Because *Bathybius* had no cellular structure, it had encouraged the belief that the properties of life appeared as soon as matter achieved the appropriate level of chemical organization. This rendered more plausible the idea that an act of chemical combination was responsible for the origin of life—an act that still might be repeated in the world today. The temporary enthusiasm for spontaneous generation soon evaporated in the 1880s, as the majority of scientists were convinced by Louis Pasteur's careful demonstration that there is no evidence of life being formed from nonliving matter. The development of cell theory and the germ plasm concept also encouraged the belief that a new living structure could arise only from a previously existing one. The continuity of nuclear material from one cell to another was seen as the essential basis of reproduction, implying that the properties of life arose from the organization of the cell. Because no amount of mere chemical activity could produce the complex structure of a cell, the spontaneous generation of life under any natural conditions seemed impossible. To evade the problem of how life first appeared on the earth, it was even supposed that spores from some unknown source in outer space originally had been responsible for "seeding" the earth (Arrhenius, 1908).

In the early twentieth century, many biologists became fascinated with the properties of colloidal suspensions, and some thought that behavior of the suspended particles somehow anticipated the activity of life itself. Oparin at first saw the origin of life as a unique event, occurring when the first cells were precipitated from a colloidal suspension. The emphasis was still on spontaneous generation by a single step, crossing at one bound the gulf between life and nonlife; the crucial advance came about through an accidental combination of chemicals. J. B. S. Haldane published a paper in 1923 in which he postulated the original formation of a viruslike structure from naturally occurring organic chemicals. The innovation in Oparin's later work was to get away from the idea of a single-step formation of life based on chance combinations. Instead, he moved to a far more sophisticated picture with no sharp dividing line marking the appearance of life. The process of chemical evolution gradually increases the level of organization, providing a complete continuity between inorganic matter and the first living cells.

Farley (1977) argues that Oparin's views changed through the influence of dialectical materialism and points out that the success of his work offers a counterweight to the disaster of Lysenkoism. If Lysenko was an opportunist, others found the dialectical approach a significant advance over older forms of materialism and a fruitful guide to research. Oparin supported Lysenko and was himself criticized by some geneticists who claimed that his theory also was intended to undermine the status of the gene as the fundamental repository of living organization. Yet in this case, the application of dialectical logic to biology produced not a travesty of science but a genuine advance. Dialectical materialism is not a mechanistic or reductionist philosophy, because it holds that entirely new laws of nature come into play as new levels of organization are reached. The dialectical law that new qualities arise out of quantitative changes at the lower level guided Oparin to the view that life must appear as the result of a gradual increase in the degree of physical complexity. In addition, the dialectic holds that the emergence of a new quality negates the qualities of the earlier state. This led Oparin to argue that once life had appeared on the earth, the process could never be repeated. Living organisms themselves would destroy the earlier stages in the process as soon as they occurred. It was thus impossible for life to be created now, just as Pasteur had shown.

Oparin's theory was presented in his *The Origin of Life* of 1936 (translation, 1938). He followed Haldane in assuming that the earth was originally a sterile planet with no free oxygen in its atmosphere. If oxygen had existed, it would have destroyed the chemicals necessary for the appearance of life. By treating oxygen as a product of living things, it was possible to uphold the claim that the evolution of life could only occur once in the planet's history. Oparin believed that the earth originally had been enveloped in a reducing atmosphere containing hydrocarbons and ammonia. Chemical reactions among these gases produced complex organic molecules that dissolved in the oceans to form a rich "primordial soup." Here the chemicals combined once again to form polymolecular open systems or coacervates, minute droplets with a definite structure, capable of adsorbing materials from their surroundings. Natural selection now came into play, allowing those coacervates with more stable structures to survive at the expense of the less stable. The more successful structures became the "prebionts," the intermediate stages before the appearance of true life in the form of the first cells. This last stage in the process was by far the most speculative part of the theory, however.

Oparin's theory was taken seriously in the West, not so much because of its dialectical origins but because a number of more imaginative biologists were moving away from simple materialism toward a more holistic view of life (a view in which behavior of the whole organism cannot be explained solely in terms of its constituent parts; see Allen, 1975*a*). In the period after World War II, the experimental opportunities to check out some stages of the process also looked promising. The most famous confirmation was the experiment performed by Stanley Miller (1953), in which an electric spark passing through a sample of the reducing atmosphere suggested by Oparin yielded amino acids—complex chemicals that form the basis of proteins. Debates have continued over the most likely course for the later stages in the process. A recent suggestion has been that clays have the appropriate property of encouraging the polymerization of molecules adsorbed into their surface. Some geneticists have objected to Oparin's gradualistic model on the grounds that the appearance of the first genes must represent the key breakthrough in the appearance of life. Now that the chemical structure of the DNA in the genes is understood, we have a better idea of the level of complexity that must be achieved. The idea that a self-replicating system must have been built up gradually has nevertheless remained popular, if only because the alternative "one step" process returns us to a situation in which a complex structure must arise suddenly out of chance collisions of much smaller molecules.

Whether the appearance of the first living forms was gradual or sudden, most biologists in the 1950s agreed that the end product represented so unlikely a combination of elements that it could only be expected after a vast period of time. As with the famous illustration of the monkey and the typewriter, it takes a large number of trials to ensure that a meaningful sequence will result from repeated acts of random combination. Another interpretation of chemical evolution, however, sees it as a way of bypassing the delays inherent in a system based purely on trial and error. Instead of starting from scratch with every attempted combination, each level of organization is preserved, once it has appeared, to form a secure foundation for the next step in the process. New evidence from geology has now shown that the development of life did not take the vast periods of time that once were thought necessary. The first primitive forms appeared very early in the earth's history, long before the sudden "explosion" of more advanced types at the beginning of the Cambrian era.

The early Darwinians had seen their evidence for Precambrian life, in the form of *Eozoön canadense,* discredited (chap. 7). For a long time, it seemed as though the advanced forms of the Cambrian indeed had appeared out of nowhere, until modern geologists began to find genuine microfossils, the remains of microscopic living cells, at ever earlier points in the Precambrian. Sedimentary rocks over three billion years old contain the remains of bacteria-like organisms (Schopf, 1978). At a somewhat later date, fossil stromatolites are found. These are composed of sediment trapped by mats of blue-green algae, structures still to be found in areas of modern oceans with high salinity. The oldest microfossils go back fully three quarters of the earth's entire history, suggesting that the formation of life took place almost as soon as the planet's surface acquired suitable conditions. The problem now is to account for the contrast between the vast period of stability after the first appearance of living cells and the comparatively sudden development of multicellular organisms in the period just before the Cambrian (Gould, 1977*c*, 1980*b*).

HUMAN EVOLUTION

At the opposite end of the evolutionary spectrum, the twentieth century has seen major changes in our conception of how the human race itself evolved. Darwin had speculated that our ancestors had separated from the apes through their adoption of an upright posture. Bipedal locomotion freed the hands for toolmaking, which then stimulated the growth of human intelligence. In the late nineteenth century, though, this insight had been largely ignored in the rush to create a theory based on the assumption that expansion of the brain was the original driving force of human evolution. Mankind was still seen as the inevitable goal of natural progress (Bowler, 1986). It could even be argued that the transition to bipedal locomotion on the open plains was a consequence of the extra intelligence that allowed our ancestors to appreciate the benefits of a move out of the forests (G. E. Smith, 1924). This implicitly teleological view of human origins only began to break down when the modern synthesis undermined the plausibility of the non-Darwinian evolutionary theories that sustained it.

When human fossils at last began to come to light, they were incorporated into preconceived ideas about our ancestry (on the major discoveries, see Leakey and Goodall, 1969; Leakey and Prost, 1971; Reader, 1981). In the 1860s, Huxley had rejected the possibility that the Neanderthal type might constitute a link between the apes and modern humans (chap. 8). More specimens of this type were discovered later in the century, and many authorities now began to ignore Huxley's warnings to create an image of the Neanderthals as an early, still apelike form of humanity. In 1891, the Dutch paleoanthropologist Eugene Dubois discovered the skull and thighbone of an even more primitive human form in Java. Dubois had gone to the East Indies to follow up Haeckel's claim that Asia, not Africa, might turn out to be the cradle of humanity. "Java man" confirmed Haeckel's and Darwin's prediction that early humans had stood upright before they acquired a modern brain capacity, although this point was again ignored by the supporters of the "brain first" theory. Dubois called his new type *Pithecanthropus erectus*, borrowing Haeckel's term for the missing link, and Haeckel enthusiastically endorsed the discovery (translation, 1898). By 1900, the conventional image of human evolution portrayed a linear progression from the apes through *Pithecanthropus* and the Neanderthals to the modern human form.

In the early twentieth century, this neat linear model was broken down to give a more sophisticated version of progressionism that could accommodate the difficulties posed by the fossils. Marcellin Boule (translation, 1923) and Arthur Keith (1915) now began to argue that the Neanderthals were too apelike to be the ancestors of modern humans: the necessary degree of development could not have been achieved in the short time indicated by the archaeological record (Hammond, 1982). Instead, the Neanderthals were dismissed as a side branch, a separate and parallel line of human development that had been exterminated by our own, more successful ancestors. Human evolution was a branching, not a linear, process—a superficially modern idea that was adapted to progressionism by arguing that all the branches were being driven by the same trend. Nature was supposed to be "experimenting" with various ways of producing the human type.

The plausibility of this model was enhanced by the notorious Piltdown fraud. In 1912, the amateur geologist Charles Dawson discovered some human remains in a gravel bed at Piltdown in Sussex. There was a large-capacity skull with a jawbone indistinguishable from that of an ape. The parts where the jaw and the skull articulated were broken off, but it was generally assumed that they belonged to the same individual. Arthur Smith Woodward of the British Museum named the new species *Eoanthropus dawsoni* and became a leading champion of the reconstruction of human evolution around the "dawn man" of Piltdown. The discovery seemed to confirm that there were several distinct branches of human development. It also supported the view that in the most important of these branches, the brain had developed faster than any other part of the body. The fraud was not exposed until 1953, when it was shown that a human skull and an ape jaw had been "planted" together at the site (Weiner, 1955). Since then, a minor literary industry has grown up around the attempt to detect the culprit (e.g., Millar, 1972; Blinderman, 1986).

Long before Piltdown was exposed, it had already become something of an anomaly. In the late 1920s, the discovery of "Pekin man" in China revealed that the *Pithecanthropus* type had been spread widely around the world. Although originally assigned to a distinct species, the Java and Pekin types are now known as variants of *Homo erectus*, the species that preceded the evolution of modern *Homo sapiens*. These discoveries might have drawn attention back to the view that the brain had expanded after the adoption of an upright posture, but their location only served to deflect interest away from what modern paleoanthropologists regard as a far more significant find. In 1924, the anatomist Raymond Dart unearthed an immature hominid skull at Taungs, South Africa, naming it *Australopithecus africanus*. He predicted that *Australopithecus* was the true ancestor of the human race and claimed that the creature had walked upright despite possessing a brain scarcely larger than an ape's. Dart's views were treated with skepticism until further discoveries in the late 1930s by Robert Broom confirmed that the Australopithecines had indeed been bipedal. Even then, progressionist ideas about human origins continued to flourish, Broom himself arguing that mankind is the goal of a divinely planned evolutionary process. The advent of the modern synthesis finally undermined the plausibility of the old theories of progress and parallelism. Modern paleoanthropologists try to understand the adaptive significance of the transition to bipedalism and the subsequent enlargement of the brain, thus returning to issues that were first sketched in by Darwin.

In the 1950s, it was assumed that *Australopithecus africanus* gave rise first to *Homo erectus* and then to *Homo sapiens*. Later discoveries have shown that the process was far more complex. Several different species of *Australopithecus* coexisted in Africa for some considerable time before dying out. In 1961, Mary Leakey discovered a much older member of the human line, *Homo habilis,* predating most of the Australopithecines. The much earlier *Australopithecus afarensis* (better known as "Lucy") has now been claimed as the common ancestor from which humans and the later Australopithecines diverged (Johanson and Edey, 1981). Paleontologists at first assumed that the main separation of the ape and hominid lines occurred as long ago as 15 million years, but recent evidence from molecular biology has reduced this considerably (Gribben and Cherfas, 1982; for a survey of modern developments, Lewin, 1984). Whatever the disagreements over details, however, human evolution has turned out to be an irregularly branching process, just as Darwinism would lead one to expect.

The latest discoveries have confirmed that the acquisition of an upright posture was the first breakthrough in human evolution but have thrown doubts on the reason given for this development. It seems clear that the trigger was a climatic change that opened up wide areas of grassland or savanna in Africa at the expense of jungles; the upright posture was adopted by the descendants of those apes that chose to explore this new environment. Darwin himself hinted that the freeing of the hands for toolmaking might have been one of the adaptive advantages that encouraged selection for bipedal locomotion. This view remained popular until quite recently. The discovery of "Lucy" has upset this explanation, because there is no evidence that *Australopithecus afarensis* used tools (Johanson and Edey, 1981). The reduction in the size of the canine teeth is another characteristic trend in human evolution that must be explained, presumably in terms of changing diet. Perhaps bipedalism offered the opportunity for carrying food in the hands, and this development was linked to the emergence of a life-style in which the individuals of a family foraged in separate directions from a temporary base (Lovejoy, 1981). Only later did tool use begin in one branch of the Australopithecines, leading to the evolution of *Homo habilis* and the establishment of intelligence as the crucial survival factor to be developed by natural selection.

Many authorities have assumed that the evolution of man proceeded by gradual transformation of whole species. In recent years, the new model of "punctuated" evolution has been applied to man (see chap. 12; Stanley, 1981; Eldredge and Tattersall, 1982). According to this interpretation, wide-ranging species are generally very stable over long periods of time, a point confirmed by the fossil record for hominids. New species are thought to appear when a small population isolated from the parent form undergoes rapid modification. The parent form is left unchanged by the "budding off" of the new species, and the two may coexist for some time thereafter.

THE IMPLICATIONS OF MODERN DARWINISM: SOCIOBIOLOGY

The biologists who created the modern synthesis were well aware of the broader implications of what they were doing. They had no wish to revive the harsh message of social Darwinism, but they were convinced that evolution offered the framework of a new world view that would replace traditional religion as a foundation for philosophy and morality. It is probably fair to say that the effort to found a new evolutionary ethic did not gain wide support while it was linked to the rather colorless humanism of the postwar years. The more popular methods of applying the principles of Darwinism to human affairs have raised all the old fears of those who see the movement as nothing more than an expression of conservative ideology (Kaye, 1986). Some students of animal behavior saw aggression as a built-in instinct that also must be present in man. More recently, the science of sociobiology has attempted to explain the evolution of all forms of behavior in terms of individualistic natural selection. Attempts to extend this approach to man have revived the social scientists' long-standing distrust of biological determinism.

Of the founders of the modern synthesis, Julian Huxley and George Gaylord Simpson made the greatest effort to extend the new Darwinism into a general world view. Huxley's *Evolution in Action* (1953) and Simpson's *The Meaning of Evolution* (1949) were intended explicitly to go beyond the technical details of the theory into a broader vision of the nature and purpose of life. They adopt a positivist philosophy in which science is the only source of knowledge and see evolution as the new foundation of morality, replacing the transcendental source of values invoked by religion. Both writers strove to get away from the mechanistic image of science by stressing the creative, opportunistic aspects of life's development within the Darwinian scheme. They also conceded that simpleminded attempts to explain society in terms of human biology were no longer plausible. It was essential to recognize that man's emergence was not preordained, because Darwinian evolution is not teleological in character; but the hope was that the process of evolution itself would teach us how best to meet future challenges. Man has become the dominant form of life on earth and is now taking charge of his own evolution. Indeed, he has become responsible for the future of all living things, and his success will be measured by the extent to which he carries on the creative heritage of nature. Huxley saw evolution as inherently progressive, leading to successively higher levels of dominant life forms; progress was measured by an increasing ability to transcend limits imposed by the environment. Freedom to realize the potentialities of life thus was seen as the highest good. Simpson was more skeptical of the progressionist interpretation but still saw man as the high point of a tendency for life to develop greater awareness of its surroundings. The drive for knowledge took on an ethical value, as something man could use for his own benefit and the benefit of all forms of life for which he is now responsible.

Not all modern evolutionists have adopted the agnostic or humanist perspective. Some have tried to incorporate the idea of evolution into an explicitly religious framework, preserving traditional spiritual values by reinterpreting the message of science. The best-known move in this direction was made by Pierre Teilhard de Chardin in his *Phenomenon of Man* (translation, 1959). Although he had some experience in paleontology, Teilhard was a Catholic priest, and the church would not allow his ideas to be published during his lifetime. To reconcile evolution and religion, he saw the development of life as a cosmic process of increasing spiritualization. The tendency of life to generate mind (noögenesis) had been greatly accelerated by the move toward "hominization," the production of man. Ultimately, the tendency of minds to unite will culminate at the "omega point" in the formation of a single hyperpersonal entity by the fusion of all human minds. Reactions to this philosophy were mixed. Julian Huxley was sufficiently impressed to write an introduction to the English translation of the *Phenomenon of Man,* while other scientists were bitterly critical of the looseness of Teilhard's borrowings from evolution theory (e.g., Medawar, 1961). Many theologians also doubted whether the prophecy of a transcendental goal for evolution and man was consistent with the Christian message (Hanson, 1970).

The association of Huxley with Teilhard suggests that evolutionary humanism and evolutionary mysticism are not that far apart in their goals. Both are forced to see some kind of purpose in evolution, although they disagree on whether that purpose should be seen as an extension of traditional spiritual values or as a new and purely naturalistic source of meaning for life. The problem in either case is that Darwinism does not really guarantee progress or at least makes it very difficult to define (Thoday, 1962; Gouge, 1967; Simpson, 1973). This is especially damaging for Teilhard's system, which quite clearly sees a goal-directed trend in evolution that is incompatible with the Darwinian mechanism. Huxley interpreted progress in a less rigid manner, more consistent with the opportunism of natural selection, yet he too was forced to argue that we can see a meaning in the end products of an apparently purposeless selection of random variations. There have been many critiques of the attempt to derive ethics from evolution (e.g., Flew, 1967; Greene, 1981). In the end, perhaps the most damaging charge is that in their efforts to twist a meaning out of the universe, the evolutionists have deceived themselves by reading their own preconceived values into nature. The principles they establish are either so vague or so tenuously linked to the logic of scientific evolution theory that their arguments could justify almost any ethical position. In the end, nature guarantees the success of whatever works out in practice—a conclusion that leads back too easily to the original, highly conservative form of social Darwinism.

Critics would argue that this danger is obvious in a number of developments associated with modern Darwinism. One is the continuation of a link with the hereditarian view of human character and with eugenics. The creation of population genetics undermined the logic of the more simpleminded claims of the early eugenicists (chap. 10), but it did not necessarily challenge the assumption that heredity is the chief determinant of human nature. R. A. Fisher was deeply committed to the eugenics movement and devoted a section of his classic book on the genetic theory of selection (1930) to social questions. Julian Huxley also supported eugenics as part of a wider program for the progress of the race, although he insisted that social reform was a necessary prerequisite of biological improvement (Bajema, 1977). There has been continuing controversy over the extent to which intelligence is determined by genetic factors, particularly whenever it has been suggested that certain groups defined by race or class are genetically inferior (Deutsch, 1968; Kamin, 1972; Clarke and Clarke, 1974; Evans and Waites, 1981). These debates have not been linked directly with evolution theory, however, and some modern evolutionists have opposed genetic determinism in man (Gould, 1981). As in an earlier epoch, there is no automatic link between evolution theory and the hereditarian philosophy, although Darwinism does have some bias in this direction.

More directly related to the selection theory is the so-called anthropology of aggression, which assumes that the evolutionary mechanism will endow all organisms including man with aggressive instincts. First suggested by Keith (1949), this view was taken up by Konrad Lorenz, the founder of the modern science of ethology, the study of animal behavior (translation, 1966). Lorenz demonstrated the existence of aggressive behavior in many animals and argued that it would be folly to assume that man was not endowed with similar instincts. Robert Ardrey (1966) suggested that the instinct to defend a piece of territory is even more basic. The "territorial imperative" has shaped the behavior of most animals, because territory is the key to food supply and successful mating. There is no reason to assume, claimed Ardrey, that man's intelligence has exempted him from the same trend, which would explain our individual behavior and our tendency toward mass warfare. Similar views were expressed in Desmond Morris's classic *Naked Ape* (1967), in which aggression was derived from the instincts of our ancestral "killer apes."

The anthropology of aggression was rejected by social scientists trained in the more liberal school of the mid-twentieth century. Their calls for reform were based on the claim that aggressive behavior is not instinctive but is the product of a deprived social environment. Many naturalists also were suspicious of the trend, as close observation began to reveal that the great apes are not, in fact, very aggressive creatures. Nor is there anything to suggest that man's early evolution should have encouraged violent instincts, because modern tribes who live by hunting do not engage in warfare with their neighbors (Leakey and Lewin, 1978). Although the anthropology of aggression was upheld by some observations of animal behavior, it did not draw on the more sophisticated view of Darwinism developed by the modern synthesis. Those who expounded this view of nature seemed obviously motivated by their desire to justify a society based on competitive individualism. A more powerful argument for biological determinism has now appeared; its scientific credentials are firmly established in the highly original application of the selection theory to animal behavior which constitutes modern sociobiology.

The original problem that led to sociobiology centers on the existence of apparently altruistic behavior in some animals. If we imagine that natural selection works solely through competition between individuals, it is difficult at first to imagine how it could give rise to an instinct that leads one organism to sacrifice itself for others. Without invoking Lamarckism, it is impossible to argue that the instinct started when the animals learned to cooperate with one another. A Darwinian explanation requires that some reproductive benefit is conferred by the behavior, and the most obvious solution is to postulate that the level of selection can be switched so that it acts among groups rather than individuals. If cooperation among its individual members makes a group more likely to succeed as a whole, then surely an instinct promoting such behavior can be built up through the fittest groups displacing those that do not develop the useful behavior. Sociobiology challenges this concept of "group selection" and seeks to explain even altruistic behavior by a more careful analysis of its reproductive consequences for the individual. It thus proposes to explain even the most complex forms of behavior in terms of the most basic interpretation of Darwinism.

Darwin himself consistently opted for selection that acted at the individual level (Ruse, 1980). He experienced some difficulty, though, in dealing with the existence of neuter castes in the Hymenoptera—the ants, bees, and wasps—whose colonies frequently contain highly specialized workers or soldiers that cannot breed. These workers exhibit an extreme form of altruism by devoting their whole lives to helping the reproduction of others. Darwin noted that selection could develop special characteristics in sterile individuals by acting on their relatives who can breed. In the case of the insect colonies, all individuals are descended from the queen, and selection can act to favor those queens whose reproductive pattern includes the production of sterile offspring that can help the whole colony to survive. Because he now supposed that selection acted among colonies, Darwin had been forced back on kind of group selection in this case. The problem of altruistic behavior was largely ignored until after the emergence of the modern synthesis. The debate that led to sociobiology was sparked by V. C. Wynne-Edwards (1962), who argued strongly for group selection to explain many aspects of animal behavior. Led by G. C. Williams (1966), many naturalists became suspicious of this approach and turned back to consider why Darwin had tried wherever possible to restrict selection to the individual level. Part of the problem with group selection lies in the difficulty of pinpointing truly altruistic behavior. An animal who gives a warning cry to his fellows on spotting the approach of a predator may seem to be acting altruistically, but he may hope to escape more easily in the resulting confusion. Even when genuinely altruistic behavior occurs, the weakness of an explanation based on group selection lies in the fact that it always can be undermined by individual selection acting to favor anyone who "cheats" on the arrangement. Such an individual gains an advantage over his fellows by refusing to reciprocate their altruism, so his characteristic will spread through the group and eradicate the altruistic instinct.

One of the most important concepts of sociobiology was pioneered by W. D. Hamilton (1964): the notion of "kin selection." This is based on a clarification of the point recognized by Darwin, that an individual can influence the course of selection not just by breeding himself but also by helping his genetic relatives to breed. Because the relatives carry some of his own genes, their success will ensure the nonbreeder some representation in the next generation. Apparently altruistic behavior thus can be explained by individual selection, which will promote the instinct to help genetic relatives in circumstances where one cannot breed oneself. Hamilton showed that Darwin need not have made an exception in the case of neuter insects, because the Hymenoptera have an unusual reproductive pattern that allows individual selection to explain the phenomenon. Males are produced from an unfertilized egg, with the result that among females, sisters are more closely related than mother and daughters. Kin selection thus favors development of sterile female castes, because a female enhances her genetic representation in the next generation best by helping to rear fertile sisters, rather than by producing her own daughters.

The technique employed here replaces the struggle for existence with a concept of selection based solely on the maximization of the individual's genetic representation in the next generation. Selection will enhance any instinct that works in this direction, because by definition, those individuals who possess the instinct will be more successful in passing on their genes. This has led Richard Dawkins (1976) to argue that the gene is really the basic unit of selection, so that organisms evolve in response to the demands of their "selfish genes." The organism is merely the gene's way of reproducing itself, and those genes corresponding to reproductively successful instincts will necessarily come to dominate the population. Dawkins warns against the danger of anthropomorphizing genes by talking of them "succeeding" or "failing" in the race to duplicate themselves—such phrases are merely shorthand descriptions of a totally mechanistic process. Some naturalists feel, nevertheless, that his approach obscures the fact that the organism, not the gene, confronts the environment and engages in the struggle for existence and reproduction. Dawkins now has replied to his critics (1982) by arguing that we need to take a much broader view of factors that contribute to successful reproduction.

The new approach to the study of behavior exposes the weakness inherent in the old idea that natural selection must automatically program organisms with aggressive instincts. This has been illustrated by John Maynard Smith (1982) through the application of games theory to the study of how selection will act on different kinds of behavior. To take the simplest possible example, imagine the two most extreme forms of behavior, which we can assign to the "hawks" and the "doves." Hawks are aggressive types who always fight their neighbors over any disputed resource, such as food or a mate, regardless of the risks that might be involved. Doves refuse to fight under any circumstances. Obviously, a population composed entirely of doves would be unstable, because a single hawk introduced by mutation would gain an immediate advantage by bullying his fellows out of their share of the resources. Thus, the hawk characteristic would be spread by natural selection. Paradoxically, however, it can be shown that a population composed entirely of hawks would also be unstable, because here a single dove introduced by mutation would actually be at an advantage. The point is that the hawks' aggressive behavior leads them to suffer frequent injury, fighting for resources that are not really worth it. The single dove escapes this risk and is thus better off than the hawks in the long run. There is a particular ratio between the proportion of hawks and doves that forms the "evolutionary stable strategy" for the species. Instead of promoting all-out aggression, selection actually works to maintain a balance of different characteristics in the population.

Clearly, an instinct that confers adaptive advantage in the traditional sense will be boosted by natural selection, but sociobiology extends selection to those areas of behavior concerned with reproduction rather than survival.

Edward O. Wilson's classic survey (1975) applies the technique at length to explain behavioral differences between males and females in various species. Because of the different level of investment they put into the growth and raising of offspring, reproductive behavior of males and females should be aimed at different goals if they are to maximize their genetic contribution to later generations. In the higher animals, the female invests a great deal of energy in reproduction, while the male can produce sperm quite easily. It is thus in the male's interest to increase the number of his matings by aggressive behavior, leaving the female in each case to raise the offspring. The female, by contrast, needs to be very careful in choosing only the fittest father for the small number of offspring she will be able to rear. The male's attitude will be rather more responsible in those species where it is essential for him to help with the rearing of the offspring—he simply cannot afford to leave for "pastures new" if this will put his existing offspring at risk. By feeding the circumstances of particular species into their model based on the power of individual selection to promote reproductively successful instincts, sociobiologists can predict the kind of behavior that should be expected. Extensive field research has shown that these predictions are extremely successful in a wide range of animal species.

The success of sociobiology in explaining animal behavior led Wilson, in particular, to argue that certain aspects of human nature might be accounted for in similar terms (1978). The development of incest taboos in almost all societies is thought to derive from an instinct created by the biological hazards of close inbreeding. More controversially, the basic differences between male and female behavior may be conditioned by exactly the same evolutionary forces that affect animal species. In Wilson's view, our sexual attitudes are biologically determined by our genes, and any society that tries to ignore this fact is doomed to failure. He concedes that human beings are governed less rigidly by their biological instincts than the other animals but insists that the genetic foundations of our behavior are strong enough to impose definite limits within which societies can flourish (see also Lumsden and Wilson, 1981).

The extension of sociobiology to man has produced a strong reaction, leading some critics to oppose the whole approach as merely a continuation of social Darwinism (for critiques, see Sahlins, 1976; Montagu, ed., 1980; Rose et al., 1984, see also the replies by Ruse, 1979*b*, 1982, and the survey edited by Caplan, 1978). The anger of social scientists is predictable: their disciplines have been built on the assumption that behavior is conditioned by the social environment, not by biological inheritance. In their anxiety to proclaim man's freedom from genetic determinism, they have tended to dismiss even animal sociobiology as a misguided application of conservative ideology to science. Because he portrays nature as based on individual competition, the sociobiologist is accused of trying to establish a foundation on which he can build his defense of racism, elitism, or sexism. In reply, it is necessary to point out that whatever its limitations when applied to man, sociobiology has been remarkably successful in explaining otherwise mysterious aspects of animal behavior. Some evolutionists who applaud sociobiology *as biology* agree that we should be suspicious of its application to man because the complexity of the human brain makes it less likely that we are dominated by instincts (e.g., Gould, 1977*c*). There is no a priori reason to expect that our instincts exert no control whatever over our behavior; indeed, it seems most unlikely that we are quite so free of our evolutionary past. What is needed is an open-minded willingness to consider the extent to which we may be governed by such instincts (Midgley, 1978; Konner, 1982). Darwinism may have drawn its inspiration from capitalist ideology, but it has proved successful as science. Those who insist that biological Darwinism must be wrong because they dislike some of its possible applications betray their own lack of objectivity.

On the question of biological determinism, the issues raised by sociobiology seem relatively straightforward in principle, however difficult it may be in practice to untangle the contributions of heredity and environment in building human character. If we *are* conditioned to some extent by our instincts, we ought to be aware of the fact, even if we choose to circumvent those instincts through the effect of education. But Wilson (1978) raises once again the deeper question of whether evolution provides a world view powerful enough to serve as a new foundation for ethics. At some points, he develops the claim that what nature has produced is necessarily good and thus should be embodied in the aims of our society (see Greene, 1981). If sociobiology explains away our urge to believe in God, then evolution itself must show us the purpose of life and the way to behave. Yet evolution works blindly by the differential success of the genes, producing mind as a kind of accident as the process strives to increase the sophistication of behavior. There are many who feel that evolution theory, however successful it may be as science, cannot stand the weight of such a superstructure. If the process of evolution shows no obvious sign of purpose, we should not use it as a guide to moral conduct, as long as we are prepared to allow that in any sense we possess free will. To know about our past is interesting and may even be practically useful, if it alerts us to difficulties that may lie in the way of certain kinds of reform. But the question of whether or not those reforms should be attempted is not one that should be decided on the basis of our past evolution. If Darwinism comes anywhere near the truth, the universe has not been designed to show us, its products, where we should go in the future. If there is an ethical message in the theory, it tells us simply that we cannot look outside ourselves for guidance. With or without the knowledge of evolution, each of us must look into his or her own conscience for a source of moral values.

12

The Modern Debates

The controversy over sociobiology was sparked by the Darwinists' self-confident extension of their theory into territory already claimed by their opponents. In most modern debates on evolution, however, Darwinians have been very much on the defensive, attempting to protect themselves against attacks from a wide range of critics. There are many different levels of disagreement. Some arguments occur within the structure of orthodox evolution theory and represent no challenge to that basic structure, even if they threaten to upset particular interpretations of the past. The development of new ideas about human evolution would fall into this category, as would Adrian Desmond's hotly debated popularization of the claim that dinosaurs were really warm-blooded (1976). If the majority of biologists came to accept this latter view (which seems unlikely at present), they certainly would have to reinterpret their views on how dinosaurs were related to birds and modern, cold-blooded reptiles—but there would be no need to reconsider the Darwinian theory of how evolution takes place. The postulation and testing of such new interpretations is a sign of evolutionism's scientific vitality, giving the lie to that often-repeated charge that the theory is only a blindly accepted dogma.

Rather more serious are the claims that modern Darwinism is an inadequate account of the evolutionary process. From Samuel Butler onward, there have always been writers from outside the scientific community willing to criticize the materialism of the selection theory. Such attacks usually end up with a plea for reconsideration of Lamarckism or of the role of growth factors in shaping evolution (recent examples include Koestler, 1967, 1972; Koestler and Smithies, 1969; Taylor, 1983). Their intention is to suggest that evolution is not just a trial-and-error process but must be directed along purposeful channels by some biological force. Scientists usually ignore nonspecialist critiques of this kind—perhaps an unwise move, given the widespread public feeling that natural selection is implausible as an explanation of organic complexity. Dawkins (1986) has now made a bold effort to enhance popular comprehension of the selection mechanism. Dawkins himself is convinced that Darwinism offers a radical but completely adequate account of how life evolved. In recent years, though, a number of working biologists have joined the ranks of those who suspect that there may be more to evolution than natural selection. In some cases, their suggestions are presented merely as additions to the synthetic theory, leaving the basic structure of modern Darwinism little changed. A few more radical biologists have now begun to support those critics who believe that the selection theory is completely inadequate.

More basic challenges to modern Darwinism come from those who wish to discard the whole idea of evolution. It is easy to forget that creationism is not the only movement dedicated to this goal. Within the academic community, there is some disagreement over whether the Darwinian approach is truly scientific. The more truculent supporters of a new technique in classification, the "transformed cladists," insist that speculation about evolutionary relationships goes completely outside the bounds of science. Many nonscientists have shown considerable interest in exotic alternatives to traditional interpretation of the earth's history, including Immanuel Velikovsky's new catastrophism and Erich von Däniken's theory of extraterrestrial interference. Even these views have less radical parallels that some scientists take seriously. The possibility that life may have arrived on the earth from outer space is advocated by Francis Crick and Fred Hoyle, both of whom clearly state their lack of confidence in the orthodox theory of chemical evolution.

Apart from its greater popularity, creationism presents a deeper threat to modern evolutionism because it is based on the claim that the origin of species must be a supernatural process. For all the talk of "scientific" creationism, this claim would appear to put the whole question outside the scope of scientific investigation. More serious still is the insistence of most creationists that the miraculous origin of life took place exactly as described in the Bible, within a period of single week, only a few thousand years ago. This entails a rejection not only of evolution but also of the whole scientific interpretation of the past established by cosmology, geology, paleontology, and archaeology. Even if one could make a plausible case for the creationist alternative, it could be argued that a movement that started with the intention of upholding a position accepted originally on religious grounds can never be truly scientific in its methods. Creationists tend to seize on the widespread criticisms of modern Darwinism as support for their case and to avoid any detailed discussion of their own alternative that would reveal its far more basic weaknesses as a comprehensive scientific theory of the past.

This chapter begins with those new ideas that, at least in principle, could be incorporated in an evolutionary view of the past and then moves on to the more radical alternatives. (For surveys of the modern debates, see Cherfas, ed., 1982; Maynard Smith, ed., 1982; Milkman, ed., 1982; Ruse, 1982; Ridley, 1985).

THE DEBATE IN BIOLOGY

In recent decades, the modern synthesis of Darwinism and genetics has been criticized as too narrowly focused, even by some biologists. Various lines of evidence have been brought forward to suggest that at least some aspects of the evolutionary process do not take place in the orthodox Darwinian manner. In some cases, as with the original theory of "punctuated equilibrium," the new idea was introduced as little more than an extension of hitherto unrecognized implications already contained within the Darwinian approach. Other theories are presented as complete alternatives to a totally inadequate selection theory. Criticism at this level is still confined to a small minority of biologists who are unable to agree among themselves over the precise nature of the new approach. The one factor uniting the critics is a feeling that the raw material of evolution must consist of something more purposeful or orderly than a flow of random micromutations. In this respect, the old tradition of non-Darwinian thinking which flourished in the nineteenth century is still alive. There are several different ways in which this alternative approach can be explored—a diversity that the critics of selectionism see as an indication of the need for a more flexible approach. Unfortunately, the inability of the critics to unite behind a single coherent alternative has left the Darwinians in a position to shrug off each individual critique as insignificant.

One of the least-known scientific critics of modern Darwinism was Leon Croizat (1958, 1964), whose work has now attracted the attention of a small but vociferous group of biologists (Nelson and Rosen, eds., 1981). Croizat's "panbiogeography" is based on the claim that the Darwinian explanation of dispersal and divergent evolution does not fit the facts of geographic distribution. Darwinians have assumed that "dominant" species spread out to take over new territory, but Croizat argued that existing distributions of related species do not correspond to their dispersal abilities. Instead, he suggested that speciation generally takes place vicariously, that is, by natural barriers splitting up the once continuous range of a widely dispersed species. In so doing, it has been claimed that he has revived issues that biogeographers have ignored since the time of Darwin (Nelson, 1978). Throughout his life, Croizat was ignored, possibly because he did not hesitate to ridicule Darwin and the orthodox modern Darwinians. His theory also required an element of non-Darwinian evolution resembling orthogenesis, since he believed that separated populations tend to evolve in a very similar way even if exposed to different environments.

Croizat's supporters now argue that his views are rendered plausible as a result of the revolution in the earth sciences leading to the acceptance of continental drift. Where traditional geology ignored the possibility of horizontal movements in the earth's crust, the new science of plate tectonics accepts drift as a natural consequence of the processes that shape the surface (Hallam, 1973; Wood, 1985). The separation of once unified landmasses is thus an integral part of the earth's history, and evolution theory must take this into account. This process could account for the distribution of species once attributed to dispersal. As yet, however, the majority of biologists do not accept that Croizat's interpretation of distribution is plausible enough to justify his wholesale attack on the Darwinian synthesis.

The element of directed evolution implicit in Croizat's thesis is the one factor linking his approach to that of the other critics of Darwinism. By far the best known of the new mechanisms is the theory of punctuated equilibrium, which began as a change of emphasis within the Darwinian camp but has gradually developed into a more comprehensive alternative to the modern synthesis. The basic issue is that of gradualism versus discontinuity: is evolution a process of slow, continuous change as supporters of the modern synthesis hold, or does it go through episodes of comparatively rapid change interspersed with long periods of stability? There are Darwinian mechanisms that could account for quite sudden changes by geological standards, and originally the punctuated equilibrium model was intended to stress the role of such events. In more recent years, though, the model has expanded into a fundamental assault on the gradualist perspective, with the aim of establishing discontinuity as a basic feature of evolution. In addition, its supporters now maintain that a hierarchical series of mechanisms can be seen operating within the evolutionary process and that the gradualism of the modern synthesis is merely the lowest level of the hierarchy. Although not integral to punctuated equilibrium theory, the concept of genetically sudden breakthroughs at key points in evolution also has reemerged. Opponents dismiss all these alternatives as unnecessary. Imperfection of the fossil record still can account for the apparently sudden introduction of new species, while the reappearance of the hopeful monster is a throwback to the wild speculations that the modern synthesis rendered superfluous.

The original proposal of the punctuated equilibrium model came in an article by Niles Eldredge and Stephen Jay Gould (1972; see also Gould and Eldredge, 1977; Eldredge, 1986). The model was the response of two paleontologists to what they perceived as excessive commitment to gradualism by the modern synthesis. Population genetics explained microevolution as a process of gradual change, and it had been widely assumed that the same process extended over greater periods of time would lead to macroevolution (production of new species, genera, etc.). G. G. Simpson's *Tempo and Mode in Evolution* (1944) had argued for applying the new Darwinism to paleontology by invoking once again the imperfection of the fossil record to account for the apparently sudden appearance of new species. Simpson and the other founders of the synthesis at first had been prepared to allow a role for rapid speciation events, but this came to be pushed aside as the gradualist metaphor grew in strength. It was held that enough examples of gradual change existed in the fossil record to show that the apparently sudden changes were an illusion, caused by lack of evidence for intermediate stages. By the 1970s, however, an increasing number of paleontologists had become dissatisfied with this approach, because many of the classic examples of gradual change had not withstood the test of modern techniques. If there were no genuine cases of gradualism in the record, then the argument for treating all cases of sudden change as the result of imperfect evidence was undermined. It might be better to reexamine the evidence in a new light, putting aside the traditional Darwinian assumption of gradualism and opting instead for a model of evolution that would allow for the sudden appearance of new forms as indicated by the fossil record.

The original argument for punctuated equilibrium was based on the application of an orthodox Darwinian idea expounded by Ernst Mayr: speciation through the separation of a "peripheral isolate" from the original population. This theory held that new species would arise most readily from small populations cut off in remote corners at the edge of the original species' range. The extreme nature of the conditions, coupled with the small size of the population, would ensure the rapid evolution of new characteristics. Such a development might take thousands of years, which is "rapid" in geological terms, although lengthy by the standards of experimental genetics. The event would be most unlikely to leave any trace in the fossil record because of its rapidity and the very small area to which the changing population was confined. In some cases, however, a new species formed in this way might reenter the area occupied by the parent species and might be better adapted to the conditions there. It would take over quickly from the parent form, spreading out over a wide area where—if it persisted long enough—it would begin to leave fossil remains. Thus, it would appear suddenly in the fossil record, giving the appearance of a totally new form with no continuous links back to the parent form.

Expressed in this way, the sudden "punctuations" marked by speciation events appear to be compatible with the orthodox Darwinism of the modern synthesis. There is a sting in their tail, however, which arises from the paleontologists' conviction that once a species comes to occupy an extensive territory, it becomes fixed in character and continues largely unchanged until it is itself replaced. This suggests that "phyletic" evolution—change within a single lineage without splitting—is not very common. An evolutionary trend thus cannot consist of a single line changing continually in a particular direction because of the action of natural selection on the individual organisms that make up the population. Instead, it must be composed of a series of speciation events, each producing fixed species; the trend will result from the consistent replacement of old species with new ones changed in the appropriate direction. The mechanism responsible for producing the trend is "species selection"—competition between species resulting in elimination of all but those best adapted to a new environment or a new way of life (Stanley, 1981). In this view, the process of speciation is not the same as the process of general macroevolution, because speciation involves selection acting on individuals, while large-scale trends depend on selection acting among species. The direction of change in speciation is not related to the macroevolutionary trend. This is partly because small populations may be unrepresentative of the original species as a result of random sampling effects. But even when changes taking place during speciation are adaptive, the fact that the population is isolated under extreme conditions means that its new characteristics are not necessarily adaptive by the standards of the parent species. Speciation thus results in essentially random production of new forms, only some of which might by chance turn out to have a general advantage over their parents which would enable them to take over the whole territory.

More recently, Gould (1980*c*, 1982) has extended the challenge to orthodox Darwinism even further. One way of understanding the direction of the new ideas is to concentrate on the alleged stability of large populations—the equilibrium that is maintained between the punctuations of speciation. Darwinism can explain why a large population exposed to a stable environment will retain the same character; "stabilizing selection" would tend to eliminate those individuals who departed too far from the norm. A Darwinian would assume, however, that if conditions change, even a large population will begin to evolve in response to individual selection. Paleontologists argue that phyletic evolution in large populations does not seem to occur, which suggests that something more rigid than stabilizing selection must be responsible for maintaining the fixed characteristic. One possibility is that "developmental constraints" may operate within the growth of the individual organism. An established pattern of embryological growth cannot be altered gradually by genetic mutation, because it has a kind of internal coherence of its own. Only in exceptional circumstances can such constraints be dismantled and a new path of development toward a different adult characteristic be established. If this is so, it may be that appearance of new

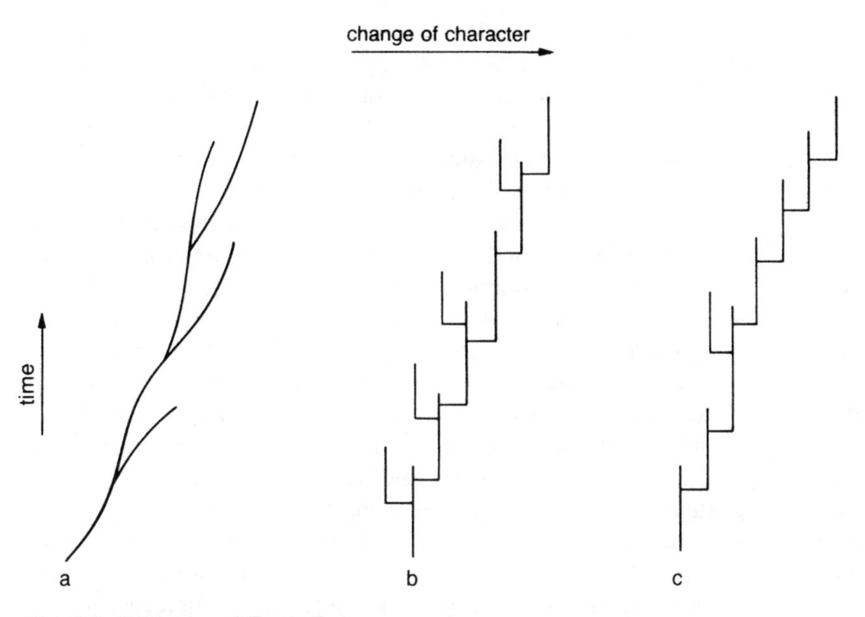

Fig. 24. Patterns of Evolution.

a) Orthodox Darwinian view of an evolutionary trend in which there is some speciation but all branches move in the same direction under pressure of natural selection acting on individual variation.

b) Punctuated-equilibrium model with species selection. Each branch remains fixed once it is formed and speciation is random with respect to the direction of the trend, which results from the preferential survival of branches to the right at the expense of branches to the left.

c) The punctuated equilibrium model with an "origin bias" influencing the trend: because of internal factors affecting variation, speciation occurs more readily to the right than to the left.

characteristics in speciation is not random. The existing course of growth may prevent all but a small number of possible changes from taking place, so that the direction of evolution is constrained to move along fixed paths. The changes taking place during speciation themselves direct the overall trend of evolution in response to internal constraints imposed by the process of embryological growth. This would give rise to directed evolution independent of the environment, somewhat along the lines of the old theories of mutation pressure and orthogenesis.

The belief that existing growth patterns can somehow control the appearance of new characteristics is known as "epigenetic" evolution. Its supporters frequently cite the long-neglected views of Richard Goldschmidt (1940; see Allen, 1974), who urged similar arguments in opposition to the emergence of the genetic theory of selection. Another idea from the same source also has gained new life: the sudden appearance of totally new characteristics through mutation. Goldschmidt's notion of the hopeful monster was rejected, because genetics seemed to show that large mutations are invariably lethal. It can be argued, however, that a comparatively small genetic change affecting early stages of the growth process might channel development along a new path that would shape it toward a new and functional goal by the integrative action of the later stages in the process. Gould (1982) suggests that such saltations might account for those occasional points at which evolution seems to have taken a totally new direction, and selection has acted to make adaptive use of the new structure once it had been formed in this way. Some advocates of epigenetic evolution see it as a factor that can be integrated into existing evolution theory (e.g., Rachootin and Thomson, 1981), while others see it more as a major alternative to Darwinism (Løvtrup, 1977; Pollard, ed., 1984; Ho and Saunders, ed., 1985).

An important by-product of this new approach is that it is no longer necessary to think that every characteristic of every organism must have an adaptive purpose. Darwinians have been accused of simpleminded "adaptationism"—following a blind assumption that because selection acts only to produce useful characteristics, then every characteristic must have a purpose (Lewontin, 1978; Gould and Lewontin, 1979). They have been led to invent hypothetical adaptive purposes for all sorts of structures, often without any good reason to suppose that the hypothesis is valid. Many apparently trivial characteristics, however, actually have been shown to be affected by natural selection. Critics insist, nevertheless, that Darwinian emphasis on adaptation has gone too far. It may well be that the shape of every organ is adapted to some function, but what about the origin of the organ's basic structure? Can every step in the creation of a totally new structure be supposed to have an adaptive advantage—of what use, for instance, is *half* a wing? Traditionally, Darwinians have responded to this question by pointing out that originally the structure may have been developed for a quite different purpose—the concept of "preadaptation." The half-formed wings of the first birds may have been used for trapping insects as food and only later exploited as a means of flying. Gould and Lewontin argue, instead, that the basic form of a new structure, its *Bauplan,* may not be adaptive at all but may result from accidental forces triggered by the saltations mentioned above. The further development of the structure also may be limited by mechanical constraints that permit only certain lines of change. Only the superficial character of the organ is molded to an adaptive purpose by selection.

Some supporters of the epigenetic approach have also expressed doubts about the notion of hard heredity derived from Weismann's theory of the germ plasm and now enshrined in the "central dogma" of molecular biology. Francis Crick and James Watson worked out the structure of DNA in 1953

(Olby, 1974) and for the first time allowed biologists to see how the information necessary to construct a new organism is transmitted by encoding in the chemical structure of the genes. Their work established the basic pattern of information flow in reproduction: the DNA from the parent cells makes RNA, which in turn manufactures the proteins from which the new organism's body is constructed. There is no feedback of information from the proteins to the DNA and hence no route by which characteristics acquired by the body can be passed on through the genes. The work of Ted Steele (1979) provoked a flurry of interest in the possibility that there might, after all, be ways in which this reverse flow of information could take place. This would, in effect, permit the reintroduction of Lamarckism.

Steele's experimental work was done on the immune system of mice, in which he claimed to have detected the inheritance of acquired tolerance to certain antigens. At the theoretical level, he postulated a trial-and-error process of adaptation within the immune system, the effects of which might be transmitted to the DNA of the germ cells by viruses. This mechanism did not, in fact, violate the principles of molecular biology, but most biologists were suspicious of Steele's claims, and attempts to reproduce his results have failed. Active support for Steele's work came from Arthur Koestler, who had earlier (1971) urged a reconsideration of Kammerer's Lamarckian experiments with the midwife toad. Unfortunately for this position, the outburst of excitement sparked by Steele's experiments in the early 1980s now seems to have died away.

Koestler's long-standing support for Lamarckism was prompted by his distrust of the selection theory's materialistic implications (1967, 1972). He preferred to see individual organisms as creative entities capable of influencing their species' evolution through purposeful changes in their behavior. Lamarckism would allow the biological effects of new habits to be inherited and hence to accumulate over many generations. According to Koestler, modern Darwinism has actively repudiated any attempt to reintroduce purposefulness into evolution. Some pioneers of the synthetic theory were certainly suspicious of attempts to "humanize" Darwinism and actively resisted the claim that new behavior patterns could influence the course of evolution (Simpson, 1953*a*). But the image of natural selection as a mechanistic, trial-and-error process has never been the only one, some biologists always preferring to adopt a more flexible approach. When Alistair Hardy (1965) agreed that innovative behavior could influence the course of evolution, he was not invoking the inheritance of acquired characteristics but a form of Darwinism modified along the lines of the old "Baldwin effect" (chap. 9). New habits mark out the direction subsequently followed by natural selection in the adaptation of bodily structures. C. H. Waddington's mechanism of "genetic assimilation" (1957) would serve a similar function, again without requiring true Lamarckism.

Support for non-Darwinian mechanisms such as Lamarckism and epigenetics has always been fueled by philosophical or moral arguments against natural selection. It is by no means clear, however, that Darwinism is necessarily as black as its opponents would paint it. Lamarckism, too, has its darker side, as we saw when discussing the racial implications of earlier theories (chap. 10). There may be genuine scientific reasons for suspecting that the synthetic theory does not offer a complete explanation of evolution, but most biologists feel that the alternatives will only command their respect if they can be dissociated from the anti-Darwinians' often one-sided pronouncements on moral issues. Many still feel that there is simply no need to postulate anything beyond natural selection.

DARWINISM NOT SCIENTIFIC?

The accusation that the Darwinians have been too eager to concoct entirely speculative explanations of the adaptation of various characteristics leads to more serious criticism of the theory. Despite the immense amount of research inspired by the modern synthesis, some writers outside the scientific community have begun to insist that Darwinism is not truly scientific. It is argued that natural selection is nothing more than an empty tautology, because in the survival of the fittest, the fittest are defined as those who survive. Led by Karl Popper, a number of philosophers have branded the Darwinian attempt to explain everything in terms of adaptation as untestable, because in practice it is always possible to cook up any number of hypotheses to explain each characteristic. In recent years, even some biologists have begun to take these charges seriously, principally the more extreme supporters of a new approach to classification known as "cladism."

The claim that Darwinism is little more than a vacuous tautology has become popular among the philosophical critics of the theory (Manser, 1965; Macbeth, 1971; Bethell, 1976). Their argument starts from the description of natural selection resulting from the "survival of the fittest"—a phrase coined by Herbert Spencer and subsequently adopted by Darwin. The problem is: how do we know which organisms are the fittest? According to critics, the biologists' only answer is that the fittest are those who do, in fact, survive longest. The "survival of the fittest" thus turns out to mean no more than the "survival of those who survive." Natural selection is reduced to a tautology, a principle that is true not because it is confirmed by the facts but because its components are defined in a way that makes its truth a logical necessity. It contains no more useful information than, say, the statement "All husbands are men," which is necessarily true since "husband" is defined as the male partner. The whole concept of natural selection amounts to nothing more than a play on words, and as such it cannot possibly constitute a workable mechanism of evolution.

Darwinians naturally reject this argument as based on a total misunderstanding of how the theory works (Ruse, 1982), and even some of those biologists skeptical of the complete Darwinian position have joined in condemnation of this particular objection (e.g., Gould, 1977*c*). Natural selection is based on the belief that the fittest do survive longer and reproduce more frequently, but fitness is defined not in terms of survival but as a measure of the organism's ability to cope with its environment by getting food, escaping predators, and so on. Those who have an advantage of this kind in the struggle for existence tend in the long run to survive at the expense of others less favorably endowed, and this is why natural selection develops adaptive characters. If the survival of the fittest were necessarily true, it would be impossible for anyone to imagine that it was false—yet there have been theories of evolution based on random factors that deny the superior reproductive power of the fit. Quite recently, it has been argued that significant amounts of evolution take place by "genetic drift," and the Darwinians had to mount a major campaign to demonstrate that survival and reproduction indeed are correlated with particular characteristics possessed only by certain individuals. The work done on melanism in the peppered moth (chap. 11) is the paradigm for a whole series of research projects that have demonstrated this point. Camouflaged individuals were shown to be missed by predators, and thus these moths survive longer than those with a conspicuous color—a clear indication that survival is linked with certain characters that confer "fitness."

Some Darwinians find it difficult to believe that anyone could take this criticism seriously, but Dawkins (1982) suggests that the problem has been created by the emergence of sociobiology, with its tendency to define fitness in terms of reproductive success. The original Darwinians knew perfectly well that fitness meant an adaptive superiority that conferred some advantage in the struggle for existence. In the modern study of reproductive behavior, however, it has become necessary to adopt a subtler approach; success is measured in terms of genetic representation in future generations. It is much easier to confuse this modern concept of reproductive fitness with sheer survival, although this does not alter the fact that success still is gained through some form of behavior that distinguishes the individual from his rivals. Brady (1979) concedes that in some cases, biologists cannot identify in practice the characteristic that actually is conferring fitness and thus have been forced to define a particular gene as "fit" because it is maintained at a high level in the population. No one doubts that some adaptive advantage is involved, although it is concealed by the sheer complexity of the situation. In these cases, the Darwinians indeed have fallen into a circular argument, although the problem is created not by the tautologous nature of natural selection but by the fact that the theory is sometimes impossible to test in practice.

This leads us to the second line of attack based on the claim that Darwinism is unfalsifiable and hence unscientific. The philosopher Karl Popper has gained his reputation through his efforts to find a criterion for distinguishing science from pseudoscience. Realizing that no generalization can be proved true by collecting positive examples, Popper has argued that science must be based not so much on the search for truth as on the detection of error (1959). A true science exposes all its hypotheses to the test of experiment by formulating them in such a way that any inconsistency with nature will be exposed as soon as possible. Scientific hypotheses are "falsifiable"—while the pseudosciences deliberately make their statements so vague that no counterinstance can ever be found. Measured by this standard, Popper insists (1974), Darwinism turns out to be untestable and hence unscientific. At best it constitutes a "metaphysical framework" for formulating properly testable theories. The accusation is based in part on the apparently speculative nature of many Darwinian explanations of how particular structures might have evolved. It is always possible, argues Popper, to come up with some sort of adaptive explanation, although in the case of extinct forms there is no way of checking whether any particular explanation is valid. These hypotheses, furthermore, always are based on individual circumstances rather than general laws, which implies that Darwinism is not capable of predicting the future course of events. This lack of predictive ability again signifies that the theory does not expose itself to proper testing.

Popper's views command great respect, but Ruse (1977, 1982) has challenged application of the philosophy to this issue. It is necessary to distinguish carefully between the causal theory of evolution and its application to particular events in the course of life's past history. The modern Darwinian mechanism has been tested on numerous occasions through analysis of the genetic structure of populations. The same point is made by Wasserman (1978), who distinguishes between the Darwinian theory of how variation is maintained in balance within a stable population—which is testable—and explanations of macroevolutionary changes—which he claims are not. The majority of Darwinians probably would feel that their theory has to explain change as well as stability or its value is impossibly restricted. Dawkins (1982) points out that different levels of selection tried out by Darwinians give very different predictions about how organisms should behave, and precisely as a result of testing these predictions, sociobiologists have rejected group selection in favor of individual selection. Ruse argues that the strength of Darwinism lies in its ability to link a wide range of phenomena into a comprehensive explanatory system. In many areas, the theory has made predictions about the general character of nature which have been substantiated.

There may be particular areas in which it seems inadequate, but many other scientific theories have been successful despite the existence of a small number of facts that apparently would falsify them. Newton's theory of gravity was accepted despite its incompatibility with the known orbit of Mercury, because in so many other areas it proved a valuable guide to research. We should not expect an active theory to provide ironclad answers to every conceivable question.

Concerning the Darwinian explanations of particular developments in the history of life, Ruse concedes that testable hypotheses cannot always be provided. He points out, though, that there are general trends in the fossil record that are consistent with the theory. It is also easy to imagine discoveries that would falsify the whole evolutionary picture of the past—human fossils in any of the more ancient geological formations, for instance. There has been some debate over whether the general trends in evolution, such as "Cope's law" of gradually increasing size, could be refined to give genuine laws of nature with predictive ability. Such laws could never be tested experimentally because of the time scale involved, but they could be checked against future discoveries in the fossil record. Even in the case of particular adaptive explanations, Ruse points out that tests are sometimes possible, because the hypotheses are based on parallels with living examples of adaptation. Popper himself has softened his position on Darwinism in response to some of these arguments (1978), but a very similar line of attack has arisen within biology itself during the last few years.

The latest source of opposition comes from a new approach to taxonomy or systematics known as cladism. The term "clade" was introduced by Julian Huxley in 1957 to denote a branch of the evolutionary tree. The new technique of classification was pioneered by Willi Hennig (translation, 1966), who insisted that the attempt to represent evolutionary relationships should concentrate on the process of branching, ignoring any changes that were not associated with an act of splitting. The name "cladism" was introduced by one of the movement's critics, Ernst Mayr, and reluctantly was accepted by Hennig's followers. Although Hennig challenged the traditional Darwinian way of linking classification with evolution, he nevertheless considered his technique a means of representing evolutionary relationships. In recent years, the more radical exponents of the new technique have maintained that representation of relationships between forms can be done without reference to evolution. These "transformed cladists" claim that the ancestor-descendant link so crucial to evolution cannot be derived from their way of expressing relationships. Outspoken critics of Darwinism, they have extended the charge that the attempt to reconstruct the past history of life is unscientific and have taken up enthusiastically some of the established arguments against natural selection.

Hennig argued that arrangement of living forms into groups should take into account only the order of branching in the evolutionary process. Clearly, orthodox evolutionary classification itself takes this into consideration, and all evolutionists recognize that the cladistic method offers a more precise way of representing some of the relationships that interest them. An evolutionist, however, also takes into account other factors, principally, the degree of change that a line of development undergoes. If a branch changes to a large extent in comparison with its ancestors, forms now lying at the end of that branch will be assigned to a higher-level taxonomic rank than those at the end of a branch that has changed little. Hennig argued that the degree of change was irrelevant, that only the branching should be used when making up natural groups. He expressed the relationships with a "cladogram" in which the branches allow genuine groupings to be displayed. Common ancestry is depicted by identifying groups linked by shared derived characteristics, that is, characteristics developed in that group alone. Shared primitive characteristics (those derived from an earlier ancestor and shared with other groups) are not taken into account. By allowing systematists to recognize the importance of shared characteristics unique to a group, cladism offers a much more precise way of identifying those characteristics most important for classification. The cladist insistence, however, that all branches derived from a single common ancestor should be assigned to a single group violates many of the traditionally established divisions within the animal kingdom. In the cladist system, the reptiles, for instance, do not form a class of their own, because they share a common ancestry with the birds and the mammals. In the traditional system, the reptile class is defined not by unique characteristics of its own but in terms of the *absence* of characteristics developed by some of its descendants; in the cladist's opinion, this is a meaningless abstraction.

Hennig's cladograms were meant to represent genuine evolutionary relationships. Nodes, at which branches leading toward modern forms divide, corresponded to ancestral forms that might be located in the fossil record. Later cladists, such as Colin Patterson (1980, 1982), have argued that all known forms, extinct and living, should be arranged along the top of the cladogram and that the lower nodes correspond only to idealized groupings of characteristics shared by individual forms. By concentrating on characteristics rather than species, the cladists have emancipated themselves from any concern about the nature of the evolutionary process. A cladogram drawn up in this way does not correspond to an evolutionary tree; in fact, it displays relationships that would be compatible with a number of evolutionary trees. At this point, transformed cladists began to argue that evolution is irrelevant for their attempt to represent the order of nature. Cladograms depict known relationships but do not tell us the evolutionary path by which those relationships have originated. Indeed, cladistic method cannot tell us how to recognize an ancestor-descendant link—it can only de-

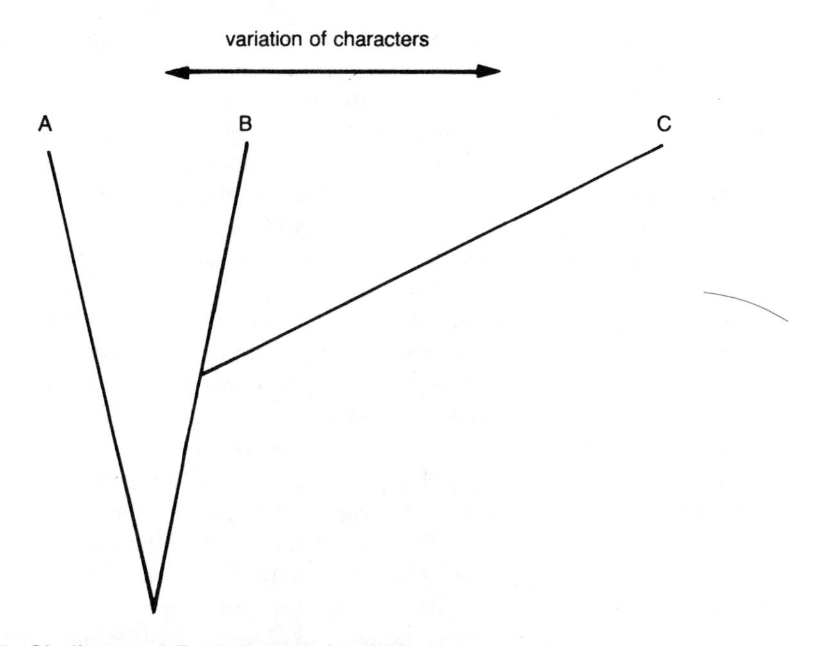

Fig. 25. Cladism and Evolutionary Classification.

The diagram represents two branching points in evolution giving rise to three forms, A, B, and C. The branch leading off to C occurs later, but in the course of its evolution, C has changed much more dramatically than A or B. In the traditional evolutionary classification, A and B would be retained in a single group, but C would be assigned to a separate group to account for the degree of change. Cladism, by contrast, links B and C together despite their differences because they share a common ancestor that A does not. To give a specific example, A might represent modern reptiles, B dinosaurs, and C birds. The evolutionist ranks birds as a separate class but links dinosaurs and modern reptiles. The cladist would regard birds and dinosaurs as the more natural grouping because they are descended from a common ancestor not shared with modern reptiles. Reptiles do not form a natural class because they can be defined only by arbitrarily excluding birds (and mammals) from the natural group formed by descent from a common ancestor.

pict the grouping of "sister" forms derived from an unknowable common ancestor. Shared characteristics can never tell us that one species is descended from another, because they may have been derived either in this way or by the descent of both species from a common ancestor. The transformed cladist thus argues that the whole effort to reconstruct the past course of evolution is based on inadequate knowledge. No study of natural relationships can ever fill in the missing information, hence many of the evolutionary trees that have been constructed over a century of research are untestable. The cladists thus have added a new line of argument to the claim that evolution theory is unscientific.

As yet, transformed cladists represent only a small minority of systematists. They have been vigorously resisted by evolutionists, who argue that for all its technical sophistication, their approach constitutes a threat to science. David Hull (1979) points out that the cladist attack is based on a methodological argument: because the technique does not allow evolutionary relationships to be investigated, evolution is unscientific. Such an attack is of dubious value since it presupposes that the present technique is the only one available. How do we know that future improvements will not allow evolutionary relationships to be incorporated into an equally rigorous method of investigation? Some aspects of the cladists' technique impose arbitrary restrictions that they have interpreted as facts of nature, for instance, the depiction of branching as always dichotomous (splitting into two). There is nothing in evolution theory that prevents multiple speciation, yet cladists insist that it does not happen because their diagrams cannot represent it. Relating the whole debate to Popper's claim that evolution is unfalsifiable, Hull points out that the argument can be applied in another way. Evolutionary trees are less certain than cladograms, but they specify more details about nature and thus are *more* exposed to potential falsification. At a practical level, evolutionists also argue that the presuppositions of Darwinism are useful in working out the way new characteristics are formed in nature (Ridley, 1982*b*).

Public controversy over cladism came to a head in 1981 with an exhibition on the origin of man prepared by the Natural History Museum in London. Exhibition literature followed the cladist line by emphasizing the uncertainty of evolutionary trees—even taking into account the important new discoveries of hominid fossils. Evolutionists responded and the debate raged for some months, principally in the form of letters to the editor of *Nature*. Because of its emphasis on splitting, cladism was linked with punctuated equilibrium theory, and both were dismissed as products of rampant Marxism. Perhaps with more justification, it also was argued that the cladists were giving encouragement to the creationists in their bid to undermine evolutionism. Transformed cladists certainly do not assert that species were miraculously created, but their claim that evolution is impossible to confirm would imply that both evolutionism and creationism are on the same scientific level. The only interest of the cladists is in setting up an abstract pattern of natural relationships; there is no concern for how the various forms were produced. Most biologists still feel that science has an obligation to propose hypotheses about the causes that may have produced the species we observe. These hypotheses refer to past events and will not be as easy to test

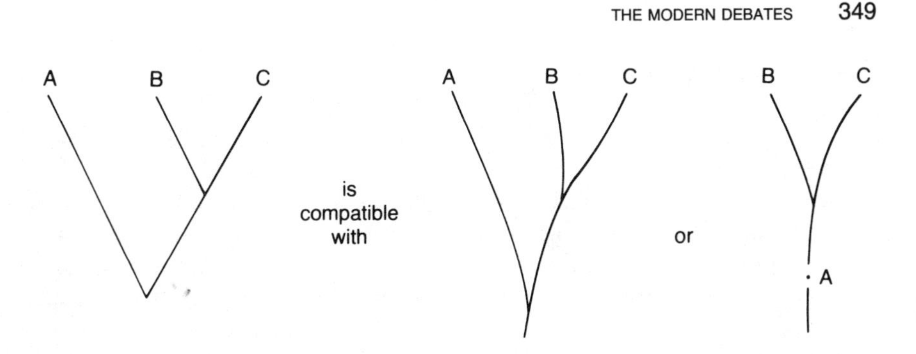

Fig. 26. Cladograms and Evolutionary Trees.

The cladogram on the left represents a group of three related species in which B and C share characteristics that are not possessed by A. The cladogram would be compatible with either of the two evolutionary trees on the right. A might be ancestral to B and C, or it might lie off on a side branch of its own. The cladogram thus cannot give information on ancestor-descendant relationships, hence the transformed cladists' rejection of evolutionary trees as unfalsifiable. This, of course, is the simplest possible case: a similar cladogram linking four forms would be compatible with twelve possible evolutionary trees (see Patterson, 1982).

as cladograms, but they *can* be tested in many indirect ways. To abandon the search would be to leave the whole question of the past history of life on earth a permanent mystery.

The cladist argument strikes far deeper than Popper's objections. Although aimed at modern Darwinism, Popper's critique would apply to any theory that sought to explain organic structures in terms of adaptation. In Hull's terminology (1979), Popper objects to setting up "evolutionary scenarios" based on hypothetical adaptive pressures. Whether Popper realizes it or not, the arguments would apply whatever the mechanism of adaptation, Darwinian or Lamarckian. Such scenarios are the most speculative product of evolution theory and will remain the most difficult to test, whatever future improvements may be made in our understanding of the mechanism of change. The cladists go beyond this to criticize the construction of evolutionary trees, and their attack thus is aimed at the whole concept of genetic relationship between species, even when no mechanism is specified. For this reason, their position tends to rank evolutionism on a par with creationism: both the existence and the lack of genetic relationship are equally untestable. Modern creationism, however, is just one of the alternatives to the orthodox scientific view of life's history; there are others that cannot be dismissed so easily as totally unscientific.

EXOTIC ALTERNATIVES

In their most extreme forms, these alternatives are certainly dismissed by the scientific community as wild speculations, intended to exploit the public's preference for spectacular ideas. Immanuel Velikovsky's catastrophist theory of the earth's history was bitterly resented by some scientists when it first appeared. Although tempers have now cooled, his views are still not taken seriously by the experts, much to the annoyance of a devoted group of supporters. Erich von Däniken's "space gods" thesis is similarly repudiated by professional archaeologists and scientists. The advocates of these unorthodox interpretations of the past invariably complain that their work is ignored by the professionals out of blind prejudice. Yet some elements of their beliefs are taken up in a more rational form by scientists of considerable reputation and have been discussed seriously in the academic literature. The willingness of a few scientists to "rock the boat" by examining the plausibility of catastrophes or extraterrestrial involvement shows that there is no absolute barrier between orthodox and heretical viewpoints.

Velikovsky's catastrophist theory of the earth's recent past was first advanced in his *Worlds in Collision* (1950) and has been expanded in a number of later books (e.g., 1952, 1955); all of them continue to be reprinted. Their starting point is the assumption that the records of great upheavals in the mythologies of many ancient peoples refer to real events. Planetary collisions are supposed to have wreaked havoc on the earth on several occasions within the last few millennia. Velikovsky backed up his ideas with geological evidence and a suggestion that electromagnetic forces might account for interplanetary movements inexplicable by standard astronomical theories. These collisions drastically altered the earth's axis of rotation, causing rapid changes of climate and the extinction of many species. It also was suggested that new species would appear very quickly as the result of mutations caused by cosmic radiation. The complete system challenged so many areas of modern science that it was greeted with skepticism and downright hostility by many professionals. Efforts were made to block the publication of Velikovsky's first book, a move that backfired because it made him a martyr in the eyes of the counterculture (De Grazia, 1966). More recently, his ideas have been debated openly, although few scientists or archaeologists have found in them a consistently satisfactory explanation of the facts (Goldsmith, 1977). Outside the academic community, Velikovsky's ideas still command wide support (Ransom, 1976; Talbott, 1976).

Velikovsky's catastrophism resembles the earlier geological theories we have studied (chaps. 2 and 5); indeed, much of his evidence is quoted directly from older texts. Some of the catastrophes mentioned refer to ancient geological periods long before the sequence of astronomical events that Velikovsky postulates. His target is obviously the uniformitarian orthodoxy that has dominated geology for a century or more. In order to make a more plausible case for catastrophes in recent times, he is prepared to turn the earth's entire history into a series of unexplained upheavals. Not surprisingly, geologists found it difficult to take his attack seriously, just as astronomers were unwilling to see the solar system as an interplanetary game of pinball. Much evidence once cited as proof of geological catastrophes can now be accounted for in terms of gradual changes. More important, the fossils indicating that parts of the earth once enjoyed different climates have now been explained by the theory of continental drift. The new geology gives a far better picture of the changing distribution of life around the earth than was possible before. Far from confirming Velikovsky's interpretation, major developments of recent years have established the principle of gradual change more firmly than ever.

At the same time, geologists have become more willing to concede that occasional catastrophic events may have punctuated the otherwise gradual development of the earth's surface. Mass extinction of the dinosaurs and many other forms at the end of the Mesozoic has long been something of a puzzle, although various efforts have been made to explain the event as the culmination of a gradual trend. It now has been suggested that the collision of an asteroid with the earth may have raised a vast dust cloud capable of producing temporary climatic effects of such severity that many forms of life would not have been able to survive. Evidence for this event is found in the presence of the element iridium in a thin layer of rock separating the upper Mesozoic and lower Cenozoic formations; iridium is far more abundant in certain kinds of meteorites (Russell, 1982; Allaby and Lovelock, 1983). Some writers argue that catastrophes might be responsible for other major changes in the earth's history. Fred Hoyle (1981) believes that the ice ages were triggered by meteorite impacts. Clube and Napier (1982) invoke periodic collisions caused by the influx of comets into the solar system from interstellar space and accept Velikovsky's claim that such an event could have occurred during human prehistory. On this view, Velikovsky's general catastrophist arguments are valid but have been obscured by his wildly improbable mechanism of planetary collisions.

Velikovsky and his followers assume that the evolution of religion was determined by the spectacular changes our ancestors observed in the heavens. A very different means of explaining the traditions that record the behavior of "gods" is advanced by Erich von Däniken, who argues that the origin of man himself, and of the earliest human civilizations, was the result of a visit by extraterrestrial beings. His *Chariots of the Gods?* (translation, 1970) aroused widespread interest with its efforts to prove that many ancient monuments could only have been built with space-age technology. Archaeologists have ignored his claims, and it is probably fair to say that his whole argument is built on a persistent refusal to credit our ancestors with the skill and patience needed to complete large projects (White, 1976). Von Däniken also claims that the human race itself is an artificial product of extraterrestrial skill at genetic engineering. In a later work (1977), he extended this argument to the whole course of evolution, postulating a series of visits by extraterrestrials to account for the sudden appearance of new types in the fossil record. He cites many of the anti-Darwinian arguments used by creationists to make a case for some form of interference with the natural course of evolution. The apparently sudden events that the creationists see as miracles are interpreted by von Däniken as genetic experiments by the "space gods."

However skeptical we may be of von Däniken's overall thesis, the possibility that an extraterrestrial civilization might have originally "seeded" the earth with the first forms of life has been supported by Francis Crick (1981). Since his pioneering role in elucidating the structure of DNA, Crick has worked on the problem of the origin of life, and he remains suspicious of the orthodox theory of chemical evolution. He argues that the complexity of life is so great that chances of even the simplest structures arising naturally on the primitive earth are too remote. To evade this difficulty, he draws on the currently popular belief among astronomers that planets suitable for life may exist in large numbers throughout the galaxy (Shklovskii and Sagan, 1966). Some of these planets are far older than the earth, and some may have conditions even more suitable for the origin of life. There is thus a chance that intelligent life with a technological civilization could have evolved within the galaxy in the remote past. If the science fiction concept of interstellar travel is impossible, such a civilization may have decided to send out the easily preserved spores of primitive organisms in order to start the process of evolution on as many planets as possible.

Along rather different lines, the astronomers Fred Hoyle and Chandra Wickramsinghe (1978) suggest that life first appeared among the vast amounts of organic matter that they believe fill interstellar space. In space, the sheer amounts of suitable chemicals would allow the extremely unlikely combinations that give rise to life to form more readily. Eventually, primitive organisms would find their way down to the earth's surface to begin the process of evolution. More recently, Hoyle and Wickramsinghe (1981) have postulated that genetic material raining down on the earth from space occasionally may have been incorporated in existing forms of life, yet another explanation of the apparently sudden episodes in evolution. They now claim (1986) that the classic *Archaeopteryx* fossils, used to support the theory of the gradual evolution of birds from reptiles, have been tampered with to add the impression of feathers. These ideas have not been taken seriously by biologists, and there is now some doubt about whether the interstellar material consists of the chemicals required by the theory.

These ideas lie at, or beyond, the limits of scientific investigation. Many scientists would reject them as totally unnecessary speculations and dismiss most of their authors as nonspecialists who have no real understanding of the areas into which they have ventured. The public, they feel, is all too easily seduced by spectacular stories of catastrophes and extraterrestrials and has no real appreciation of how little these speculations agree with orthodox theories at the level of specialist investigation. It is one thing to pick out difficulties with existing ideas but quite another to come up with something that is really better. The appearance of these alternatives in the literature, nevertheless, raises some important points. It shows, for a start, that radical ideas not only can circulate but also can command the support of reputable scientists. However suspicious the profession may be of popular authors out to found a new cult, specialists are sometimes willing to explore parallel hypotheses that might be expressed in a more reasonable form. New ideas have a chance to be considered seriously, even if their prospect of success is slight. The apparent conservatism of scientists comes not from short-sighted dogmatism but from deep knowledge of just how many barriers a new idea must overcome before it shows any superiority over well-accepted theories. The opponents of Darwinism have always dismissed the theory as a blindly accepted dogma, its weaknesses ignored by a brainwashed biological profession. Yet we already have seen (chap. 9) that Darwinism has been challenged seriously before. The wealth of alternatives being considered both inside and outside science today shows quite clearly that there is no academic conspiracy to protect modern evolutionism.

Every scientist knows that existing theories cannot solve all the problems confronting them, but it will be difficult to find an alternative theory that does not face even greater difficulties. The complexity of the situation is illustrated by the wide range of unorthodox ideas that have been suggested. Critics who lambast professional scientists for ignoring the critics' own pet theories are themselves often guilty of failing to acknowledge the other alternatives under discussion, a point worth bearing in mind as we turn to the most active source of anti-Darwinism in the modern world: creationism. Even if "creation science" were shown to be a legitimate subject deserving equal time in the schools, could not the same case be made for all the ideas mentioned above? Some of them also enjoy the support of many ordinary people and at least a few professional scientists. Should they not also be given equal time to present their case? It is one thing to mention alternatives to orthodox theories, but to give each of them an equal share in the school curriculum would leave hardly any time to teach the interpretation accepted by the vast majority of working scientists.

CREATIONISM

Religious opposition to teaching evolution in the schools is not a new phenomenon in America. The belief that evolution was the spearhead of an atheistic attack on religion and morality was widespread in the "Bible belt" states over fifty years ago. Efforts to ban teaching evolution in the schools culminated in 1925 with the passing of Tennessee's notorious Butler Act. It was under this act that John Thomas Scopes was prosecuted for teaching evolution in the high school at Dayton, Tennessee. In the much-publicized "monkey trial," Scopes was prosecuted by the fundamentalist politician William Jennings Bryan and defended by the agnostic lawyer Clarence Darrow (Ginger, 1958; Scopes, 1967; de Camp, 1968; Settle, 1972). In the end, Scopes was convicted and fined, but the resulting publicity exposed fundamentalists to national ridicule. It often has been assumed that the event was a triumph for evolutionists, with the Butler Act and similar legislation in other states gradually sinking into oblivion. In fact, fear of controversy forced textbook publishers to eliminate direct references to evolution, and the subject hardly was mentioned in the schools for several decades afterward (Grabiner and Miller, 1974). One noted biologist eventually complained that "one hundred years without Darwin are enough" (Muller, 1959), and a movement began to introduce evolution as an integral part of the science curriculum in American schools. Modern creationism is a direct response to this challenge (Numbers, 1982).

The new creationism has changed its tactics and no longer seeks to get evolution banned from the schools. Instead, it attempts to show that evolution theory has so many weaknesses that a creationist interpretation of the facts is just as plausible. The demand is now that "creation science" be given equal time alongside evolution in school biology classes (Nelkin, 1977). Efforts have been made to persuade several states to pass the appropriate legislation, but so far the American Civil Liberties Union has been successful in blocking these moves by arguing that creationism is based on religious principles and thus violates the constitutional separation of church and state. Equal-time legislation in Arkansas was declared unconstitutional in 1982 amid much public debate, and a more sophisticated bill in Louisiana has now been struck down by the Supreme Court. A similar movement is at work in some parts of Canada, while in Britain, a smaller band of creationists conducts guerilla warfare against evolutionism.

The creationists make no secret of their distaste for the "atheistic" philosophy of evolution, but they present their case in an ostensibly scientific form that does not refer explicitly to the Genesis story. Some of their publications are aimed at the general public, while others are meant as alternative textbooks in the schools (Whitcomb and Morris, 1961; Watchtower Bible and Tract Society, 1967; Moore and Slusher, 1970; Gish, 1972; Morris,

1974; Kofahl and Segraves, 1975; Parker, 1980). Their argument is based on an attempt to show that biologists have greatly exaggerated the success of evolution theory. In fact, the theory faces a number of major scientific problems, which would all disappear if the possibility were accepted instead that species are miraculously created. Many arguments are directed against natural selection, which creationists see as the heart of evolutionary materialism because of its emphasis on a trial-and-error sorting of random mutations. They argue that such a haphazard process could never generate the complex structures of living things, and they have revived the argument from design in the form once employed by Paley. The fossil record is brought in on the assumption that gradual development of new forms must have left evidence in the rocks if it occurred at all. Instead, they argue, we see only the sudden appearance of new types and classes. Because they believe that man can have no connection with the apes, creationists are particularly anxious to discredit the fossil intermediates found recently in Africa.

The creationists are adept at turning the scientists' own words against them. If any evolutionist has expressed doubts about the theory's application to a particular problem, creationists quote these doubts as though they expressed an overall verdict on the theory itself. If evolution is really scientific, they argue, its supporters should have no doubts to express, because they should have a complete explanation for everything. Creationists have gleefully seized on recent debates in which evolution has been branded as untestable and hence unscientific. Because these claims have been made by reputable philosophers and biologists, they appear to give independent support to the creationist position. If evolution itself is untestable, then miraculous creation cannot be dismissed for postulating events outside the scope of direct investigation.

Until recently, evolutionists have concentrated largely on defending themselves against these criticisms. They argued that creationists have greatly exaggerated the problems facing the theory in order to claim that the origin of life will never be explained in terms of natural causes. Many of the creationist objections are based on outdated or distorted interpretations of both the theory and the evidence. Far from being on the brink of collapse, the evolutionary theory is seen by its supporters as based on a wide range of supporting evidence, and they insist that to invoke miracles at this stage is to abandon over a century of successful scientific research. The creationist image of natural selection has been grossly oversimplified, so that it is easier to discredit before an audience of nonspecialists. Like most opponents of Darwinism, creationists portray it as a selection based on individual mutations, most of which are harmful. This ignores the modern concept of the gene pool as a reservoir of genetic variability that allows a wide flexibility of response to environmental challenges. Recent fossil discoveries, such as those in the Precambrian, are ignored; and classic examples of intermediate forms are presented in a distorted manner to conceal their true character. Finally, in response to the modern accusation that evolution is unfalsifiable, it can be pointed out that even Popper has recanted many of his earlier claims, while the cladists represent only a handful of biologists with a very narrow definition of science.

As the creationist challenge has become mixed up in the general debate over the nature of evolutionism, biologists have begun a more active campaign to defend the theory (Eldredge, 1982; Godfrey, 1983; Halstead, 1983; Kitcher, 1982; Montagu, 1982; Newell, 1982; Ruse, 1982; Futuyma, 1982). In part, these books present a more detailed, factual response to creationist attacks, but increasingly evolutionists are going on the offensive by exposing the weaknesses of creationism itself. They no longer content themselves with reiterating the evidence for evolution but go on to show how difficult it is to reconcile the facts with the bizarre alternative creationists propose. This tactic has been made necessary by the very one-sided argument creationists have adopted. A striking fact about their literature is the amount of space it devotes to attacking evolution, rather than expounding the creation alternative. Lengthy arguments of varying degrees of plausibility are developed against evolution, and the reader is left to assume that if evolution is in such a poor state, creation must surely be preferable. Yet the details are never supplied of how a scientist might set up a comprehensive explanation based on the Genesis story. This remarkably successful technique has put evolutionists on the defensive without exposing creationists to any equivalent danger of attack.

The importance of getting creationist theory into the open to be properly examined may seem obvious, but this move raises deeper questions about the nature of science which may force evolutionists to reassess their position. Creationist tactics have succeeded because they could exploit the oversimplified public image of science as a process of fact-gathering. If evolution is scientific, then it must be a "fact," and its supporters should be able to prove that their explanation of every aspect of the history of life is adequate. If it can be shown that there are any gaps in the case for evolution, any questions that cannot be answered, then it is not an established fact and should not be taught as such in the schools. Inevitably, with a theory of this magnitude, there are some areas where the evolutionist cannot make a satisfactory response to this line of attack. We do not have a complete theory of the origin of life, and there are indeed some gaps in the fossil record where we cannot see how evolution occurred. Once these points are conceded, the creationist feels himself in a position to dismiss the whole evolutionary approach as inadequate—at least when he is addressing an audience of people who have been led to believe that evolution is an established fact.

The evolutionists themselves may have helped to build the trap they have been led to by this line of argument. All too often they have claimed that evolution is a demonstrable fact and have traded on the general reputation of scientists for infallibility. Yet, clearly, evolution is not a "fact" in the sense that the man in the street understands the word. Without a time machine, we cannot *prove* that birds evolved from reptiles; we can only show that the known fossil record is consistent with this belief. Nor can we prove that natural selection is the mechanism responsible for the whole development of life on earth, which is why alternatives—such as punctuated equilibrium—are being considered by some biologists. The creationists have hailed these new alternatives to the Darwinian synthesis as a sign that the whole evolutionary program is beginning to break down. The public, remembering the confidence with which biologists once proclaimed that they had solved all the problems, can be forgiven for taking this claim seriously. Yet proliferation of new hypotheses is actually a sign of vitality within the basic evolutionary perspective. It shows that new ideas are being considered and new lines of research opened. To understand why this ferment of activity is a healthy sign and why creationism cannot hope to match it with research initiatives of its own, the public must be given a more realistic idea of how science works. They must also be shown the difference between facts and theories and given a better understanding of the true status of scientific knowledge.

To make these points, however, evolutionists must be willing to concede that they do not have a completely adequate theory. A scientific theory is not a fact: it is a working hypothesis of great explanatory power employed because it continues to guide research along fruitful lines. No theory can be expected to have all the answers to problems that confront it, and hence no theory can ever be secure against criticism. The very idea of scientific research depends on the existence of unsolved problems and thus on the admission that the present state of knowledge is inadequate. If ordinary people can be persuaded to adopt this more sophisticated view of scientific activity, they will realize that evolutionists cannot be expected to answer all the questions put to them. They will also realize that a decision to abandon evolutionism in favor of creationism could only be justified if creationism shows that it can be elaborated into a comprehensive explanatory system that can serve as a better guide to research. It may sound paradoxical, but the only way creationists can be forced to expose their own extraordinary theory to close analysis is if evolutionists concede that they cannot prove their own case. It is only by admitting the lack of certainty inherent in any theory that we can shift the argument to a new level where creationists can no longer get away with merely attacking evolution but are forced to present their own alternative for detailed scientific scrutiny.

The creationists argue that it is only fair to give their alternative equal time in the schools, but if this demand were accepted, it would seem equally reasonable to insist that they spend the time expounding the substance of their own beliefs. Because they have concentrated so heavily on attacking evolution, it is not always clear exactly what they do stand for. At first sight, it would seem that their only concern is to establish the possibility that the miraculous creation of new species could have occurred at some point in the past. Much of their literature takes the fossil record for granted and implies that it reveals the sudden appearance of new types in a manner more consistent with creation than with evolution. This would represent a position very similar to that of nineteenth-century catastrophists (chap. 5)—accepting the sequence of geological periods in the earth's history and arguing merely for the sudden creation of new forms at certain points in time. Understood in this way, the attempt to turn creationism into science implies only that biologists should explore the basic notion of miracles within an otherwise orthodox version of the earth's past. Yet when we expose the details of their proposals, we find that the majority of modern creationists adopt a much more rigid position, one in which the creation is interpreted in literal biblical terms—as a single event taking place only a few thousand years ago. To turn this into a scientific theory would require the elaboration of an alternative interpretation of all facts encompassed by those sciences dealing with the past history of the earth and the universe. If this position could be established, evolution would disappear almost as a casual by-product of the massive reduction in the time scale of the earth's history. For this reason alone, arguments against evolution are quite irrelevant in the general debate between the scientific and the biblical interpretations of the past.

Can the basic idea of divine creation be turned into a workable scientific theory of the origin of species? It must certainly be admitted that miraculous creation is *compatible* with what we know. There is no way of disproving the claim that new species were created at certain points in the earth's past. Anyone who believes in the existence of a God who could and would interfere with the normal operations of His universe thus is free to assert that He created all (or perhaps only some) of the forms that appear in the fossil record and in the world today. However, there is no way of *proving* that miracles occurred, or even that they can occur, because supernatural events of this kind do not appear to have taken place during the period of recorded human history. Creationists argue that evolution theory is so unconvincing that it is more sensible to reject the whole idea of a natural explanation for the origin of life; but this is a purely negative argument. If we concede that present knowledge is imperfect but insist that research is capable of improving our level of understanding, there is no guarantee that the future will not see the outstanding problems of evolutionism resolved. Neither side can offer proof, so the question must be decided on the basis of which approach offers the best hope of increasing our knowledge of how life is distributed around the globe and through the fossil record. At this level, it proves very difficult for the concept of miraculous creation to function as an effective scientific theory.

Consider first the question of the mechanism by which new species are formed. Perhaps the modern synthesis of genetics and natural selection is deficient in some respects, but biologists are already devising new mechanisms to supplement Darwinism, and their research is guided by the effort to determine which hypotheses work out best in practice. It seems obvious that creationism can offer no parallel research projects at this level, because miracles are by definition supernatural events in which the laws of nature are transcended in a manner inexplicable to the human observer. If we accept this mode of origin for new species, we commit ourselves to the belief that we shall never understand what actually happens in the process. Creationism is a counsel of despair as far as this aspect of biology is concerned. It encourages the scientist to abandon all research into the question because he has no hope of answering it in rational terms. Such a negative policy can hardly be described as part of an active scientific program. In fact, no area of science would be able to progress if every difficulty were treated as an excuse to give up research because the problem might be impossible to solve. Creationism is not a scientific approach to the origin of species; it is a call to give up scientific research into that particular question.

To begin to illustrate this point, it is perfectly conceivable that only *some* species were miraculously created, while others were produced by natural evolution. Unless we invoke revelation—which a scientific creationist is not supposed to do—we have no reason to suppose that every distinct form was individually created. Even hard-line creationists admit that the Galápagos finches were formed by natural divergence but argue that they all are still the same "kind" of bird. Yet biologists classify these finches not only into separate species but even separate genera, which surely gives natural evolution a fairly wide scope. A few more-sophisticated creationists (e.g., Shute, 1961) argue that only major steps in the history of life required miracles, that smaller developments were filled in by natural means. But if we adopt this view, how do we determine at what point a miracle must be postulated? Is a miracle required to introduce a new order, a new class, or perhaps only a new phylum? The only way of finding out where this limit lies is to keep trying to extend the natural explanation as far as possible, because it is impossible to work actively with the concept of miracle. It thus can be argued that even someone who believes in the possibility of miraculous creation should conduct his research on the basis of evolution theory. If he adopts any other policy, he runs the risk of giving up too soon and postulating a miracle where a little more research might have revealed a natural explanation.

It also would seem that creation can offer no guide to the study of the fossil record or the geographic distribution of species around the world.

These fields would be reduced to a purely descriptive level, because if we appeal to miracles, we can never explain why a particular species appeared at the time or place it is found. Why are there discontinuities at some points in the fossil record and regular trends elsewhere? The creationist can only point to the inscrutable will of God. He can never predict what new fossils might be found to link known forms, because he is predisposed to assume that any gap in the record indicates a genuine discontinuity in the advance of life. If an intermediate eventually turns up, the creationist merely adds another miracle to his list. Nor can he offer any natural explanation of why species exist in the locations they presently occupy: each was formed wherever the Creator chose. Why were kangaroos and other marsupials created in Australia, instead of placental mammals, especially as the climate of that continent has proved eminently suitable for mammals introduced by man from Europe? No rational answer can be given to such a question. The whole program of research that has tried to explain geographic distribution in terms of isolation, migration, and evolution would have to be abandoned if we admitted the possibility that species could appear anywhere at the whim of the Creator.

It thus can be argued that even the most flexible form of creationism does not offer a suitable basis for a scientific theory of the origin of species. It cannot be proved wrong, but it can offer no guide to research and limits rather than extends our understanding. Then what of the fundamentalist version of creationism so popular today, in which the acts of creation are supposed to have taken place exactly as described in Genesis? Can any legitimate scientific procedure lend support to the claim that the earth and all its inhabitants were formed in a short period of time only a few thousand years ago? This would require not only rejection of evolution theory but also complete rewriting of the sciences of archaeology, paleontology, geology, cosmology, and even physics itself. It is difficult to believe that the creationists themselves have any idea of the magnitude of the task they have set themselves. They have to provide workable theories to cover this whole vast range of knowledge, theories that can function at least as well as those they wish to replace. Instead, they seem limited to a few pinpricks of criticism aimed at the orthodox theories, along with vague suggestions for alternatives that fall apart as soon as they are subjected to close analysis.

It is impossible here to cover all the areas that creationism must try to cope with. Let us take a single example: the sequence of geological formations and the fossils they contain. The creationists are forced to deny that this sequence is the product of a vast period of earth history. Instead, they assert that all sedimentary rocks were laid down during the great flood and that the fossils are the remains of creatures drowned in the catastrophe. Distribution of fossils in the rocks is caused by the fact that primitive forms such as trilobites were killed very early, while more advanced creatures were able to climb up the mountains and only later were drowned. No working geologist has taken this idea seriously since the early eighteenth century (chap. 2), and even Werner's more sophisticated Neptunist theory was abandoned soon after 1800. Even geologists with sincere religious beliefs were forced to concede that the evidence pointed to a sequence of periods in which the rocks were laid down, accompanied by vast movements of the earth's crust. There are many cases in which earlier rocks have been twisted and eroded before newer rocks were laid down on top of them. Very ancient rocks are frequently found at higher elevations than younger rocks. There are intrusions by molten rocks into the sedimentary formations, sometimes on an enormous scale. Occasionally, we find strata composed of volcanic ash, not waterborne sediment. All of these facts are consistent with the modern view of the earth's history and are impossible to reconcile with the idea of a single episode of deposition from a universal ocean.

The creationist explanation of the fossil record runs into similar difficulties. Many forms known only from the more ancient rocks were aquatic creatures, so why did some of them not survive long enough in the flood to be entombed in later deposits? In all the turmoil of the deluge, would we expect *any* regular sorting of the dead bodies before they sank to the bottom, bearing in mind that corpses can float for a considerable time after death? How are we to explain the occurrence of fossil footprints of dinosaurs and other terrestrial creatures? Were they able to walk around on the bottom of the great ocean as sediment was accumulating? It is very easy to throw out the casual suggestion that the rocks could have been laid down in a deluge; but it is far more difficult to substantiate this claim in the face of the kind of detailed criticism that creationists eagerly apply to evolution theory. Unless they can come up with a theory that provides satisfactory answers to these questions, they are not in a position to insist that their views are plausible as science. They must provide an explanatory system at least as comprehensive and as practically useful as the existing body of earth science. When oil and mining companies will hire geologists trained in the flood theory because they can locate minerals more skillfully than their orthodox counterparts, then creationists will be able to say that they have a better interpretation of the earth's past.

Finally, what are we to make of the story of Noah's ark, which the creationists are forced to take seriously as the only means by which terrestrial life could survive a universal deluge? The fact that so many species were left to perish is itself a warning that we cannot take the biblical story literally, because God commanded Noah to take specimens of all living things onto the ark. Apart from the difficulty of keeping all the animals in a confined space for a considerable time, what happened after the ark grounded? How did the carnivores live, except by killing off the sole survivors of the prey species? How did the heavier animals move around at all, considering that most of the earth was covered by thousands of feet of fresh sediment, which surely must have taken a considerable time to harden? To adapt an earlier example, why did marsupials such as the kangaroos make only for Australia, while no placental mammals were able to reach that continent? How did all the animals travel vast distances from the ark's resting place? Imagine, for instance, the koala, which is hardly adapted for long-distance walking and has a very specialized diet, trekking all the way from Mount Ararat to Australia! The fact that many creationists insist on incorporating such a highly implausible episode into their position is clear evidence that their approach is based on the Bible rather than on an objective consideration of the facts.

In view of these difficulties, creationists might begin to reconsider their demand for equal time in the schools, if it were made clear that their time must be devoted to detailed consideration of their own proposals, rather than an attack on evolutionism. Even if it were agreed that school biology classes should deal with the problems encountered by evolutionists, these should not be presented to imply that a simple leap to creationism was the only alternative. Imaginative science teachers might welcome the opportunity to discuss other ideas being debated by researchers. There are, however, valid pedagogical reasons for concentrating on the strong points of existing theories when teaching students who have to pass examinations. All scientific theories have some areas of weakness at the frontiers of research, but we do not force junior students to grapple with these problems even before they have learned to apply the theory in areas where it is successful. The creationist demand thus raises very important issues about how science should be taught in the schools. Should we retain the present method of teaching accepted theories as "facts" in order to simplify the pressures on the students? Or should we admit more openly that science is a process of discovery in which research actually changes our understanding of how things work? The teaching profession must resolve this debate in its own terms, but there is no excuse for imposing the more flexible approach by law in the single area of evolution theory. Creationists may object that the moral issues here are so important that students must be made aware of a less materialistic alternative to Darwinism. Even accepting the idea that ethical questions should determine how science is taught, though, there is no reason why biblical creationism should be presented as though it were the only conceivable alternative. Once we begin to discuss the broader issues, all possibilities deserve to be mentioned—and it is difficult to see why biblical creationism should be given more time than the others.

The most disturbing aspect of modern creationism is not its implausibility but its open admission that the whole theory is based on the Book of Genesis. For all their talk of "scientific creationism," members of the Creation Research Society are obliged to declare their belief in the literal truth of divine revelation. It is difficult to see how such a deliberate commitment to a particular interpretation in advance of any investigation can be seen as part of a scientific program. There is no possibility of change or development in the creationist scheme, because for its adherents the truth is already known on extrascientific grounds. In practice, this gives their opponents another weapon to use against them, and great success has been obtained in some public debates by simply pointing out that the sacred record does not offer a consistent story of how creation occurred. Biblical scholars now recognize that Genesis is a compilation from different sources (see Asimov, 1981), which give different accounts of the creation and the deluge. Was man created first or last (compare Genesis I and II)? Did Noah take two or seven of each "clean" species on the ark (Genesis VI and VII)? The vast majority of religious thinkers realize that such discrepancies are inevitable in a record that was assembled for the purpose of religious and moral instruction, not as a science textbook. Whatever the merits of the text for its intended purpose, it cannot be used as the basis for a detailed theory of the earth's history, because it is inconsistent both within itself and with the facts of nature.

The link with Genesis has allowed creationism to be challenged as a disguised form of religion, specifically, of fundamentalist Protestantism. If this is accepted, then its teaching in the schools would indeed violate the American constitution. But our reasons for rejecting the demand for equal time must not be based on narrow, legalistic grounds. They must reflect a distaste for seeing the attitudes and values of a particular religion taught in schools where many of the students do not come from backgrounds in which this religion is accepted. Many Christian denominations do not believe that the Bible must be taken literally on scientific matters, and they condemn fundamentalists for their oversimplification of the relationship between God and man. In America, Canada, and Britain, large numbers of immigrants from non-Christian countries would quite rightly oppose any attempt to teach their children that Genesis is supported by science. Some of the world's great religions have no problems with evolution, because they do not include the notion of a single act of creation. Others have their own particular difficulties, as does fundamentalist Islam. For three years, I taught in a Muslim country (Malaysia) and encountered the resentment of some Muslim students who saw evolution as incompatible with the word of God. Yet for them divine revelation meant the Koran, not the Bible, which has its own rather different account of the creation of man. In a pluralistic society, we cannot accede to the demands of those who see Genesis as the only source of divinely inspired knowledge about the creation. Must Koranic creationism also be given equal time in some of our schools, or should we learn from this the need to keep at arm's length any group that claims to have prior knowledge of what science *must* discover?

The creationists are quite right to oppose evolution if it is taught as the basis for a completely materialistic philosophy. The time is past when scientific experts could appeal to a spurious infallibility to back up their own moral preconceptions. But a return to the literal word of the Bible cannot solve the dilemmas of modern life, any more than it can solve the scientific problems of evolutionism. Creationists attack Darwinism because its trial-and-error mechanism seems particularly damaging to the argument from design. Yet in political terms, Darwinism always has been linked with the far right—which is also where most of the fundamentalist sects align themselves. Lamarckism is the mechanism of evolution that has been hailed as a means of retaining a role for design, but it has a long history of association with the political left. The basic idea of evolution, therefore, cannot be identified with any particular philosophy or ideology. Where one observer sees only the blind struggle for existence, another sees a divinely preordained ascent toward perfection. Even Darwinism has not always been branded as atheistic. Moore (1979) points out that in Darwin's own time, a few strict Calvinists welcomed the selection theory, because they were suspicious of the argument from design and believed that a universe governed by so harsh a law of development was a fit place for sinful man. Any system that assumes that nature gives us direct information about the purpose of life runs the risk of limiting human freedom by lending support to the claim that we *must* follow the prescribed rules—or face the consequences in this world or the next. Philosophers such as John Dewey and William James have pointed out (chap. 8) that Darwinism can be interpreted to encourage freedom, precisely because it does not predict a goal toward which nature and man must inevitably progress.

Bibliography

Despite its size, this is by no means an exhaustive bibliography. Its aim is to provide an outline of the most easily accessible primary sources and a good introduction to the secondary literature. Sarjeant (1980) is a comprehensive bibliography covering the development of geology and paleontology, with much additional material on evolution theory. Wherever possible, I have included English translations of works published originally in a foreign language and modern reprints of books published before 1900. Facsimile reprints of primary sources are cited under the original date of publication, but modern editions of older works are listed under the later date. In general, the original publisher is not given for books printed before 1920. Some books are issued by a different publisher in Britain and America, and because I have worked on both sides of the Atlantic, it is a matter of chance which is listed here.

ADAMS, FRANK DAWSON. 1938. *The Birth and Development of the Geological Sciences*. Baltimore: reprinted New York, Dover, 1954.

ADAMS, MARK B. 1968. "The Foundations of Population Genetics: Contributions of the Chetverikov School, 1924–34." *J. Hist. Biology* 1: 23–39.

———. 1970. "Toward a Synthesis: Population Concepts in Russian Evolutionary Thought." *J. Hist. Biology* 3: 107–209.

———. 1979. "From 'Gene Fund' to 'Gene Pool': On the Evolution of Evolutionary Language." *Stud. Hist. Biology* 3: 241–285.

ADANSON, MICHEL. 1763. *Familles de plantes*. Paris: reprinted Lehre, Cramer.

———. 1779. "Sur les changements des espèces dans les plantes." *Hist. Acad. Roy. des Sci.*, 71–77.

ADELMANN, HOWARD B. 1966. *Marcello Malpighi and the Evolution of Embryology*. 5 vols. Ithaca, N.Y.: Cornell University Press.

AGASSIZ, ELIZABETH CARY. 1885. *Louis Agassiz: His Life and Correspondence*. 2 vols. London.

AGASSIZ, LOUIS. 1833–43. *Recherches sur les poissons fossiles*. 5 vols and plates. Neuchatel.

———. 1842. "On the Succession and Development of Organized Beings at the Surface of the Terrestial Globe." *Edinburgh New Phil. J*. 33: 388–399.

———. 1863. *Methods of Study in Natural History*. New York: reprinted New York, Arno Press.

———. 1962. *Essay on Classification*. Edited by Edward Lurie. Cambridge, Mass.: Harvard University Press.

———. 1967. *Studies on Glaciers*. Translated by Albert V. Carozzi. New York: Hafner.

———, and GOULD, A. A. 1848. *Principles of Zoology*. Boston: reprinted New York, AMS Press.

ALEXANDER, H. G. 1956. *The Leibniz-Clarke Correspondence*. Manchester: Manchester University Press.

ALEXANDER, SAMUEL. 1920. *Space, Time and Deity*. 2 vols. London: Macmillan.

ALLABY, MICHAEL, and LOVELOCK, JAMES. 1983. *The Great Extinction*. London: Secker and Warburg.

ALLAN, MEA. 1967. *The Hookers of Kew, 1785–1911*. London: Joseph.

———. 1977. *Darwin and His Flowers: The Key to Natural Selection*. New York: Tapplinger.

ALLEN, D. C. 1978. *The Naturalist in Britain: A Social History*. Harmondsworth, Middlesex: Penguin Books.

ALLEN, GARLAND E. 1968. "Thomas Hunt Morgan and the Problem of Natural Selection." *J. Hist. Biology* 1: 113–139.

———. 1969*a*. "Thomas Hunt Morgan and the Emergence of a New American Biology." *Quart. Rev. Biol*. 44: 168–188.

———. 1969*b*. "Hugo De Vries and the Reception of the Mutation Theory." *J. Hist. Biology* 2: 55–87.

———. 1974. "Opposition to the Mendelian-Chromosome Theory: The Physiological and Developmental Genetics of Richard Goldschmidt." *J. Hist. Biology* 7: 49–92.

———. 1975*a*. *Life Science in the Twentieth Century*. New York: Wiley.

———. 1975*b*. "Genetics, Eugenics and Class Struggle." *Genetics* 79: 29–45.

———. 1976. "Genetics, Eugenics and Society: Internalists and Externalists in Contemporary History of Science." *Social Studies of Science* 6: 105–122.

———. 1978. *Thomas Hunt Morgan: The Man and his Science*. Princeton. Princeton University Press.

———. 1986. "The Eugenics Records Office at Cold Spring Harbor, 1910–1940." *Osiris* n.s. 2: 225–264.

ANDERSON, LORRAINE. 1976. "Bonnet's Taxonomy and Chain of Being." *J. Hist. Ideas* 37: 45–58.

———. 1982. *Charles Bonnet and the Order of the Known*. Dordrecht. D. Reidel.

APPEL, TOBY A. 1980. "Henri de Blainville and the Animal Series: A Nineteenth-Century Chain of Being." *J. Hist. Biology* 13: 291–319.

———. 1987. *The Cuvier-Geoffroy Debate: French Biology in the Decades before Darwin*. Oxford: Oxford University Press.

APPLEMAN, PHILIP, ed. 1970. *Darwin: A Norton Critical Edition*. New York: Norton.

ARDREY, ROBERT. 1966. *The Territorial Imperative: A Personal Inquiry into the Animal Origins of Property and Nations*. New York: Athenaeum.

ARGYLL, GEORGE DOUGLAS CAMPBELL, DUKE OF. 1867. *The Reign of Law*. London. 5th ed., London, 1868.

———. 1868. *Primeval Man: And Examination of Some Recent Speculations*. London.

———. 1898. *Organic Evolution Cross-examined*. London.

ARGYLL, DOWAGER DUCHESS OF, ed. 1906. *George Douglas Campbell, Eighth Duke of Argyll, K.G., K.T., (1823–1900): Autobiography and Memoirs*. 2 vols. London.

ARONSON, LESTER R. 1975. "The Case of *The Case of the Midwife Toad*." *Behavior Genetics* 5: 115–125.

ARRHENIUS, SVANTE. 1908. *Worlds in the Making: The Evolution of the Universe*. Translated by H. Borns. New York and London: Harper.

ASHFORTH, ALBERT. 1969. *Thomas Henry Huxley*. Boston: Twayne Publishers.

ASIATIC EXCLUSION LEAGUE. 1907–13. *Proceedings*. San Francisco: reprinted New York, Arno, 1977.

ASIMOV, ISAAC. 1981. *In the Beginning*. London: New English Library.

AYALA, F. J., and VALENTINE, J. W. 1979. *Evolving: The Theory and Processes of Organic Evolution*. Menlo Park, Calif.: Benjamin Cummings.

BABBAGE, CHARLES. 1838. *The Ninth Bridgewater Treatise: A Fragment*. 2d ed., London: reprinted London, Cass, 1968.

BAER, KARL ERNST VON. 1828. *Über Entwickelungsgeschichte der Thiere: Beobachtung und Reflexion. Erster Theil*. Königsburg: reprinted Brussels, Culture et Civilization, 1967.

BAGEHOT, WALTER. 1872. *Physics and Politics: Or Thoughts on the Application of the Principles of "Natural Selection" and "Inheritance" to Political Society*. London: reprinted Farnborough, Gregg International, 1971.

BAILEY, EDWARD B. 1967. *James Hutton: The Founder of Modern Geology*. Amsterdam, London and New York: Elsevier Publishing Co.

BAJEMA, CARL JAY. 1977. *Eugenics: Then and Now*. Stroudsburg, Pa.: Dowden, Hutchinson and Ross Inc.

BAKER, J. A. 1952. *Abraham Trembley of Geneva: Scientist and Philosopher*. London: Arnold.

BAKER, KEITH, M. 1975. *Condorcet: From Natural Philosophy to Social Mathematics*. Chicago: University of Chicago Press.

BALDWIN, JAMES MARK. 1902. *Development and Evolution: Including Psychophysical Evolution, Evolution by Orthoplasy, and the Theory of Genetic Modes*. New York.

BANNISTER, ROBERT C. 1970. "'The Survival of the Fittest is Our Doctrine': History or Histrionics?" *J. Hist. Ideas* 31: 377–398.

———. 1979. *Social Darwinism: Science and Myth in Anglo-American Social*

Thought. Philadelphia: Temple University Press.

BANTON, MICHAEL. 1987. *Racial Theories*. Cambridge: Cambridge University Press.

BARASH, DAVID P. 1977. *Sociobiology and Behavior*. New York, Oxford and Amsterdam: Elsevier.

BARBER, W. H. 1955. *Leibniz in France from Arnauld to Voltaire: A Study in French Reactions to Leibnizianism, 1670–1769*. Oxford: Clarendon Press.

BARBOUR, IAN C. 1966. *Issues in Science and Religion*. Englewood Cliffs, N.J.: Prentice-Hall.

———. 1968. *Science and Religion: New Perspectives on the Dialogue*. New York: Harper & Row.

BARKER, A. D. 1969. "An Approach to the Theory of Natural Selection." *Philosophy* 44: 271–290.

BARLOW, NORA. 1946. *Charles Darwin and the Voyage of the Beagle*. New York: Philosophical Library.

———. 1967. *Darwin and Henslow: The Growth of an Idea*. London: John Murray.

BARNES, BARRY, and SHAPIN, STEPHEN. 1979. *Natural Order: Historical Studies of Scientific Culture*. Beverly Hills and London: Sage Publications.

BARNETT, S. A., ed. 1962. *A Century of Darwin*. Reprinted London: Mercury Books.

BARRETT, PAUL H. 1974. "The Sedgwick-Darwin Geological Tour of North Wales." *Proc. Am. Phil. Soc*. 118: 146–164.

BARTHÉLEMY-MADAULE, MADELEINE. 1982. *Lamarck the Mythical Precursor: A Study of the Relations between Science and Ideology*. Cambridge, Mass.: MIT Press.

BARTHOLOMEW, MICHAEL. 1973. "Lyell and Evolution: An Account of Lyell's Response to the Prospect of an Evolutionary Ancestry for Man." *Brit. J. Hist. Sci*. 6: 261–303.

———. 1975. "Huxley's Defence of Darwinism." *Annals of Science* 32: 525–535.

———. 1976. "The Non-Progress of Non-Progressionism: Two Responses to Lyell's Doctrine." *Brit. J. Hist. Sci*. 9: 166–174.

———. 1979. "The Singularity of Lyell." *Hist. of Science* 17: 276–293.

BARZUN, JACQUES. 1958. *Darwin, Marx, Wagner: Critique of a Heritage*. 2d ed., Garden City, N.Y.: Doubleday.

———. 1965. *Race: A Study in Superstition*. New York: Harcourt Brace.

BASALLA, GEORGE, et al., eds. 1970. *Victorian Science*. Garden City, N.Y.: Doubleday.

BATES, HENRY WALTER. 1862. "Contributions to an Insect Fauna of the Amazon Valley: Lepidoptera: Heliconidae." *Trans. Linn. Soc. Lond*. 23: 495–515.

———. 1863. *The Naturalist on the River Amazons*. 2 vols. London.

BATESON, BEATRICE. 1928. *William Bateson, F.R.S.: Naturalist*. Cambridge: Cambridge University Press.

BATESON, WILLIAM. 1894. *Materials for the Study of Variation: Treated with Especial Regard to Discontinuity in the Origin of Species*. London.

———. 1902. *Mendel's Principles of Heredity*. Cambridge.

———. 1914. President's Address. *Report of the British Association for the Advancement of Science*, 3–28.

BAYNE, PETER. 1871. *The Life and Letters of Hugh Miller*. 2 vols. London.

BEATTY, JOHN. 1982. "What's in a Word? Coming to Terms in the Darwinian Revolution." *J. Hist. Biology* 15: 215–239.

BEDDALL, BARBARA G. 1968. "Wallace, Darwin and the Theory of Natural Selection." *J. Hist. Biology* 1: 261–323.

———. 1969. *Wallace and Bates in the Tropics: An Introduction to the Theory of Natural Selection*. London: Macmillan.

———. 1972. "Wallace, Darwin and Edward Blyth: Further Notes on the Development of Evolution Theory." *J. Hist. Biology* 5: 153–158.

———. 1973. "Notes for Mr. Darwin: Letters to Charles Darwin from Edward Blyth at Calcutta: A Study in the Process of Discovery." *J. Hist. Biology* 6: 69–95.

BEER, GILLIAN. 1983. *Darwin's Plots: Evolutionary Narrative in Darwin, George Eliot, and Nineteenth-Century Fiction*. London: Routledge and Kegan Paul.

BELL, P. R., ed. 1959. *Darwin's Biological Work: Some Aspects Reconsidered*. New York: Wiley.

BENNETT, J. H., ed. 1983. *Natural Selection, Heredity, and Eugenics*. Oxford: Oxford University Press.

BENTHAM, JEREMY. 1838–43. *The Works of Jeremy Bentham*. 11 vols. London: reprinted New York, Russell and Russell, 1962.

BERG, LEO S. 1926. *Nomogenesis: Or Evolution Determined by Law*. Translated by J. N. Rostovtsov, introduced by D'Arcy Wentworth Thomson. London: reprinted with a preface by Theodosius Dobzhansky, Cambridge, Mass., MIT Press, 1969.

BERGSON, HENRI. 1911. *Creative Evolution*. Trans. by Arthur Mitchell. New York.

BETHELL, T. 1976. "Darwin's Mistake." *Harper's Magazine* 252: 70–75.

BIBBY, HAROLD CECIL. 1959. *T. H. Huxley: Scientist, Humanist and Educator*. London: Watts.

———. 1972. *Scientist Extraordinary: The Life and Scientific Career of Thomas Henry Huxley, 1825–1895*. Oxford and New York: Pergamon Press.

BILBERG, ISAAC. 1752. "Oeconomia naturae." *In* Linnaeus, ed., *Amoenitates Academicae* 1749–90, 2: 1–58. Translation, "The Economy of Nature" in Benjamin Stillingfleet, ed., *Miscellaneous Tracts relating to Natural History*. London, 1762. Pp. 39–126.

BLACKER, C. 1952. *Eugenics: Galton and After*. Cambridge, Mass.: Harvard University Press.

BLAISDELL, MURIEL. 1982. "Natural Theology and Nature's Disguises." *J. Hist. Biology* 15: 163–189.,

BLANCKAERT, CL., et al. 1979. "Les néo-lamarckiens français." *Révue de synthèse* 3d ser., nos. 95–96, 283–468.

BLINDERMAN, CHARLES. 1986. *The Piltdown Inquest*. Buffalo, N.Y.: Prometheus Books.

BLIXT, S. 1975. "Why Didn't Gregor Mendel Find Linkage?" *Nature* 256: 206.

BLUMENBACH, JOHANN FRIEDRICH. 1865. *On the Natural Varieties of Mankind*. Translated by Thomas Bendyshe. London: reprinted New York, Bergman, 1969.

BLUNT, WILFRED. 1971. *The Compleat Naturalist: A Life of Linnaeus*. With the assistance of William T. Stearn. New York: Viking Press.

BOCK, KEITH. 1955. "Darwin and Social Theory." *Phil. Sci.* 22: 123–134.

BOEHME, JACOB. 1912. *The Signature of All Things*. London: Everyman.

BÖLSCHE, WILHELM. 1906. *Haeckel: His Life and Work*. Trans. Joseph McCabe. London: T. Fisher Unwin.

BOLT, CHRISTINE. 1971. *Victorian Attitudes to Race*. London: Routledge and Kegan Paul.

BONNET, CHARLES. 1779. *Oeuvres d'histoire naturelle et de philosophie*. 19 vols. Neuchatel.

BOULE, MARCELLIN. 1923. *Fossil Man: Elements of Human Palaeontology*. Edinburgh: Oliver and Boyd.

BOURDIER, FRANK. 1969. "Geoffroy Saint Hilaire versus Cuvier: The Campaign for Paleontological Evolution." *In* Schneer, ed., *Toward a History of Geology*. Pp. 36–61.

BOWLER, PETER J. 1971. "Preformation and Pre-existence in the Seventeenth Century: A Brief Analysis." *J. Hist. Biology* 4: 221–244.

———. 1973. "Bonnet and Buffon: Theories of Generation and the Problem of Species." *J. Hist. Biology* 6: 259–281.

———. 1974*a*. "Evolutionism in the Enlightenment." *History of Science* 12: 159–183.

———. 1974*b*. "Darwin's Concepts of Variation." *J. Hist. Medicine* 29: 196–212.

———. 1975. "The Changing Meaning of 'Evolution.'" *J. Hist. Ideas* 36: 95–114.

———. 1976*a*. *Fossils and Progress: Paleontology and the Idea of Progressive Evolution in the Nineteenth Century*. New York: Science History Publications.

———. 1976*b*. "Malthus, Darwin and the Concept of Struggle." *J. Hist. Ideas* 37: 631–650.

———. 1976*c*. "Alfred Russel Wallace's Concepts of Variation." *J. Hist. Medicine* 31: 17–29.

———. 1977*a*. "Darwinism and the Argument from Design: Suggestions for a Reevaluation." *J. Hist. Biology* 10: 29–43.

———. 1977*b*. "Edward Drinker Cope and the Changing Structure of Evolutionary Theory." *Isis* 68: 249–265.

———. 1978. "Hugo De Vries and Thomas Hunt Morgan: The Mutation Theory and the Spirit of Darwinism." *Annals of Science* 35: 55–73.

———. 1979. "Theodor Eimer and Orthogenesis: Evolution by Definitely Directed Variation." *J. Hist. Medicine* 34: 40–73.

———. 1983. *The Eclipse of Darwinism: Anti-Darwinian Evolution Theories in the Decades around 1900*. Baltimore: Johns Hopkins University Press.

———. 1984. "E. W. MacBride's Lamarckian Eugenics and Its Implications for the Social Construction of Scientific Knowledge." *Annals of Science* 41: 245–260.

———. 1985. "Scientific Attitudes to Darwinism in Britain and America." *In* D. Kohn, ed., *The Darwinian Heritage*. Pp. 641–682.

———. 1986. *Theories of Human Evolution: A Century of Debate, 1844–1944*. Baltimore: Johns Hopkins University Press. Oxford: Basil Blackwell.

———. 1988. *The Non-Darwinian Revolution: Reinterpreting a Historical Myth*. Baltimore: Johns Hopkins University Press.

BOX, JOAN FISHER. 1978. *R. A. Fisher: The Life of a Scientist*. New York: Wiley.

BRACKMAN, ARNOLD C. 1980. *A Delicate Arrangement: The Strange Case of Charles Darwin and Alfred Russel Wallace*. New York: Times Books.

BRADY, RONALD H. 1979. "Natural Selection and the Criteria by which a Theory Is Judged." *Systematic Zoology* 28: 600–621.

BRENT, PETER. 1981. *Charles Darwin: A Man of Enlarged Curiosity*. New York: Harper and Row.

BREWSTER, SIR DAVID. 1855. *Memoirs of the Life, Writings and Discoveries of Sir Isaac Newton*. 2 vols. London: reprinted New York, Johnson Reprint Corporation, 1965.

BROBERG, GUNNAR, ed. 1980. *Linnaeus: Progress and Prospects in Linnaean Research*. Stockholm: Almquist and Wiksell International.

BROCK, W. H., and MACLEOD, R. M. 1976. "The Scientists' Declaration: Reflections on Science and Belief in the Wake of *Essays and Reviews*." *Brit. J. Hist. Sci.* 9: 39–66.

BRONGNIART, ADOLPHE. 1828*a*. *Prodome d'une histoire des végétaux fossiles . . .* Paris and Strasburg.

———. 1828*b*. "Considerations sur la nature de la végétation qui couvrait la surface de la terre aux diverses époques de la formation de son écorce." *Ann. des sci. nat.* 15: 225–258.

———. 1829. "General Considerations on the Nature of the Vegetation which Covered the Earth at the Different Epochs of the Formation of Its Crust." *Edinburgh New Phil. J.* 6: 349–371.

BRONN, HEINRICH GEORG. 1858. *Untersuchungen über die Entwickelungsgesetze der organischen Welt*. Stuttgart.

———. 1859. "On the Laws of Evolution of the Organic World during the Formation of the Crust of the Earth." *Ann. & Mag. of Nat. Hist.*, 3d series, 4: 81–89, 175–184.

BRONOWSKI, JACOB. 1975. *Science and Human Values*. New York & London: Harper Colophon.

BROOKE, JOHN HEDLEY. 1977. "Richard Owen, William Whewell and the *Vestiges*." *Brit. J. Hist. Sci.* 10: 132–145.

———. 1985. "The Relations between Darwin's Science and His Religion." *In* Durant, ed., *Darwinism and Divinity*. Pp. 40–75.

BROOKS, JOHN LANGDON. 1983. *Just before the Origin: Alfred Russel Wallace's Theory of Evolution*. New York: Columbia University Press.

BROWNE, JANET. 1980. "Darwin's Botanical Arithmetic and the 'Principle of Divergence,' 1854–1858." *J. Hist. Biology* 13: 53–89.

———. 1983. *The Secular Ark: Studies in the History of Biogeography*. New Haven: Yale University Press.

BRUNET, PIERRE. 1929. *Maupertuis*. 2 vols. Paris: Blanchard.

BRUSH, STEPHEN G. 1978. *The Temperature of History: Phases of Science and Culture in the Nineteenth Century*. New York: Burt Franklin Inc.

BUCKLAND, WILLIAM. 1820. *Vindiciae Geologicae: Or the Connexion of Geology and Religion Explained*. Oxford.

———. 1823. *Reliquiae Diluvianae: Or Observations of the Organic Remains contained in Caves, Fissures and Diluvial Gravel, and other Geological Phenomena, attesting the Action of a Universal Deluge*. London: reprinted New York, Arno Press, 1977.

———. 1836. *Geology and Mineralogy Considered with Reference to Natural The-*

ology. 2 vols. London.
BUFFETAUT, ERIC. 1986. *A History of Vertebrate Paleontology*. Bekenham: Croom Helm.
BUFFON, GEORGES LOUIS LECLERC, COMTE DE. 1749–67. *Histoire naturelle, générale et particulière*. 15 vols. Paris.
———. 1774–1789. *Histoire naturelle: Supplément*. 7 vols. Paris.
———. 1785. *Natural History*. 9 vols. Translated by William Smellie, 2d ed. London.
———. 1962. *Les époques de la nature*. Edited by Jacques Roger. Paris: Muséum d'histoire naturelle.
———. 1964. *Oeuvres philosophiques*. Edited by J. Piveteau, Paris.
———. 1976. "The 'Critical Discourse' to Buffon's *Histoire Naturelle*: The First Complete English Translation." By John Lyon, *J. Hist. Biology* 9: 133–181.
———. 1981. *From Natural History to the History of Nature: Readings from Buffon and His Critics*. Ed. Philip R. Sloan and J. Lyon. London and Notre Dame: University of Notre Dame Press.
BURCHFIELD, JOE D. 1974. "Darwin and the Dilemma of Geological Time." *Isis* 65: 301–321.
———. 1975. *Lord Kelvin and the Age of the Earth*. New York: Science History Publications.
BURKHARDT, FREDERICK, et al., eds. 1985. *A Calendar of the Correspondence of Charles Darwin*. New York and London: Garland Publishing.
BURKHARDT, RICHARD W., JR. 1970. "Lamarck, Evolution and the Politics of Science." *J. Hist. Biology* 3: 275–296.
———. 1972. "The Inspiration of Lamarck's Belief in Evolution," *J. Hist. Biology* 5: 413–438.
———. 1977. *The Spirit of System: Lamarck and Evolutionary Biology*. Cambridge, Mass., and London: Harvard University Press.
BURNET, THOMAS. 1691. *The Sacred Theory of the Earth*. London: reprinted with an introduction by Basil Willey, London, Centaur Press, 1965.
BURR, CLINTON STODDARD. 1922. *America's Race Heritage*. New York: reprinted New York, Arno, 1977.
BURROW, J. W. 1966. *Evolution and Society: A Study in Victorian Social Theory*. Cambridge: Cambridge University Press.
BURSTYN, H. C. 1975. "If Darwin Wasn't the *Beagle*'s Naturalist, Why Was He on Board?" *Brit. J. Hist. Sci*. 8: 62–69.
BURY, J. B. 1932. *The Idea of Progress: An Inquiry into Its Growth and Origins*. Reprinted New York: Dover, 1955.
BUSS, ALLAN R. 1976. "Galton and the Birth of Differential Psychology and Eugenics: Social, Political and Economic Forces." *J. Hist. Behavioural Sci*. 12: 47–58.
BUTLER, SAMUEL. 1879. *Evolution, Old and New: Or the theories of Buffon, Dr. Erasmus Darwin, and Lamarck, as Compared with that of Mr. Charles Darwin*. London.
———. 1916. *Life and Habit*. 2d ed. London.
———. 1920*a*. *Unconscious Memory*. 3d ed. London.

———. 1920*b*. *Luck, or Cunning, as the Main Means of Organic Modification?* 2d ed. London.

BYNUM, WILLIAM F. 1973. "The Anatomical Method, Natural Theology, and the Function of the Brain." *Isis* 64: 445–468.

———. 1975. "The Great Chain of Being." *History of Science* 13: 1–28.

———. 1984. "Charles Lyell's *Antiquity of Man* and Its Critics." *J. Hist. Biol.* 17: 153–187.

CABANIS, PIERRE. 1981. *On the Relations between the Physical and the Moral Aspects of Man*. Edited by George Mora. Baltimore: Johns Hopkins University Press.

CAHN, THEOPHILE. 1962. *La vie et l'oeuvre d'Etienne Geoffroy Saint Hilaire*. Paris: Presses Universitaires de France.

CALHOUN, WILLIAM PATRICK. 1902. *The Caucasian and the Negro in the United States*. Columbia. S.C.: reprinted New York, Arno, 1977.

CALLOT, EMILE. 1965. *La philosophie de la vie au XVIII*[e] *siècle*. Paris: Marcel Rivière.

———. 1971. *La philosophie biologique de Goethe*. Paris: Marcel Rivière.

CAMPBELL, JOHN ANGUS. 1974. "Nature, Religion and Emotional Response: A Reconsideration of Darwin's Affective Decline." *Victorian Studies* 18: 154–174.

CAMPER, PETRUS. 1779. "Account of the Organs of Speech of the Orang Outang." *Phil. Trans. Roy. Soc.* 69: 155–156.

CANGUILHEM, GEORGES. 1960. "Du développement à l'évolution au XIXe siècle." *Thales* 2: 3–68.

———. 1970. "Les concepts de 'lutte pour l'existence' et de 'selection naturelle' en 1858: Charles Darwin et Alfred Russel Wallace." *In* Canguilhem, ed., *Etudes d'histoire et de philosophie des sciences*. 2d ed. Paris: Vrin.

———, ed. 1970. "Georges Cuvier . . ." *Rev. Hist. Sci. Appl.* 23: 7–92.

CANNON, WALTER F. 1960*a*. "The Uniformitarian-Catastrophist Debate." *Isis* 51: 38–55.

———. 1960*b*. "The Problem of Miracles in the 1830s." *Victorian Studies* 4: 5–32.

———. 1961*a*. "The Bases of Darwin's Achievement: A Reevaluation." *Victorian Studies* 5: 109–132.

———. 1961*b*. "The Impact of Uniformitarianism: Two Letters from John Herschel to Charles Lyell, 1836–37." *Proc. Am. Phil. Soc.* 105: 301–314.

———. 1961*c*. "John Herschel and the Idea of Science." *J. Hist. Ideas* 20: 215–239.

———. 1964. "Scientists and Broad Churchmen: An Early Victorian Intellectual Network." *J. Brit. Studs.* 4: 65–88.

CANNON, SUSAN F. 1978. *Science in Culture: The Early Victorian Period*. New York: Science History Publications.

CAPLAN, ARTHUR L., ed. 1978. *The Sociobiology Debate*. New York: Harper & Row.

CARLSON, ELOF AXEL. 1966. *The Gene: A Critical History*. Philadelphia: Saunders.

CAROZZI, ALBERT V. 1969. "De Maillet's *Telliamed* (1748): An Ultra-Neptunian Theory of the Earth." *In* Schneer, ed., *Toward a History of Geology*. Pp. 80–91.

CARPENTER, WILLIAM BENJAMIN. 1851. *Principles of Physiology: General and Comparative*. 3d ed. London.

———. 1888. *Nature and Man: Essays Scientific and Philosophical. With an Intro-*

ductory Memoir by J. Erstlin Carpenter. London: reprinted Farnborough, Gregg, 1970.

CARTER, G. S. 1957. *A Hundred Years of Evolution*. New York: Macmillan.

CASSIRER, ERNST. 1945. *Rousseau, Kant, Goethe: Two Essays*. Translated by James Gutman et al. Princeton: Princeton University Press.

———. 1951. *The Philosophy of the Enlightenment*. Translated by Fritz Koelln and James Pettegrove. Princeton: Princeton University Press.

CASTLE, W. E. 1911. *Heredity in Relation to Evolution and Animal Breeding*. New York: Appleton.

CHADWICK, OWEN. 1966. *The Victorian Church*. London: A. C. Black.

———. 1975. *The Secularization of the European Mind in the Nineteenth Century*. Cambridge: Cambridge University Press.

CHAMBERS, ROBERT. 1844. *Vestiges of the Natural History of Creation*. London: reprinted with an introduction by Sir Gavin De Beer, Leicester, Leicester University Press. 4th ed. London, 1845. 12th ed. with an introduction by Alexander Ireland, London, 1884.

———. 1846. *Explanations: A Sequel to the Vestiges of the Natural History of Creation*. 2d ed. London.

CHAPMAN, ROGER G. 1982. *Charles Darwin 1809–1882; a Centennial Commemorative*. Wellington, New Zealand: Nova Pacifica.

CHERFAS, JEREMY, ed. 1982. *Darwin Up to Date*. London: IPC Magazines.

CHETVERIKOV, S. S. 1961. "On Certain Aspects of the Evolutionary Process from the Standpoint of Modern Genetics." Translated by M. Barker, edited by I. M. Lerner. *Proc. Am. Phil. Soc*. 105: 167–195.

CHURCHILL, FREDERICK B. 1968. "August Weismann and a Break from Tradition." *J. Hist. Biology* 1: 91–112.

———. 1974. "William Johannsen and the Genotype Concept." *J. Hist. Biology* 7: 5–30.

———. 1976. "Rudolph Virchow and the Pathologist's Criterion for the Inheritance of Acquired Characteristics." *J. Hist. Medicine* 31: 117–148.

———. 1986. "Weismann, Hydromedusae, and the Biogenetic Imperative: A Reconsideration." *In* T. J. Horder et al., eds., *A History of Embryology*. Pp. 7–34.

CLARK, J. W., and HUGHES, T. M. 1890. *The Life and Letters of the Reverend Adam Sedgwick*. 2 vols. Cambridge: reprinted Farnborough, Gregg, 1970.

CLARK, LINDA L. 1984. *Social Darwinism in France*. University, Ala.: University of Alabama Press.

CLARK, RONALD W. 1969. *JBS: The Life and Work of J. B. S. Haldane*. New York: Coward-McCann.

CLARKE, A. M., and CLARKE, A. D. B., eds. 1974. *Mental Deficiency: The Changing Outlook*. London: Macmillan.

CLUBE, VICTOR, and NAPIER, BILL. 1982. *The Cosmic Serpent*. London: Faber.

CLUTTON-BROCK, T. H., and HARVEY, PAUL H. 1978. *Readings in Sociobiology*. San Francisco: W. H. Freeman.

COCK, A. G. 1973. "William Bateson, Mendelism, and Biometry." *J. Hist. Biology* 6: 1–36.

COLBERT, E. H. 1971. *Men and Dinosaurs: The Search in Field and in Laboratory*. Reprinted Harmondsworth: Penguin Books.

COLE, F. J. 1930. *Early Theories of Sexual Generation*. Oxford and New York: Oxford University Press.

COLEMAN, WILLIAM. 1962. "Lyell and the Reality of Species." *Isis* 53: 325–338.

———. 1964. *Georges Cuvier, Zoologist: A Study in the History of Evolution Theory*. Cambridge, Mass.: Harvard University Press.

———. 1965. "Cell, Nucleus, and Inheritance: A Historical Study." *Proc. Am. Phil. Soc.* 109: 124–158.

———. 1966. "Science and Symbol in the Turner Frontier Hypothesis." *Am. Hist. Rev.* 72: 22–49.

———. 1970. "Bateson and Chromosomes: Conservative Thought in Science." *Centaurus* 15: 228–315.

———. 1971. *Biology in the Nineteenth Century: Problems of Form, Function and Transmutation*. New York: Wiley.

———. 1973. "Limits of the Recapitulation Theory: Carl Freidrich Kielmeyer's Critique of the Presumed Parallelism of Earth History, Ontogeny and the Present Order of Organisms." *Isis* 64: 341–350.

———. 1976. "Morphology between Type Concept and Descent Theory." *J. Hist. Medicine* 31: 149–175.

COLLINGWOOD, R. G. 1945. *The Idea of Nature*. Oxford: Clarendon Press.

COLLINS, JAMES. 1959. "Darwin's Impact on Philosophy." *Thought* 34: 185–248.

COLP, RALPH, JR. 1974. "The Contacts between Karl Marx and Charles Darwin." *J. Hist. Ideas* 35: 329–338.

———. 1977. *To be an Invalid: The Illness of Charles Darwin*. Chicago: University of Chicago Press.

———. 1986. "Confessing a Murder: Darwin's First Revelations about Transmutation." *Isis* 77: 9–32.

COMTE, AUGUST. 1975. *August Comte and Positivism: The Essential Writings*. Ed. Gertrud Lenzer. New York: Harper & Row.

CONDILLAC, ETIENNE BONNOT DE. 1756. *An Essay on the Origins of Human Knowledge*. Reprinted with an introduction by James H. Stam. New York: AMS Press, 1974.

CONDORCET, MARIE JEAN ANTOINE NICHOLAS. 1955. *Sketch for a Historical Picture of the Progress of the Human Mind*. Translated by J. Barraclough. London: Weidenfeld and Nicholson.

CONYBEARE, W. D., and PHILLIPS, WILLIAM. 1822. *Outlines of the Geology of England and Wales*. London: reprinted New York, Arno Press, 1977.

CONRY, YVETTE. 1972. *Correspondence entre Charles Darwin et Gaston de Saporta: precedé d'une histoire de la paléobotanique en France*. Paris: Presses Universitaires de France.

———. 1974. *L'introduction du Darwinisme en France au XIXe siècle*. Paris: Vrin.

COOTER, ROGER. 1985. *The Cultural Meaning of Popular Science: Phrenology and the Organization of Consent in Nineteenth-Century Britain*. Cambridge: Cambridge University Press.

COPE, EDWARD DRINKER. 1887. *The Origin of the Fittest: Essays in Evolution*. Reprinted with Cope, *The Primary Factors of Organic Evolution*. New York: AMS Press, 1974.

———. 1896. *The Primary Factors of Organic Evolution*. Chicago.

COPLESTON, FREDERICK. 1963. *A History of Philosophy*. vol. 7. *Fichte to Nietzsche*. London: Burns and Oates.

———. 1966. *A History of Philosophy*. vol. 8. *Bentham to Russell*. London: Burns and Oates.

CORNELL, JOHN F. 1983. "From Creation to Evolution: Sir William Dawson and the Idea of Design in the Nineteenth Century." *J. Hist. Biol.* 16: 137–170.

———. 1984. "Analogy and Technology in Darwin's Vision of Nature." *J. Hist. Biol.* 17: 303–344.

CORSI, PIETRO. 1978. "The Importance of French Transformist Ideas for the Second Volume of Lyell's *Principles of Geology*." *Brit. J. Hist. Sci.* 11: 221–244.

COUNT, E. W. 1950. *This Is Race*. New York: Schuman.

COWAN, RUTH SCHWARTZ. 1972*a*. "Francis Galton's Contributions to Genetics." *J. Hist. Biology* 5: 389–412.

———. 1972*b*. "Francis Galton's Statistical Ideas: The Influence of Eugenics." *Isis* 63: 509–528.

———. 1977. "Nature and Nurture: The Interplay of Biology and Politics in the Work of Francis Galton." *Stud. Hist. Biol.* 1: 133–208.

CRAVENS, HAMILTON. 1978. *The Triumph of Evolution: American Scientists and the Heredity-Environment Controversy, 1900–1941*. Philadelphia: University of Pennsylvania Press.

CRICK, FRANCIS. 1981. *Life Itself: Its Origin and Nature*. New York: Simon and Schuster.

CROCKER, LESTER G. 1959. "Diderot and Eighteenth-Century French Transformism." *In* Glass et al., eds., *Forerunners of Darwin*. Pp. 114–143.

———. 1963. *Nature and Culture: Ethical Thought in the French Enlightenment*. Baltimore: Johns Hopkins University Press.

CROIZAT, LEON. 1958. *Panbiogeography*. Caracas: the author. 3 vols.

———. 1964. *Space, Time, Form: The Biological Synthesis*. Caracas: the author.

CUVIER, GEORGES. 1805. *Leçons d'anatomie comparée*. 5 vols. Paris: reprinted Brussels, Culture et Civilisation. 1969.

———. 1812*a*. *Recherches sur les ossemens fossiles de quadrupèdes* . . . 4 vols. Paris: reprinted Brussels, Culture et Civilisation, 1969.

———. 1812*b*. "Sur un nouveau rapprochement à établir entre les classes qui composent le règne animal." *Ann. Mus. Nat. Hist.* 12: 73–84.

———. 1813. *An Essay on the Theory of the Earth*. Translated by Robert Kerr, with notes by Robert Jameson. Edinburgh: reprinted Farnborough, Gregg, 1971. 3d ed. Edinburgh, 1817, reprinted New York: Arno Press, 1977.

———. 1817. *Le règne animal distribué d'après son organization* . . . 4 vols. Paris: reprinted Brussels, Culture et Civilisation, 1969.

———. 1825. *Discours sur les révolutions de la surface du globe* . . . 2d ed. Paris: reprinted Brussels, Culture et Civilization, 1969.

———. 1863. *The Animal Kingdom Arranged after Its Organization* . . . New ed. London: reprinted Millwood N.Y., Kraus Reprints.

———. and BRONGNIART, ALEXANDRE. 1825. *Description geologiques des environs de Paris*. 2d ed. Paris: reprinted Brussels, Culture et Civilisation, 1969.

DANIEL, GLYN. 1975. *A Hundred and Fifty Years of Archaeology*. London: Duckworth.

DANIELS, GEORGE. 1968. *Darwinism Comes to America*. Waltham, Mass.: Blaisdell Publishing Co.

DÄNIKEN, ERICH VON. 1970. *Chariots of the Gods? Unsolved Mysteries of the Past*. Translated by Michael Heron. New York: G. P. Putnam's Sons.

———. 1977. *According to the Evidence: My Proof of Man's Extraterrestrial Origins*. Translated by Michael Heron. London: Souvenir Press.

DARDEN, LINDLEY. 1976. "Reasoning in Scientific Change: Charles Darwin, Hugo De Vries, and the Rediscovery of Segregation." *Stud. Hist. & Phil. Sci.* 7: 127–169.

———. 1977. "William Bateson and the Promise of Mendelism." *J. Hist. Biology* 10: 87–106.

DARLINGTON, CYRIL D. 1961. *Darwin's Place in History*. New York: Macmillan.

DARWIN, CHARLES ROBERT. 1839. *Journal of Researches into the Geology and Natural History of the Various Countries Visited by H.M.S. Beagle*. London: reprinted Brussels, Culture et Civilisation, 1969. 1896 ed. reprinted New York: AMS Press, 1972.

———. 1842. *The Structure and Distribution of Coral Reefs*. London: reprinted Brussels, Culture et Civilisation, 1969. 1897 ed. reprinted New York: AMS Press, 1972.

———. 1845. *Journal of Researches into the Natural History and Geology of the Various Countries Visited by H.M.S. Beagle*. London: reprinted London, Everyman.

———. 1851–53. *A Monograph on the Sub-Class Cirripedia*. 2 vols. London: reprinted New York, Johnson Reprint Corporation, 1964.

———. 1859. *On the Origin of Species by Means of Natural Selection: Or the Preservation of Favoured Races in the Struggle for Life*. London, reprinted with an introduction by Ernst Mayr. Cambridge, Mass.: Harvard University Press, 1964.

———. 1862. *On the Various Contrivances by which British and Foreign Orchids are Fertilised by Insects* . . . London. 1896 ed. reprinted New York: AMS Press, 1972.

———. 1865. *The Movements and Habits of Climbing Plants*. London: reprinted Brussels, Culture et Civilisation, 1969. 1893 ed. reprinted New York: AMS Press, 1972.

———. 1868. *The Variation of Animals and Plants under Domestication*. 2 vols. London: reprinted Brussels, Culture et Civilisation, 1969. 1896 ed. reprinted New York: AMS Press, 1972.

———. 1871. *The Descent of Man and Selection in Relation to Sex*. 2 vols. London: reprinted Brussels, Culture et Civilisation, 1969. 1895 ed. reprinted New York: AMS Press, 1972.

———. 1872. *The Expression of the Emotions in Man and the Animals*. London: reprinted Brussels, Culture et Civilisation, 1969. 1896 ed. reprinted New York: AMS Press, 1972.

———. 1873. *Insectivorous Plants*. London: reprinted, Brussels, Culture et Civilisation, 1969. 1893 ed. reprinted New York: AMS Press, 1972.

———. 1877. *The Different Forms of Flowers on Plants of the Same Species*. London: reprinted Brussels, Culture et Civilisation, 1969. 1896 ed. reprinted New

York: AMS Press, 1972.

———. 1880. *The Power of Movement in Plants*. London: reprinted Brussels, Culture et Civilisation, 1969.

———. 1881. *The Formation of Vegetable Mould through the Action of Earthworms*. London: reprinted Brussels, Culture et Civilisation, 1969. 1896 ed. reprinted New York: AMS Press, 1972.

———. 1933. *Charles Darwin's Diary of the Voyage of H.M.S. Beagle*. Edited from the original manuscript by Nora Barlow. New York: reprinted New York, Krause Reprints, 1969.

———. 1958. *The Autobiography of Charles Darwin: With the Original Omissions Restored*. Edited by Nora Barlow. New York: Harcourt Brace.

———. 1959*a*. "Darwin's Journal." Edited by Sir Gavin de Beer. *Bull. Brit. Mus. (Nat. Hist.), Historical Series* 2: no. 1.

———. 1959*b*. *The Origin of Species . . . A Variorum Text*. Edited by Morse Peckham. Philadelphia: University of Pennsylvania Press.

———. 1960–61. "Darwin's Notebooks on the Transmutation of Species." Edited by Sir Gavin de Beer. *Bull. Brit. Mus. (Nat. Hist.), Historical Series* 2: nos. 2–6.

———. 1963. "Darwin's Ornithological Notes." Edited by Nora Barlow. *Bull. Brit. Mus. (Nat. Hist.), Historical Series* 2: no. 7.

———. 1967. "Darwin's Notebooks . . . Pages Excised by Darwin." Edited by Sir Gavin de Beer et al. *Bull. Brit. Mus. (Nat. Hist.), Historical Series* 3: no. 5.

———. 1974. Notebooks on Man. Edited by Paul H. Barrett. *In* Howard E. Gruber, *Darwin on Man: A Psychological Study of Scientific Creativity*. New York: E. P. Dutton.

———. 1975. *Charles Darwin's Natural Selection: Being the Second Part of His Big Species Book Written from 1856 to 1858*. Edited by Robert C. Stauffer. London: Cambridge University Press.

———. 1977. *The Collected Papers of Charles Darwin*. Edited by Paul H. Barrett. 2 vols. Chicago: University of Chicago Press.

———. 1979. *The Red Notebook of Charles Darwin*. Edited by Sandra Herbert. Ithaca, New York: Cornell University Press.

———. 1980. *Metaphysics, Materialism, and the Evolution of Mind: The Early Writings of Charles Darwin*. Edited by Paul H. Barrett. Chicago: University of Chicago Press.

———. 1981. *A Concordance to Darwin's Origin of Species, First Edition*. Edited by Paul H. Barrett et al. Ithaca, N.Y.: Cornell University Press.

———, 1984, 1986. *The Correspondence of Charles Darwin*. Edited by Frederick Burkhardt and Sydney Smith. I (1821–1836). II (1837–1843). Cambridge: Cambridge University Press.

———. 1987. *Charles Darwin's Theoretical Notebooks (1836–1844)*. Edited by Paul H. Barrett et al. Cambridge: Cambridge University Press.

———, and HENSLOW, J. 1962. *Darwin and Henslow: The Growth of an Idea: Letters, 1831–60*. Edited by Nora Barlow. Berkeley and Los Angeles: University of California Press.

———, and WALLACE, ALFRED RUSSEL. 1958. *Evolution by Natural Selection*. With a foreword by Sir Gavin de Beer. Cambridge: Cambridge University Press.

DARWIN, ERASMUS. 1791. *The Botanic Garden: A Poem in Two Parts. Part One:*

Containing the Economy of Nature, Part Two: The Loves of the Plants. 2 vols. London: reprinted Elmsford, N.Y., Pergamon (British Book Center).
———. 1794–96. *Zoonomia: Or the Laws of Organic Life*. 2 vols. London: reprinted New York, AMS Press, 1974.
———. 1803. *The Temple of Nature*. London: reprinted Elmsford, N.Y., Pergamon (British Book Center).
———. 1968. *The Essential Erasmus Darwin*. Edited by Desmond King-Hele. London: McGibbon and Kee.
———. 1981. *The Letters of Erasmus Darwin*. Edited by Desmond King-Hele. Cambridge: Cambridge University Press.
DARWIN, FRANCIS, ed. 1887. *The Life and Letters of Charles Darwin*. 3 vols. London: reprinted New York, Johnson Reprint Corporation, 1969.
———. 1903. *More Letters of Charles Darwin*. 2 vols. London: reprinted New York, Johnson Reprint Corporation, 1972.
DAUDIN, HENRI. 1926. *Etudes d'histoire des sciences naturelles*. I. *De Linné à Jussieu: Méthodes de la classification et l'idée de série en botanique et en zoologie*. II. *Cuvier et Lamarck: Les classes zoologiques et l'idée de série animale*. Paris: Alcan.
DAVIES, GORDON L. 1969. *The Earth in Decay: A History of British Geomorphology, 1578–1878*. New York: Science History Publications.
DAWKINS, R. 1976. *The Selfish Gene*. Oxford: Oxford University Press.
———. 1982. *The Extended Phenotype: The Gene as the Unit of Selection*. Oxford and San Francisco: W. H. Freeman.
———. 1986. *The Blind Watchmaker*. Harlow: Longman Scientific and Technical.
DAWSON, SIR JOHN WILLIAM. 1890. *Modern Ideas of Evolution*. London: reprint edited by William R. Shea and John F. Cornell, New York, Prodist.
DEAN, DENNIS R. 1975. "James Hutton on Religion and Geology." *Annals of Science* 32: 187–193.
———. 1981. "The Age of the Earth Controversy: Beginnings to Hutton." *Annals of Science* 38: 435–456.
DE BEER, GAVIN. 1930. *Embryology and Evolution*. Oxford: Oxford University Press.
———. 1940. *Embryos and Ancestors*. Oxford: Oxford University Press.
———. 1963. *Charles Darwin: Evolution by Natural Selection*. London: Nelson.
———. 1964. "Mendel, Darwin and Fisher." *Notes Roy. Soc. London* 19: 192–226.
DE CAMP, L. S. 1968. *The Great Monkey Trial*. Garden City, N.Y.: Doubleday.
DE GRAZIA, A. 1966. *The Velikovsky Affair*. New Hyde Park, N.Y.: University Books.
DELAGES, YVES, and GOLDSMITH, MARIE. 1912. *The Theories of Evolution*. Translated by André Tridon. New York.
DELAIR, JUSTIN B., and SARJEANT, WILLIAM. 1975. "The Earliest Discoveries of Dinosaurs." *Isis* 66: 5–25.
DELUC, JEAN ANDRÉ. 1790–91. "Letters to Dr. James Hutton, F. R. S. Ed., on His Theory of the Earth." *Monthly Review* 2d series, 2: 206–227 and 582–601, 3: 573–586 and 4: 564–585.
———. 1798. *Lettres sur l'histoire physique de la terre, addressés à M. le Professeur Blumenbach*. Paris.
DE MAILLET, BENOÎT. 1968. *Telliamed, or Conversations between an Indian Phi-*

losopher and a French Missionary on the Diminution of the Sea . . . Trans. Albert V. Carozzi. Urbana: University of Illinois Press.

De Marrais, Robert. 1974. "The Double-Edged Effect of Sir Francis Galton: A Search for the Motives of the Biometrician-Mendelian Debate." *J. Hist. Biology* 7: 141–174.

Dennert, E. 1904. *At the Deathbed of Darwinism*. Translated by E. V. O'Hara and John H. Peschges. Burlington, Iowa.

Derham, William. 1713. *Physico-Theology: Or a Demonstration of the Being and Attributes of God from His Works of Creation*. London.

Descartes, René. 1964–73. *Oeuvres*. Edited by Charles Adam and Paul Tannery. Paris: Libraire Philosophique J. Vrin.

Desmond, Adrian. 1976. *The Hot-Blooded Dinosaurs: A Revolution in Paleontology*. New York: Dial Press.

———. 1979. "Designing the Dinosaur: Richard Owen's Response to Robert Edmond Grant." *Isis* 70: 224–234.

———. 1982. *Archetypes and Ancestors in Victorian London, 1850–1875*. London: Blond and Briggs; Chicago: University of Chicago Press.

———. 1984. "Robert E. Grant: The Social Predicament of a Pre-Darwinian Evolutionist." *J. Hist. Biol.* 17: 189–223.

———. 1985. "Richard Owen's Reaction to Transmutation in the 1830s." *Brit. J. Hist. Sci.* 18: 25–50.

———. 1987. "Artisan Resistance and Evolution in Britain, 1819–1848." *Osiris*, n.s. 3: 72–110.

Deutsch, M., et al., eds. 1968. *Social Class, Race and Psychological Development*. New York: Holt, Rinehart and Winston.

De Vries, Hugo. 1904. *Species and Varieties: Their Origin by Mutation*. Edited by D. T. MacDougal. Chicago. New edition 1910.

———. 1910*a*. *Intracellular Pangenesis*. Translated by C. Stuart Gager. Chicago.

———. 1910*b*. *The Mutation Theory: Experiments and Observations on the Origin of Species in the Vegetable Kingdom*. Translated by J. B. Farmer and A. D. Darbyshire. 2 vols. London.

Dewey, John. 1910. *The Impact of Darwin on Philosophy: And Other Essays in Contemporary Thought*. Reprinted Bloomington: Indiana University Press, 1965.

Dexter, Ralph W. 1979. "The Impact of Evolutionary Theories on the Salem Group of Agassiz Zoologists (Morse, Hyatt, Packard, Putnam)." *Essex Institute Historical Collections* 115: 144–171.

D'Holbach, Paul Henri Thiry, baron. 1821. *Système de la nature: Ou des lois du monde physique et du monde moral*. Reprinted with an introduction by Yvon Belavel. 2 vols. Hildesheim: Georg Olms, 1966.

———. 1868. *The System of Nature: Or Laws of the Moral and Physical World*. Translated by H. D. Robinson. Reprinted New York: Burt Franklin, 1970.

Diderot, Denis. 1875–77. *Oeuvres complètes*. Edited by J. Assezat. 20 vols. Paris.

———. 1964. *Oeuvres philosophiques*. Ed. Paul Vernière. Paris: Classiques Garnier.

———. 1966. *D'Alembert's Dream and Rameau's Nephew*. Translated by L. W. Tancock. Harmondsworth, Middlesex: Penguin Books.

Di Gregorio, Mario A. 1982. "The Dinosaur Connection: A Reinterpretation of T. H. Huxley's Evolutionary View." *J. Hist. Biology* 15: 397–418.

———. 1984. *T. H. Huxley's Place in Natural Science*. New Haven: Yale University Press.

Dobzhansky, Theodosius. 1937. *Genetics and the Origin of Species*. New York: Columbia University Press.

———. 1980. *The Roving Naturalist: Travel Letters of Theodosius Dobzhansky*. Ed. Bentley Glass. Philadelphia: American Philosophical Society.

———. 1981. *Dobzhansky's Genetics of Natural Populations. I-XLIII*. Edited by R. C. Lewontin et al. New York: Columbia University Press.

———, et al., 1977. *Evolution*. San Francisco: W. H. Freeman.

Dott, R. H., Jr. 1969. "James Hutton and the Concept of a Dynamic Earth." *In* Schneer, ed., *Toward a History of Geology*. Pp. 122–141.

Drummond, Henry. 1894. *The Ascent of Man*. New York.

Duncan, David, ed. 1911. *The Life and Letters of Herbert Spencer*. Reissue, London.

Dunn, Leslie Clarence. 1965. *A Short History of Genetics*. New York: McGraw-Hill.

Dupree, A. Hunter. 1959. *Asa Gray*. Cambridge, Mass.: Harvard University Press.

Durant, John R. 1979. "Scientific Naturalism and Social Reform in the Thought of Alfred Russel Wallace." *Brit. J. Hist. Sci.* 12: 31–58.

———, ed. 1985. *Darwinism and Divinity: Essays on Evolution and Religious Belief*. Oxford: Basil Blackwell.

East, Edward M. 1910. "A Mendelian Interpretation of Variation that is Apparently Continuous." *Am. Naturalist* 44: 65–82.

———. 1926. *Mankind at the Crossroads*. New York and London: reprinted New York, Arno, 1977.

Eddy, J. H., Jr. 1984. "Buffon, Organic Alterations, and Man." *Stud. Hist. Biol.* 7: 1– 46.

Egerton, Frank N. 1968. "Studies of Animal Population from Lamarck to Darwin." *J. Hist. Biology* 1: 225–259.

———. 1970*a*. "Humboldt, Darwin and Population." *J. Hist. Biology* 3: 325–360.

———. 1970*b*. "Refutation and Conjectures: Darwin's Response to Sedgwick's Attack on Chambers." *Stud. Hist. & Phil. Sci.* 1: 176–183.

———. 1971. "Darwin's Method or Methods." *Stud. Hist. & Phil. Sci.* 2: 281–286.

———. 1973. "Changing Concepts of the Balance of Nature." *Quart. Rev. Biology* 58: 322–350.

Eimer, Gustav Heinrich Theodor. 1890. *Organic Evolution as the Result of the Inheritance of Acquired Characters According to the Laws of Organic Growth*. Translated by James T. Cunningham. London.

———. 1898. *On Orthogenesis and the Impotence of Natural Selection in Species Formation*. Translated by J. M. McCormack. Chicago.

Eiseley, Loren. 1958. *Darwin's Century: Evolution and the Men Who Discovered It*. New York: Doubleday.

———. 1959. "Charles Darwin, Edward Blyth, and the Theory of Natural Selection." *Proc. Am. Phil. Soc.* 103: 94–158.

ELDREDGE, N. 1982. The *Monkey Business; a Scientist Looks at Creationism*. New York: Washington Square Press.

———. 1985. *Unfinished Synthesis: Biological Hierarchies and Modern Evolutionary Thought*. New York and Oxford: Oxford University Press.

———. 1986. *Time Frames: The Rethinking of Darwinian Evolution and the Theory of Punctuated Equilibria*. London: Heinemann.

ELDREDGE, N., and GOULD, S. J. 1972. "Punctuated Equilibria: An Alternative to Phyletic Gradualism." *In* T. J. M. Schopf, ed., *Models in Paleobiology*. San Francisco: Freeman, Cooper and Co. Pp. 82–115.

ELDREDGE, N., and TATTERSALL, I. 1982. *Myths of Human Evolution*. New York: Columbia University Press.

ELIE DE BEAUMONT, LÉONCE. 1830. "Recherches sur quelque-unes des révolutions de la surface du globe." *Ann. des sci. nat*. 18: 5–26, 284–416; 19: 5–99, 177–240.

ELLEGÅRD, ALVAR. 1957. "The Darwinian Theory and Nineteenth-Century Philosophies of Science." *J. Hist. Ideas* 18: 362–393.

———. 1958. *Darwin and the General Reader: The Reception of Darwin's Theory of Evolution in the British Periodical Press, 1859–1872*. Göteburg: Acta Universitatis Gothenburgensis.

EMMET, DOROTHY. 1932. *Whitehead's Philosophy of Organism*. New ed. London: Macmillan, 1966.

EVANS, BRIAN, and WAITES, BERNARD. 1981. *IQ and Mental Testing: An Unnatural Science and Its Social History*. London: Macmillan.

EVANS, L. T. 1984. "Darwin's Use of the Analogy between Artificial and Natural Selection." *J. Hist. Biol*. 17: 113–140.

EYLES, JOAN M. 1969. "William Smith: Some Aspects of His Life and Work." *In* Schneer, ed., *Toward a History of Geology*. Pp. 142–158.

FAIRCHILD, HENRY PRATT. 1926. *The Melting Pot Mistake*. Boston: reprinted New York, Arno, 1977.

FANCHER, RAYMOND E. 1983. "Francis Galton's African Ethnography and Its Role in the Development of His Psychology." *Brit. J. Hist. Sci*. 16: 67–79.

FARBER, PAUL LAWRENCE. 1972. "Buffon and the Problem of Species." *J. Hist. Biology* 5: 259–284.

———. 1975. "Buffon and Daubenton: Divergent Traditions in the *Histoire naturelle*." *Isis* 66: 63–74.

———. 1982. "The Transformation of Natural History in the Nineteenth Century." *J. Hist. Biology* 15: 145–152.

FARRALL, LYNDSAY A. 1979. "The History of Eugenics: A Bibliographical Review." *Annals of Science* 36: 111–123.

FARLEY, JOHN. 1974. "The Initial Reaction of French Biologists to Darwin's *Origin of Species*." *J. Hist. Biology* 7: 275–300.

———. 1977. *The Spontaneous Generation Controversy: From Descartes to Oparin*. Baltimore: Johns Hopkins University Press.

———. 1982. *Gametes and Spores: Ideas about Sexual Reproduction, 1750–1914*. Baltimore: Johns Hopkins University Press.

FAY, MARGARET A. 1978. "Did Marx Offer to Dedicate *Capital* to Darwin?" *J. Hist. Ideas* 39: 133–146.

FELLOWS, OTIS T. 1955. "Voltaire and Buffon: Clash and Conciliation." *Symposium* 10: 222–235.

———, and MILLIKEN, STEPHEN. 1972. *Buffon*. Boston: Twayne.

FEUER, LEWIS S. 1975. "Is the Darwin-Marx Correspondence Authentic?" *Annals of Science* 32: 1–12.

FICHMAN, MARTIN. 1977. "Wallace's Zoogeography and the Problem of Land Bridges." *J. Hist. Biology* 10: 45–63.

———. 1981. *Alfred Russel Wallace*. Boston: Twayne.

———. 1984. "Ideological Factors in the Dissemination of Darwinism in England, 1860–1900." *In* E. Mendelsohn, ed., *Transformation and Tradition in the Sciences: Essays in Honor of I. Bernard Cohen*. Cambridge: Cambridge University Press. Pp. 471–485.

FICHTE, JOHANN GOTTLEIB. 1956. *The Vocation of Man*. Edited by Roderick M. Chisholm. New York: Bobbs Merrill.

FISHER, R. A. 1918. "The Correlation between Relatives on the Supposition of Mendelian Inheritance." *Trans. Roy. Soc. Edinb*. 52: 399–433.

———. 1930. *The Genetical Theory of Natural Selection*. Oxford: Clarendon Press, reprinted New York: Dover, 1958.

———. 1936. "Has Mendel's Work been Rediscovered?" *Annals of Science* 1: 115–137.

FISKE, JOHN. 1874. *Outlines of Cosmic Philosophy*. Boston: reprinted with an introduction by David W. Noble, New York, Johnson Reprint Corporation, 1969.

FLEMING, DONALD. 1961. "Charles Darwin: The Anaesthetic Man." *Victorian Studies* 4: 219–236.

FLEW, A. G. N. 1967. *Evolutionary Ethics*. London: Macmillan.

FONTENELLE, BERNARD DE. 1688. *A Plurality of Worlds*. Translated by John Glanville. Reprinted London: Centaur Press, 1929.

FORCE, JAMES E. 1984. *William Whiston: Honest Newtonian*. Cambridge: Cambridge University Press.

FORD, E. B. 1931. *Mendelism and Evolution*. London: Methuen.

FORREST, D. 1974. *Francis Galton: The Life and Work of a Victorian Genius*. New York: Tapplinger.

FOTHERGILL, PHILLIP G. 1952. *Historical Aspects of Organic Evolution*. London: Hollis and Carter.

FOUCAULT, MICHEL. 1966. *Les Mots et les choses*. Paris: Gallimard. Translated as:

———. 1970. *The Order of Things: The Archaeology of the Human Sciences*. New York: Pantheon Books.

FOX, R., ed. 1976. "Lyell Centenary Issue: Papers Delivered at the Charles Lyell Centenary Symposium, London, 1975." *Brit. J. Hist. Sci*. 9: part 2.

FRÄNGSMYR, TORE, ed. 1984. *Linnaeus: The Man and His Work*. Berkeley, Los Angeles, London: University of California Press.

FREDERICKSON, GEORGE. 1971. *The Black Image in the White Mind: The Debate on Afro-American Character and Destiny, 1817–1914*. New York: Harper & Row.

FREEMAN, DEREK. 1974. "The Evolutionary Theories of Charles Darwin and Herbert Spencer." *Current Anthropology* 15: 211–237.

FREEMAN, R. B. 1976. *The Works of Charles Darwin: An Annotated Bibliographical Handlist*. Folkestone: Dawson.

———. 1979. *Charles Darwin: A Companion*. Folkestone: Dawson.

FROGGATT, P., and NEVIN, N. C. 1971*a*. "The 'Law of Ancestral Heredity' and the Mendelian-Ancestrian Controversy in England, 1889–1900." *J. Med. Genetics* 8: 1–36.

———. 1971*b*. "Galton's 'Law of Ancestral Heredity': Its Impact on the Early Development of Human Genetics." *Hist. Sci.* 10: 1–27.

FUTUYMA, D. 1979. *Evolutionary Biology*. Sunderland, Mass.: Sinauer.

———. 1982. *Science on Trial: The Case for Evolution*. New York: Pantheon.

GAISSINOVITCH, A. E. 1980. "The Origins of Soviet Genetics and the Struggle with Lamarckism, 1922–1929." *J. Hist. Biology* 13: 1–52.

GALE, BARRY G. 1972. "Darwin and the Concept of the Struggle for Existence: A Study in the Extra-Scientific Origins of Scientific Ideas." *Isis* 63: 321–344.

———. 1982. *Evolution without Evidence: Charles Darwin and the Origin of Species*. Albuquerque: University of New Mexico Press. London: Harvester.

GALLAGHER, IDELLA J. 1970. *Morality in Evolution: The Moral Philosophy of Henri Bergson*. The Hague: Martinus Nijhoff.

GALTON, FRANCIS. 1883. *Inquiries into Human Faculty and Its Development*. London.

———. 1889. *Natural Inheritance*. London.

———. 1892. *Hereditary Genius*. Rev. ed. London: reprinted ed. C. D. Darlington, Cleveland: Meridian Books, 1962.

———. 1908. *Memories of My Life*. London.

GASKING, ELIZABETH. 1959. "Why Was Mendel's Work Ignored?" *J. Hist. Ideas* 20: 60–84.

———. 1967. *Investigations into Generation, 1651–1828*. London: Hutchinson.

GASMAN, DANIEL. 1971. *The Scientific Origins of National Socialism: Social Darwinism in Ernst Haeckel and the Monist League*. New York: American Elsevier.

GAY, PETER. 1966, 1969. *The Enlightenment: An Interpretation*. Vol. 1. *The Rise of Modern Paganism*. Vol. 2. *The Science of Freedom*. New York: Alfred A. Knopf.

GEIKIE, SIR ARCHIBALD, ed. 1875. *Life of Sir Roderick I. Murchison*. 2 vols. London: reprinted Farnborough, Gregg, 1972.

———. 1897. *The Founders of Geology*. London: reprinted New York, Dover, 1962.

GEISON, GERALD. 1969. "Darwin and Heredity: The Evolution of his Hypothesis of Pangenesis." *J. Hist. Medicine* 24: 375–411.

GEOFFROY SAINT HILAIRE, ETIENNE. 1818–22. *Philosophie anatomique: Les organes résPiratoires sous le rapport de la détermination et de l'identité de leur pièces osseuses*. 2 vols. Paris: reprinted Brussels, Culture et Civilisation, 1968.

———. 1833. "Le degré d'influence du monde ambiant pour modifier les formes animales. . . . " *Mem. Acad. Roy. des Sciences* 12: 63–92.

GEOFFROY SAINT HILAIRE, ISIDORE. 1847. *Vie, traveaux et doctrine scientifique d'Etienne Geoffroy Saint Hilaire*. Paris and Strasburg: reprinted Brussels, Culture et Civilization.

GEORGE, WILMA. 1964. *Biologist-Philosopher: A Study of the Life and Writings of Alfred Russel Wallace*. New York: Abelard-Schuman.

GERSTNER, PATSY. 1968. "James Hutton's *Theory of the Earth* and His Theory of Matter." *Isis* 59: 26–31.
———. 1970. "Vertebrate Paleontology: An Early-Nineteenth-Century Transatlantic Science." *J. Hist. Biology* 3: 137–148.
———. 1971. "The Reaction to James Hutton's Use of Heat as a Geological Agent." *Brit. J. Hist. Sci.* 5: 353–362.
GHISELIN, MICHAEL T. 1969. *The Triumph of the Darwinian Method*. Berkeley and Los Angeles: University of California Press.
———. 1971. "The Individual in the Darwinian Revolution." *New Literary History* 3: 113–134.
———. 1974. *The Economy of Nature and the Evolution of Sex*. Berkeley and Los Angeles: University of California Press.
———, and JAFFE, L. 1973. "Phylogenetic Classification in Darwin's *Monograph on the Sub-Class Cirripedia*." *Systematic Zoology* 22: 132–140.
GILBERT, SCOTT F. 1978. "The Embryological Origins of the Gene Theory." *J. Hist. Biol.* 11: 307–351.
GILLESPIE, NEAL C. 1977. "The Duke of Argyll, Evolutionary Anthropology, and the Art of Scientific Controversy." *Isis* 68: 40–54.
———. 1979. *Charles Darwin and the Problem of Creation*. Chicago and London: University of Chicago Press.
———. 1987. "Natural History, Natural Theology, and Social Order: John Ray and the 'Newtonian Ideology.'" *J. Hist. Biol.* 20: 1–50.
GILLISPIE, CHARLES COULSTON. 1951. *Genesis and Geology: A Study in the Relations of Scientific Thought, Natural Theology and Social Opinions in Great Britain, 1790–1850*. Reprinted New York: Harper, 1959.
———. 1956. "The Formation of Lamarck's Evolutionary Theory." *Arch. Internat. Hist. Sci.* 9: 323–338.
———. 1959. "Lamarck and Darwin in the History of Science." *In* Glass et al., eds., *Forerunners of Darwin*. Pp. 265–291.
———. 1960. *The Edge of Objectivity: An Essay in the History of Scientific Ideas*. Princeton: Princeton University Press.
———, ed. in chief. 1970–80. *Dictionary of Scientific Biography*. 16 vols. New York: Charles Scribner's Sons.
GINGER, RAY. 1958. *Six Days or Forever*. Boston: Beacon Press.
GINSBERG, MORRIS. 1953. *The Idea of Progress: A Revaluation*. Reprinted Westport, Conn.: Greenwood Press, 1972.
GISH, DUANE T. 1972. *Evolution: The Fossils Say No!* San Diego: Creation Life Publishers.
GLASS, BENTLEY. 1959*a*. "Maupertuis, Pioneer of Genetics." *In* Glass et al., eds. *Forerunners of Darwin*. Pp. 51–83.
———. 1959*b*. "Heredity and Variation in the Eighteenth-Century Concept of the Species." *In* Glass et al., eds., *Forerunners of Darwin*. Pp. 144–172.
GLASS, BENTLEY, et al., eds. 1959. *Forerunners of Darwin, 1745–1859*. Baltimore: Johns Hopkins University Press.
GLICK, THOMAS F., ed. 1974. *The Comparative Reception of Darwinism*. Austin and London: University of Texas Press.
GODE VON AESCH, A. 1941. *Natural Science in German Romanticism*. New York:

Columbia University Press.

GODFREY, L. R., ed. 1983. *Scientists Confront Creationism*. New York: W. W. Norton.

GOLDSCHMIDT, RICHARD. 1940. *The Material Basis of Evolution*. New Haven: Yale University Press. Reprinted Paterson, N.J.: Pageant Books, 1960.

GOLDSMITH, DONALD, ed. 1977. *Scientists Confront Velikovsky*. Ithaca, N.Y.: Cornell University Press.

GOUGE, T. A. 1967. *The Ascent of Life*. Toronto: University of Toronto Press.

GOULD, STEPHEN JAY. 1968. "*Trigonia* and the Origin of Species." *J. Hist. Biology* 1: 41–56.

———. 1970. "Dollo on Dollo's Law: Irreversibility and the Status of Evolutionary Laws." *J. Hist. Biology* 3: 189–212.

———. 1974*a*. "On Biological and Social Determinism." *Hist. Sci.* 12: 212–220.

———. 1974*b*. "The Origin and Function of 'Bizarre' Structures: Antler Size and Skull Size in the 'Irish Elk,' *Megaloceros giganteus*." *Evolution* 28: 191–220.

———. 1977*a*. "The Eternal Metaphors of Paleontology." *In* A. Hallam, ed., *Patterns of Evolution*. Amsterdam: Elsevier. Pp. 1–26.

———. 1977*b*. *Ontogeny and Phylogeny*. Cambridge, Mass.: Harvard University Press.

———. 1977*c*. *Ever Since Darwin: Reflections in Natural History*. New York: W. W. Norton.

———. 1979. "Agassiz's Marginalia in Lyell's *Principles*, or the Perils of Uniformity and the Ambiguity of Heroes." *Stud. Hist. Biology* 3: 119–138.

———. 1980*a*. "G. G. Simpson, Paleontology, and the Modern Synthesis." *In* Mayr and Provine, eds., *The Evolutionary Synthesis*. Pp. 153–172.

———. 1980*b*. *The Panda's Thumb: More Reflections on Natural History*. New York: Norton.

———. 1980*c*. "Is a New and General Theory of Evolution Emerging?" *Paleobiology* 6: 119–130. Reprinted in Maynard Smith, ed., *Evolution Now*. Pp. 129–145.

———. 1981. *The Mismeasure of Man*. New York: Norton.

———. 1982. "The Meaning of Punctuated Equilibrium and Its Role in Validating a Hierarchical Approach to Macroevolution." *In* Roger Milkman, ed., *Perspectives on Evolution*. Pp. 83–104. Abridged in Cherfas, ed., *Darwin Up to Date*. Pp. 26–30.

———. 1983. "The Hardening of the Modern Synthesis." *In* M. Grene, ed., *Dimensions of Darwinism*. Pp. 71–93.

———. 1987. *Time's Arrow, Time's Cycles: Myth and Metaphor in the Discovery of Geological Time*. Cambridge, Mass.: Harvard University Press.

GOULD, STEPHEN J., and ELDREDGE, NILES. 1977. "Punctuated Equilibria: The Tempo and Mode of Evolution Reconsidered." *Paleobiology* 3: 115–151.

———, and LEWONTIN, R. C. 1979. "The Spandrels of San Marco and the Panglossian Paradigm: A Critique of the Adaptationist Programme." *Proc. Roy. Soc. London* B, 205: 581–598.

GRABINER, J. V., and MILLER, P. D. 1974. "Effects of the Scopes Trial: Was It a Victory for Evolutionists?" *Science* 185: 832–837.

GRANT, MADISON. 1918. *The Passing of the Great Race*. New York: reprinted New York, Arno, 1977.

GRAY, ASA. 1963. *Darwiniana: Essays and Reviews Pertaining to Darwinism*. Edited by A. Hunter Dupree. Cambridge, Mass.: Harvard University Press.

GRAYSON, DONALD K. 1983. *The Establishment of Human Antiquity*. New York: Academic Press.

GREENAWAY, A. P. 1973. "The Incorporation of Action into Associationism: The Psychology of Alexander Bain." *J. Hist. Behavioural Sci*. 9: 42–52.

GREENE, JOHN C. 1959*a*. *The Death of Adam: Evolution and Its Impact on Western Thought*. Ames: Iowa State University Press.

———. 1959*b*. "Biology and Social Theory in the Nineteenth Century: Auguste Comte and Herbert Spencer." *In* M. Clagett, ed., *Critical Problems in the History of Science*. Madison: University of Wisconsin Press. Pp. 419–466. Reprinted in Greene, *Science, Ideology, and World View*. Pp. 60–94.

———. 1961. *Darwin and the Modern World View*. Baton Rouge: Louisiana State University Press.

———. 1971. "The Kuhnian Paradigm and the Darwinian Revolution in Natural History." *In* Duane H. D. Roller, ed., *Perspectives in the History of Science and Technology*. Pp. 3–25. Reprinted in Greene, *Science, Ideology, and World View*. Pp. 30–59.

———. 1975. "Reflections on the Progress of Darwin Studies." *J. Hist. Biology* 8: 243–273.

———. 1977. "Darwin as a Social Evolutionist." *J. Hist. Biology* 10: 1–27. Reprinted in Greene, *Science, Ideology, and World View*. Pp. 95–127.

———. 1981. *Science, Ideology, and World View: Essays in the History of Evolutionary Ideas*. Berkeley, Los Angeles, London: University of California Press.

GREENE, MOTT T. 1982. *Geology in the Nineteenth Century: Changing Views of a Changing World*. Ithaca, N.Y.: Cornell University Press.

GREENWOOD, DAVYDD J. 1984. *The Taming of Evolution: The Persistence of Nonevolutionary Views in the Study of Humans*. Ithaca, N.Y.: Cornell University Press.

GRENE, MARJORIE, ed. 1983. *Dimensions of Darwinism: Themes and Counterthemes in Twentieth-Century Evolutionary Theory*. Cambridge: Cambridge University Press.

GREGORY, FREDERICK. 1977. *Scientific Materialism in Nineteenth-Century Germany*. Dordrecht: D. Reidel.

GRIBBEN, JOHN, and CHERFAS, JEREMY. 1982. *The Monkey Puzzle: A Family Tree*. London: Bodley Head.

GRINNELL, GEORGE. 1974. "The Rise and Fall of Darwin's First Theory of Transmutation." *J. Hist. Biology* 7: 259–274.

———. 1985. "The Rise and Fall of Darwin's Second Theory." *J. Hist. Biol*. 18: 50–70.

GRUBER, HOWARD E. 1968. "Who Was the *Beagle*'s Naturalist?" *Brit. J. Hist. Sci*. 4: 266–282.

———. 1974. *Darwin on Man: A Psychological Study of Scientific Creativity . . . together with Darwin's Early and Unpublished Notebooks*. New York: E. P. Dutton.

———, and GRUBER, V. 1962. "The Eye of Reason: Darwin's Development during the *Beagle* Voyage." *Isis* 53: 186–200.

GRUBER, JACOB W. 1960. *A Conscience in Conflict: The Life of St. George Jackson*

Mivart. New York: Columbia University Press.
GUEDON, JEAN-CLAUDE. 1977. "Michel Foucault: The Knowledge of Power and the Power of Knowledge." *Bull. Hist. Med.* 51: 245–277.
GUETTARD, JEAN ETIENNE. 1752. "Sur quelques montanes de France qui ont été volcans." *Hist. Acad. Roy. des Sci.* (1752): 1–8.
GULICK, ADDISON. 1932. *Evolutionist and Missionary: John Thomas Gulick*. Chicago: University of Chicago Press.
GULICK, JOHN T. 1888. "Divergent Evolution through Cumulative Segregation." *J. Linn. Soc. (Zool.)* 20: 189–274.
GURALNICK, STANLEY M. 1972. "Geology and Religion before Darwin: The Case of Edward Hitchcock, Theologian and Geologist." *Isis* 63: 529–543.
GUYENOT, EMILE. 1941. *Les Sciences de la vie aux XVIIe et XVIIIe siècles: L'idée d'évolution*. Paris: Albin Michel.
HABER, FRANCIS C. 1959. *The Age of the World: Moses to Darwin*. Baltimore: Johns Hopkins University Press.
HAECKEL, ERNST. 1876. *The History of Creation: Or the Development of the Earth and Its Inhabitants by the Action of Natural Causes. A Popular Exposition of the Doctrine of Evolution in General and of that of Darwin, Goethe and Lamarck in Particular*. 2 vols. New York.
———. 1879. *The Evolution of Man: A Popular Exposition of the Principal Points of Human Ontogeny and Phylogeny*. New York.
———. 1898. *The Last Link: Our Present Knowledge of the Descent of Man*. London.
HAGBERG, KNUT. 1953. *Carl Linnaeus*. Translated by Alan Blair. New York: Dutton.
HALDANE, J. B. S. 1932. *The Causes of Evolution*. London: reprinted Ithaca, N.Y., Cornell University Press, 1966.
———. 1938. *Heredity and Politics*. London: Allen and Unwin.
HALÉVY, ELIE. 1955. *The Growth of Philosophic Radicalism*. Translated by Mary Morris. Boston: Beacon Press.
HALL, G. STANLEY. 1904. *Adolescence: Its Psychology and Its Relation to Physiology, Anthropology, Sociology, Sex, Crime, Religion and Education*. 2 vols. New York.
HALL, THOMAS S. 1968. "On Biological Analogs of Newtonian Paradigms." *Philosophy of Science* 35: 6–27.
———. 1969. *Ideas of Life and Matter*. 2 vols. Chicago and London: University of Chicago Press.
HALLAM, ANTHONY. 1973. *A Revolution in the Earth Sciences: From Continental Drift to Place Tectonics*. Oxford: Clarendon Press.
———. 1983. *Great Geological Controversies*. Oxford: Oxford University Press.
HALLER, ALBRECHT VON. 1755. *A Dissertation on the Sensible and Irritable Parts of Animals*. Reprint edited by Owsei Temkin. Baltimore: Johns Hopkins University Press, 1936.
———. 1786. *First Lines in Physiology*. Reprint introduced by Lester S. King. London and New York: Johnson Reprint Corporation, 1966.
HALLER, JOHN S. 1975. *Outcasts from Evolution: Scientific Attitudes of Racial Inferiority, 1859–1900*. Urbana: University of Illinois Press.
HALLER, Mark H. 1963. *Eugenics: Hereditarian Attitudes in American Thought*.

New Brunswick, N.J.: Rutgers University Press.

HALLEY, EDMUND. 1724–25. "Some considerations about the cause of the universal deluge, laid before the Royal Society on the 12th of December 1694" and "Some farther thoughts upon the same subject. . . ." *Phil. Trans. Roy. Soc.* 33: 118–125.

HALLIDAY, R. J. 1971. "Social Darwinism: A Definition." *Victorian Studies* 14: 389–405.

HALSTEAD, L. B. 1983. *Hunting the Past: Fossils, Rocks, Tracks and Trails: The Search for the Origin of Life*. London: Hamish Hamilton.

HAMILTON, W. D. 1964. "The Genetical Evolution of Social Behavior, I and II." *J. Theoretical Biology* 7: 1–32.

HAMMOND, MICHAEL. 1980. "Anthropology as a Weapon of Social Combat in Late Nineteenth-Century France." *J. Hist. Behavioral Sci.* 16: 118–132.

———. 1982. "The Expulsion of the Neanderthals from Human Ancestry: Marcellin Boule and the Social Context of Scientific Research." *Social Studies of Science* 12: 1–36.

HAMPSON, NORMAN. 1968. *The Enlightenment*. Harmondsworth, Middlesex: Penguin Books.

HANKINS, THOMAS L. 1970. *Jean d'Alembert: Science in the Enlightenment*. Oxford: Clarendon Press.

HANKS, LESLIE. 1966. *Buffon avant l'Histoire naturelle*. Paris: Presses Universitaires de France.

HANSON, ANTHONY, ed. 1970. *Teilhard Reassessed: A Symposium of Critical Studies in the Thought of Pierre Teilhard de Chardin*. London: Darton, Longmann and Todd

HARDY, A. C. 1965. *The Living Stream; a Restatement of Evolution Theory*. London: Collins.

HARMSEN, H. 1955. "The German Sterilization Act of 1933." *Eugenics Review* 46: 227–232.

HARRIS, MARVIN. 1968. *The Rise of Anthropological Theory: A History of Theories of Culture*. New York: Thomas Y. Crowell.

HARRISON, JAMES. 1972. "Erasmus Darwin's Views on Evolution." *J. Hist. Ideas* 32: 247–264.

HARTLEY, DAVID. 1749. *Observations upon Man: His Frame, His Duty and His Expectations*. London: reprinted Gainsville, Fla.: Scholars' Reprint Corporation, 1966.

———. 1775. *Hartley's Theory of the Human Mind*. Edited by Joseph Priestley. London: reprinted New York, AMS, 1972.

HARTMANN, JOHANN. 1756. "Plantae Hybridae." In Linnaeus, ed., *Amoenitates Academicae* 3: 28–62.

HARWOOD, JONATHAN. 1984. "The Reception of Morgan's Chromosome Theory in Germany." *Medizin Historisches Journal* 9: 3–32.

———. 1985. "Genetics and the Evolutionary Synthesis in Interwar Germany." *Annals of Science* 42: 279–301.

HASTINGS, HESTER. 1936. "Man and Beast in French Thought of the Eighteenth Century," *Johns Hopkins Studies in Romance Literature and Languages* 27 (1): 19–63.

HATCH, ELVIN. 1973. *Theories of Man and Culture*. New York: Columbia University Press.

HAZARD, PAUL. 1953. *The European Mind, 1680–1715*. Trans. J. Lewis May. London: Hollis and Carter.

———. 1963. *European Thought in the Eighteenth Century: from Montesquieu to Lessing*. Translated by J. Lewis May. Cleveland and New York: Meridian Books.

HEGEL, G. W. F. 1953. *Reason in History: a General Introduction to the Philosophy of History*. Translated by Robert S. Hartman. New York: Bobbs Merrill.

HEIM, ROGER, ed. 1952. *Les grands naturalists français: Buffon*. Paris: Museum d'histoire naturelle.

HEIMANN, P. M., and MCGUIRE, J. E. 1971. "Newtonian Forces and Lockean Powers: Concepts of Matter in Eighteenth Century Thought." *Hist. Stud. Phys. Sci.* 3: 233–306.

HELFAND, M. S. 1977. "T. H. Huxley's 'Evolution and Ethics': The Politics of Evolution and the Evolution of Politics." *Victorian Studies* 20: 159–177.

HELVÉTIUS, CLAUDE ADRIEN. 1810. *A Treatise on Man: His Intellectual Faculties and His Education*. Translated by W. Hooper. Reprinted New York: Franklin, 1969.

HEMPEL, CARL. 1966. *Philosophy of Natural Science*. Englewood Cliffs, N.J.: Prentice-Hall.

HENFREY, ARTHUR, and HUXLEY, T. H., eds. 1853. *Scientific Memoirs: Selected from the Transactions of Foreign Academies of Science and from Foreign Journals: Natural History*. London: reprinted New York, Johnson Reprint Corporation, 1966.

HENKIN, LEO J. 1963. *Darwinism in the English Novel, 1860–1910*. New York: Russell and Russell.

HENNIG, WILLI. 1966. *Phylogenetic Systematics*. Trans. D. Dwight Davis and Rainer Zangerl. Urbana: University of Illinois Press.

HENSLOW, GEORGE. 1888. *The Origin of Floral Structures Through Insect and Other Agencies*. London.

———. 1895. *The Origin of Plant Structures by Self-Adaptation to the Environment*. London.

HERBERT, SANDRA. 1971. "Darwin, Malthus and Selection." *J. Hist. Biology* 4: 209–218.

———. 1974–77. "The Place of Man in the Development of Darwin's Theory of Transmutation. Part I, to July 1837." *J. Hist. Biology* 7: 217–258. ". . . Part II." *Ibid.*, 10: 155–227.

HERDER, JOHANN GOTTFRIED VON. 1968. *Reflections on the Philosophy of the History of Mankind*. Translated by T. O. Churchill, edited by Frank E. Manuel. Chicago: University of Chicago Press.

HERSCHEL, Sir J. F. W. 1830. *A Preliminary Discourse on the Study of Natural Philosophy*. London, reprinted with an introduction by Michael Partridge, New York: Johnson Reprint Corporation, 1966.

HEYER, PAUL. 1982. *Nature, Human Nature, and Society: Marx, Darwin, Biology and the Human Sciences*. Westport, Conn.: Greenwood Press.

HILL, EMITA. 1968. "Materialism and Monsters in *Le rêve de d'Alembert*." *Diderot Studies* 10: 69–94.

HIMMELFARB, GERTRUDE. 1959. *Darwin and the Darwinian Revolution*. Reprinted New York: Norton.
HO, MAE-WAN, and SAUNDERS, PETER T., eds. 1984. *Beyond Neo-Darwinism: An Introduction to the New Evolutionary Paradigm*. London: Academic Press.
HOBBES, THOMAS. 1957. *Leviathan: Or the Matter, Forme and Power of a Commonwealth, Ecclesiastical and Civil*. Ed. Michael Oakeshott. Oxford: Blackwell.
HODGE, M. J. S. 1971. "Lamarck's Science of Living Bodies." *Brit. J. Hist. Sci*. 5: 323–352.
———. 1972. "The Universal Gestation of Nature: Chambers' *Vestiges* and *Explanations*." *J. Hist. Biology* 5: 127–152.
———. 1977. "The Structure and Strategy of Darwin's 'Long Argument.'" *Brit. J. Hist. Sci*. 10: 237–246.
———. 1982. "Darwin and the Laws of the Animate Part of the Terrestrial System (1835–1837): On the Lyellian Origins of His Zoonomical Explanatory Programme." *Stud. Hist. Biol*. 6: 1–106.
———. 1985. "Darwin as a Lifelong Generation Theorist." *In* D. Kohn, ed., *The Darwinian Heritage*. Pp. 207–243.
HODGE, M. J. S., and KOHN, D. 1985. "The Immediate Origins of Natural Selection." *In* D. Kohn, ed., *The Darwinian Heritage*. Pp. 185–206.
HOFSTADTER, RICHARD. 1959. *Social Darwinism in American Thought*. Revised edition, New York: George Braziller.
HOLT, NILES R. 1971. "Ernst Haeckel's Monist Religion." *J. Hist. Ideas* 32: 265–280.
HOOKE, ROBERT. 1705. *The Posthumous Works of Robert Hooke*. London: reprinted New York, Johnson, 1969.
———. 1977. *Lectures and Discourses of Earthquakes and Subterraneous Eruptions*. Reprinted New York: Arno Press, 1977.
HOOKER, JOSEPH DALTON. 1860. "On the Origination and Distribution of Vegetable Species: Introductory Essay to the Flora of Tasmania." *Am. J. Sci*. 2d series, 29: 1–25, 305–326.
HOOYKAAS, R. 1957. "The Parallel between the History of the Earth and the History of the Animal World." *Arch. Internat. Hist. Sci*. 10: 3–18.
———. 1959. *Natural Law and Divine Miracle: The Principle of Uniformity in Geology, Biology and History*. Leiden: Brill.
———. 1966. "Geological Uniformitarianism and Evolution." *Arch. Internat. Hist. Sci*. 19: 3–19.
———. 1970. *Catastrophism in Geology: Its Scientific Character in Relation to Actualism and Uniformitarianism*. Amsterdam and London: North Holland Publishing Co.
HORDER, T. J., WITKOWSKI, J. A., and WYLIE, C. C., eds. 1986. *A History of Embryology*. Cambridge: Cambridge University Press.
HOSKIN, M. A. 1964. *William Herschel and the Construction of the Heavens*. New York: W. W. Norton, Inc.
HOYLE, FRED. 1981. *Ice*. London: Hutchinson.
HOYLE, FRED, and WICKRAMSINGHE, CHANDRA. 1978. *Lifecloud: The Origin of Life in the Universe*. London: J. M. Dent.
———. 1981. *Evolution from Space*. London: J. M. Dent.
———. 1986. *Archaeopteryx: The Primordial Bird*. Swansea: Christopher Davies.

HULL, DAVID. 1973*a*. "Charles Darwin and Ninteenth Century Philosophies of Science." *In* N. Giere and R. S. Westfall, eds., *Foundations of Scientific Thought: The Nineteenth Century*. Bloomington: Indiana University Press.

———. 1973*b*. *Darwin and His Critics: The Reception of Darwin's Theory of Evolution by the Scientific Community*. Cambridge, Mass.: Harvard University Press.

———. 1978. "Sociobiology: a Scientific Bandwagon or a Travelling Medicine Show?" *In* M. S. Gregory, et al., eds., *Sociobiology and Human Nature*. San Francisco: Jossey-Bass. Pp. 136–163.

———. 1979. "The Limits of Cladism." *Systematic Zoology* 28: 416–440.

———. et al. 1978. "Plank's Principle: Do Younger Scientists Accept New Scientific Ideas with Greater Alacrity than Older Scientists?" *Science* 202: 717–723.

HUMBOLDT, ALEXANDER VON. 1814–29. *Personal Narrative of Travels to the Equinoctal Regions of the New Continent During the Years 1799–1804*. 7 vols. Translated by Helen Maria Williams. London: reprinted New York, AMS Press, 1966.

HUME, DAVID. 1888. *A Treatise on Human Nature*. Edited by A. Selby Bigge. Oxford: Clarendon Press.

———. 1942. *Dialogues Concerning Natural Religion*. Edited by Norman Kemp Smith. New York: Bobbs Merrill.

———. 1962. *On Human Nature and the Understanding*. Edited by Anthony Flew. New York: Collier.

HUNTINGTON, ELLSWORTH. 1925. *The Character of Races*. New York: reprinted New York, Arno, 1977.

HUTTON, JAMES. 1795. *Theory of the Earth, with Proofs and Illustrations*. 2 vols. Edinburgh: reprinted Weinheim/Bergstr., H. R. Engelmann (J. Cramer) and Codiocote, Herts., Wheldon and Wesley, 1960.

HUXLEY, JULIAN S. 1932. *Problems of Relative Growth*. London: Methuen.

———, ed. 1940. *The New Systematics*. Oxford: Oxford University Press.

———. 1953. *Evolution in Action; Based on the Pattern Foundation Lectures Delivered at Indiana University in 1951*. London: Chatto & Windus.

———, ed. 1961. *The Humanist Frame*. London: Allen and Unwin.

———. 1963. *Evolution: The Modern Synthesis*. New ed., London: Chatto & Windus.

———. 1970. *Memories*. London: Allen and Unwin.

HUXLEY, LEONARD. 1900. *The Life and Letters of Thomas Henry Huxley*. 2 vols. London: reprinted Farnborough, Gregg International, 1969.

———. 1918. *The Life and Letters of Sir Joseph Dalton Hooker*. . . . 2 vols. London.

HUXLEY, T. H. 1854. "Vestiges of the Natural History of Creation." *Brit. & Foreign Med. Chirurg. Rev*. 13: 332–343.

———. 1863. *Man's Place in Nature*. London: Reprinted in Huxley, *Collected Essays*, vol. 7.

———. 1887. "The Reception of the 'Origin of Species.'" In F. Darwin, ed. *Life and Letters of Charles Darwin*. 2: 179–204.

———. 1888. *American Addresses: With a Lecture on the Study of Biology*. New York.

———. 1893–94. *Collected Essays*. 9 vols. London: reprinted Hildesheim, Georg

Olms, 1970.
———. 1893. *Darwiniana*. Collected Essays, vol. 2. London: reprinted New York, AMS Press, 1970.
———. 1894. *Evolution and Ethics*. Collected Essays, vol. 9. Reprinted New York: AMS Press, 1970.
HYATT, ALPHEUS. 1866. "On the Parallelism between the Different Stages of Life in the Individual and Those in the Entire Group of the Molluscous Order Tetrabranchiata." *Mem. Boston Soc. Nat. Hist.* 1: 193–209.
———. 1880. "Genesis of the Tertiary Species of Planorbis at Steinheim." *Boston Soc. Nat. Hist., Anniversary Memoir*. Abstracted *Proc. A.A.A.S.* 1880, 527–550 and *Am. Naturalist* 16 (1882): 441–453.
———. 1884. "Evolution of the Cephalopods." *Science* 3: 122–127 and 145–149.
———. 1889. *Genesis of the Arietidae*. Washington.
ILTIS, HUGO. 1932. *Life of Mendel*. Translated by Eden and Cedar Paul. Reprinted New York: Hafner, 1966.
INGOLD, TIM. 1987. *Evolution and Social Life*. Cambridge: Cambridge University Press.
IRVINE, WILLIAM. 1955. *Apes, Angels and Victorians: The Story of Darwin, Huxley and Evolution*. London: reprinted Cleveland, Meridian Books, 1959.
JAKI, STANLEY. 1978*a*. *The Road of Science and the Way to God*. Chicago: University of Chicago Press.
———. 1978*b*. *Planets and Planetarians: A History of Theories of the Origin of Planetary Systems*. Edinburgh: Scottish University Press.
JAMES, WILLIAM. 1901–05. *Principles of Psychology*. 2 vols. London.
JAMESON, ROBERT. 1804–08. *A System of Mineralogy*. 3 vols. Edinburgh.
———. 1808. *The Wernerian Theory of the Neptunian Origin of Rocks*. Reprint of vol. 3 of *System of Mineralogy*. Introduced by Jesse M. Sweet. New York: Hafner/London: Collier Macmillan, 1976.
JENKIN, FLEEMING. 1867. "The Origin of Species." *North British Review* 46: 277–318.
JOHANNSEN, WILHELM. 1955. "Concerning Heredity in Populations and in Pure Lines." Translated by Harold Gall and Elga Putsch. In *Selected Readings in Biology for Natural Sciences*. Chicago: University of Chicago Press. Pp. 172–215.
JOHANSON, D. C., and EDEY, M. A. 1981. *Lucy: The Beginnings of Humankind*. New York: Simon and Schuster.
JONES, GRETA. 1980. *Social Darwinism and English Thought: The Interaction between Biological and Social Theory*. London: Harvester Press.
———. 1986. *Social Hygiene in Twentieth-Century Britain*. London: Croom Helm.
JONES, RICHARD FOSTER. 1965. *Ancients and Moderns: A Study of the Rise of the Scientific Movement in Seventeenth-Century England*. 2d ed., Berkeley and Los Angeles: University of California Press.
JORAVSKY, D. 1970. *The Lysenko Affair*. Cambridge, Mass.: Harvard University Press.
JORDAN, Z. A. 1967. *The Evolution of Dialectical Materialism*. New York: St. Martin's Press.
JORDANOVA, L. 1984. *Lamarck*. Oxford: Oxford University Press.

JORDANOVA, L., and PORTER, R. S., eds. 1979. *Images of the Earth: Essays in the History of the Environmental Sciences*. Chalfont St. Giles, Bucks.: British Society for the History of Science.

KAMIN, LEON. 1972. *The Science and Politics of I.Q*. Potomac, Md.: Erlbaum.

KAMMERER, PAUL. 1923. "Breeding Experiments on the Inheritance of Acquired Characters." *Nature* 111: 637–640.

———. 1924. *The Inheritance of Acquired Characteristics*. New York: Boni and Liveright.

KANT, IMMANUEL. 1934. *Critique of Pure Reason*. Translated by J. M. D. Meiklejohn. London: Everyman.

———. 1963. *On History*. Edited by Lewis White Beck. New York: Bobbs Merrill.

———. 1969. *Universal Natural History and Theory of the Heavens*. Introduction by Milton K. Munitz. Ann Arbor: University of Michigan Press.

KAYE, HOWARD L. 1986. *The Social Meaning of Modern Biology: From Social Darwinism to Sociobiology*. New Haven: Yale University Press.

KEITH, SIR ARTHUR. 1915. *The Antiquity of Man*. London.

———. 1949. *A New Theory of Human Evolution*. New York: Philosophical Library.

———. 1955. *Darwin Revalued*. London: Watts.

KELLOGG, VERNON L. 1907. *Darwinism Today: A Discussion of Present Day Scientific Criticism of the Darwinian Selection Theories . . .* New York and London.

KELLY, ALFRED. 1981. *The Descent of Darwinism: The Popularization of Darwinism in Germany, 1860–1914*. Chapel Hill: University of North Carolina Press.

KELLY, SUZANNE. 1969. "Theories of the Earth in Renaissance Cosmologies." *In* Schneer, ed., *Toward a History of Geology*. Pp. 214–225.

KENNEDY, JAMES G. 1978. *Herbert Spencer*. Boston: Twayne Publishers.

KERKUT, G. A. 1960. *The Implications of Evolution*. Oxford and New York: Pergamon Press.

KETTLEWELL, H. B. D. 1955. "Selection Experiments on Industrial Melanism in the Lepidoptera." *Heredity* 9: 323–342.

———. 1973. *The Evolution of Melanism*. Oxford: Clarendon Press.

KEVLES, DANIEL. 1985. *In the Name of Eugenics: Genetics and the Uses of Human Heredity*. New York: Knopf.

KEYNES, RICHARD DARWIN. 1979. *The Beagle Record: Selections from the Original Pictorial Records and Written Accounts of the Voyage of H.M.S. Beagle*. Cambridge, London and New York: Cambridge University Press.

KIMURA, MOTOO. 1983. *The Neutral Theory of Molecular Evolution*. Cambridge: Cambridge University Press.

KIMURA, M., and OHTA, T. 1971. *Theoretical Aspects of Population Genetics*. Princeton: Princeton University Press.

KINCH, MICHAEL PAUL. 1980. "Geographical Distribution and the Origin of Life: The Development of Early Nineteenth-Century British Explanations." *J. Hist. Biology* 13: 91–119.

KING, LESTER S. 1964. "Stahl and Hoffmann: A Study in Eighteenth Century Animism." *J. Hist. Medicine* 19: 118–130.

———. 1967. "Basic Concepts of Eighteenth Century Animism." *Am. J. Psychiatry* 124: 797–802.

KING-HELE, DESMOND. 1963. *Erasmus Darwin*. New York: Scribner.

KIRWAN, RICHARD. 1799. *Geological Essays*. London, reprinted New York: Arno Press, 1977.
KITCHER, PHILIP. 1982. *Abusing Science: The Case against Creationism*. Cambridge, Mass.: MIT Press.
KITTS, DAVID B. 1977. "Karl Popper, Univerifiability, and Systematic Zoology." *Systematic Zoology* 26: 185–194.
KNIGHT, ISOBEL F. 1968. *The Geometric Spirit: The Abbé de Condillac and the French Enlightenment*. New Haven: Yale University Press.
KOESTLER, ARTHUR. 1967. *The Ghost in the Machine*. New York: Macmillan.
———. 1971. *The Case of the Midwife Toad*. London: Hutchinson.
———. 1972. *The Roots of Coincidence*. London: Hutchinson.
———, and SMITHIES, J. R., eds. 1969. *Beyond Reductionism: New Perspectives in the Life Sciences*. Reprinted Boston: Beacon Press, 1971.
KOFAHL, R. E., and SEGRAVES, K. L. 1975. *The Creation Explanation: A Scientific Alternative to Evolution*. Wheaton, Ill.: Harold Shaw.
KOHN, DAVID. 1980. "Theories to Work By: Rejected Theories, Reproduction, and Darwin's Path to Natural Selection." *Stud. Hist. Biol.* 4: 67–170.
———. 1981. "On the Principle of Diversity." *Science* 213: 1105–1108.
———. ed., 1985. *The Darwinian Heritage: A Centennial Retrospect*. Princeton: Princeton University Press.
KÖLREUTER, JOSEPH GOTTLEIB. 1761–66. *Vorläufige Nachricht von einigen Beobachtungen nebst Forsetzungen 1, 2 & 3*. Reprinted Leipzig: Ostwaldt's Klassiker der Exacten Wissenschaften, 1893.
KONNER, MELVIN. 1982. *The Tangled Wing: Biological Constraints on the Human Spirit*. New York: Holt, Rinehart & Winston.
KOTTLER, MALCOLM JAY. 1974. "Alfred Russel Wallace, the Origin of Man, and Spiritualism." *Isis* 65: 145–192.
———. 1978. "Charles Darwin's Biological Species Concept and Theory of Geographic Speciation: The Transmutation Notebooks." *Annals of Science* 35: 275–297.
———. 1979. "Hugo De Vries and the Rediscovery of Mendel's Laws." *Annals of Science* 36: 517–538.
———. 1980. "Darwin, Wallace, and the Origin of Sexual Dimorphism." *Proc. Am. Phil. Soc.*: 124: 203–226.
———. 1985. "Charles Darwin and Alfred Russel Wallace: Two Decades of Debate over Natural Selection." *In* D. Kohn, ed., *The Darwinian Heritage*. Pp. 367–432.
KRAUSE, ERNST. 1879. *Erasmus Darwin*. Translated by W. S. Dallas, with a preliminary notice by Charles Darwin. London: reprinted Farnborough, Gregg, 1971.
KROEBER, A. L. 1917. "The Superorganic." *Am. Anthropologist* n.s. 19: 163–213.
KROPOTKIN, PETER. 1902. *Mutual Aid: A Factor in Evolution*. London: reprinted with an introduction by Ashley Montagu, Boston, Extending Horizon Books.
KUBRIN, D. 1967. "Newton and the Cyclical Cosmos: Providence and the Mechanical Philosophy." *J. Hist. Ideas* 28: 325–346.
KUHN, THOMAS S. 1962. *The Structure of Scientific Revolutions*. Chicago: University of Chicago Press. Reprinted 1969.
LACK, DAVID. 1947. *Darwin's Finches*. Cambridge: Cambridge University Press.

LAKATOS, IMRE, and MUSGRAVE, ALAN. 1970. *Criticism and the Growth of Knowledge*. Cambridge: Cambridge University Press.

LAMARCK, JEAN BAPTISTE PIERRE ANTOINE DE MONET, Chevalier de. 1815–22. *Histoire naturelle des animaux sans vertèbres*. 6 vols. Paris: reprinted Brussels, Culture et Civilisation, 1969.

———. 1873. *Philosophie zoologique: ou exposition des considérations relatives à l'histoire naturelle des animaux* . . . Edited by Charles Martin. 2 vols. Paris.

———. 1914. *Zoological Philosophy*. Translated by Hugh Elliot. London: reprinted New York, Hafner, 1963.

———. 1964. *Hydrogeology*. Translated by Albert V. Carozzi. Urbana: University of Illinois Press.

LA METTRIE, JULIEN OFFRAY DE. 1774. *Oeuvres philosophiques*. Berlin, 2 vols.

———. 1960. *L'homme machine*. Edited by Aram Vartanian. Princeton: Princeton University Press.

LANHAM, URL. 1973. *The Bone Hunters*. New York: Columbia University Press.

LAPLACE, PIERRE SIMON, MARQUIS DE. 1830. *The System of the World*. 2 vols. Translated by H. H. Harte. Dublin and London.

LARSON, JAMES L. 1971. *Reason and Experience: The Representation of Natural Order in the Work of Carl von Linné*. Berkeley, Los Angeles, London: University of California Press.

LAUDAN, RACHEL. 1977. "Ideas and Organization in British Geology: A Case Study in Institutional History." *Isis* 67: 527–538.

———. 1982. "The Role of Methodology in Lyell's Science." *Stud. Hist. & Phil. Sci.* 13: 215–249.

———. 1987. *From Mineralogy to Geology: The Foundations of a Science, 1650–1830*. Chicago: University of Chicago Press.

LAWRENCE, PHILIP. 1977. "Heaven and Earth: The Relation of the Nebular Hypothesis to Geology." In Wolfgang Yourgrau & Allen D. Breck, eds., *Cosmology, History, and Theology*. New York: Plenum. Pp. 253–281.

———. 1978. "Charles Lyell versus the Theory of Central Heat: A Reappraisal of Lyell's Place in the History of Geology." *J. Hist. Biology* 11: 101–128.

LEAKEY, L. S. B., and GOODALL, VANNE MORRIS. 1969. *Unveiling Man's Origins: Ten Decades of Thought about Human Evolution*. Cambridge, Mass.: Schenkman Publishing Co.

———, and PROST, JACK and JOSEPHINE, eds. 1971. *Adam or Ape: A Sourcebook of Discoveries about Early Man*. Cambridge, Mass.: Schenkman Publishing Co.

LEAKEY, RICHARD, and LEWIN, ROBERT. 1977. *Origins: What New Discoveries Reveal about the Emergence of our Species and its Possible Future*. London: Macdonald and Jane's.

LE CONTE, JOSEPH. 1899. *Evolution: Its Nature, Its Evidences and Its Relation to Religious Thought*. 2d ed. New York: reprinted New York: Kraus, 1970.

LECOURT, DOMINIQUE. 1977. *Proletarian Science? The Case of Lysenko*. Introduced by Louis Althusser. London: NLB Books.

LE MAHIEU, D. L. 1976. *The Mind of William Paley: A Philosopher of His Age*. Lincoln, Nebraska and London: University of Nebraska Press.

LENOIR, TIMOTHY. 1978. "Generational Factors in the Origin of *Romantische Naturphilosophie*." *J. Hist. Biol.* 11: 57–100.

———. 1982. *The Strategy of Life: Teleology and Mechanics in Nineteenth-Century German Biology*. Dordrecht: D. Reidel.

LESCH, JOHN E. 1975. "The Role of Isolation in Evolution: George J. Romanes and John T. Gulick." *Isis* 66: 483–503.

LEWIN, ROGER. 1984. *Human Evolution: An Illustrated Introduction*. Oxford: Basil Blackwell.

LEWONTIN, R. C. 1974. *The Genetic Basis of Evolutionary Change*. New York: Columbia University Press.

———. 1978. "Adaptation." *Sci. American* 239: 156–169.

LEWONTIN, RICHARD, and LEVINS, RICHARD. 1976. "The Problem of Lysenkoism." *In* Hilary and Steven Rose, eds., *The Radicalization of Science: Ideology of/in the Natural Sciences*. London: Macmillan. Pp. 32–64.

LIMOGES, CAMILLE. 1970. *Le selection naturelle: étude sur le première construction d'un concept*. Paris: Presses Universitaires de France.

———. 1976. "Natural Selection, Phagocytosis and Preadaptation: Lucien Cuénot, 1886–1901." *J. Hist. Medicine* 31: 176–214.

LINNAEUS, CAROLUS (Karl von Linné). 1735. *Systema Naturae*. Reprinted ed. M. S. J. Engel-Ledeboer and H. Engel, Nieuwkoop: B. De Graff, 1964.

———. 1736. *Fundamenta Botanica*. Reprinted Munich: W. Frisch, 1968.

———, ed. 1749–90. *Amoenitates Academicae, seu Dissertationes variae physicae, medicae et botanicae*. 10 vols. Leiden/Amsterdam.

———. 1751. *Philosophia Botanica*. Reprinted Codicote, Herts/New York: Wheldon and Wesley/Stechert Hafner Service Agency, 1966.

———. 1753. *Species Plantarum*. 2 vols. Reprinted London: Ray Society, 1955–59.

———. 1754. *Genera Plantarum*. 5th ed., reprinted Weinheim: J. Cramer, 1960.

———. 1758–59. *Systema Naturae*. 10th ed., reprinted vol. 1, London: Ray Society, 1956; vol. 2, Weinheim: J. Cramer, 1964.

———. 1760. *Dissertation de Sexu Plantarum*. Reprinted in Linnaeus, ed. *Amoenitates Academicae* 10: 100–131.

———. 1781. "On the increase of the Habitable Earth." Translated in F. J. Brand, *Select Dissertations from the Amoenetates Academicae: A Supplement to Mr. Stillingfleet's Tracts relating to Natural History*. London.

LIVINGSTONE, DAVID. 1987. *Darwin's Forgotten Defenders: The Encounter between Evangelical Theology and Evolutionary Thought*. Edinburgh: Scottish Universities Press. Grand Rapids, Mich.: Eerdmans.

LOCKE, JOHN. 1960. *Two Treatises of Government*. Edited by Peter Laslett. Cambridge: Cambridge University Press.

———. 1975. *An Essay Concerning Human Understanding*. Edited by P. Nidditch. Oxford: Clarendon Press.

LOEWENBERG, BERT JAMES. 1959. *Darwin, Wallace and the Theory of Natural Selection: Including the Linnean Society Papers*. Cambridge, Mass.: Arlington Books.

———. 1965. "Darwin and Darwin Studies." *History of Science* 4: 15–54.

———. 1969. *Darwin Comes to America: 1859–1900*. Philadelphia: Fortress Press.

LORENZ, KONRAD. 1966. *On Aggression*. Translated by Marjorie Kerr Wilson. New York: Harcourt, Brace and World.

LOVEJOY, ARTHUR O. 1936. *The Great Chain of Being: A Study in the History of*

an Idea. Reprinted New York: Harper, 1960.
———. 1959*a*. "Buffon and the Problem of Species." *In* Glass et al., eds., *Forerunners of Darwin*. Pp. 84–113.
———. 1959*b*. "Herder: Progressionism without Transformism." *In* Glass et al., eds., *Forerunners of Darwin*. Pp. 207–221.
———. 1959*c*. "The Argument for Organic Evolution before the *Origin of Species*, 1830–1858." *In* Glass et al., eds., *Forerunners of Darwin*. Pp. 356–414.
———. 1959*d*. "Schopenhauer as an Evolutionist." *In* Glass et al., eds., *Forerunners of Darwin*. Pp. 415–437.
———. 1959*e*. "Recent Criticism of the Darwinian Theory of Recapitulation: Its Ground and Its Initiator." *In* Glass et al., eds., *Forerunners of Darwin*. Pp. 438–458.
LOVEJOY, C. O. 1981. "The Origin of Man." *Science* 211: 341–350.
LØVTRUP, SØREN, 1977. *The Phylogeny of the Vertebrata*. London & New York: John Wiley.
LUBBOCK, JOHN. 1870. *The Origin of Civilization and the Primitive Condition of Man*. New York.
LUCAS, J. R. 1979. "Wilberforce and Huxley: A Legendary Encounter." *Historical Journal* 22: 313–330.
LUDMERER, KENNETH M. 1972. *Genetics and American Society: A Historical Appraisal*. Baltimore: Johns Hopkins University Press.
LUMSDEN, C. J., and WILSON, E. O. 1981. *Genes, Mind and the Coevolutionary Process*. Cambridge, Mass.: Harvard University Press.
LURIE, EDWARD. 1960. *Louis Agassiz: A Life in Science*. Chicago: University of Chicago Press.
———. 1959–60. "Louis Agassiz and the Idea of Evolution." *Victorian Studies* 3: 87–108.
LYELL, CHARLES. 1830–33. *Principles of Geology: Being an Attempt to Explain the Former Changes of the Earth's Surface by Reference to Causes Now in Operation*. 3 vols. London: reprinted with an introduction by M. J. S. Rudwick, Lehre, Cramer/Codicote, Herts., Wheldon and Wesley/ New York, Strechert-Hafner, 1970.
———. 1845. *Travels in North America in the Years 1841–42*. 2 vols. New York: reprinted New York, Arno Press, 1977.
———. 1851. Presidential Address. *Quart. J. Geol. Soc. Lond.* 7: 25–76.
———. 1863. *Geological Evidences of the Antiquity of Man: With Remarks on Theories of the Origin of Species by Variation*. London. 4th edition, London, 1873, reprinted New York: AMS Press, 1973.
———. 1970. *Sir Charles Lyell's Journals on the Species Question*. Edited by Leonard J. Wilson. New Haven and London: Yale University Press.
LYELL, MRS. K. M., ed. 1881. *The Life, Letters and Journals of Sir Charles Lyell*. 2 vols. London: reprinted Farnborough, Gregg, 1970.
MACBETH, NORMAN. 1971. *Darwin Retried: An Appeal to Reason*. Boston: Gambit Inc.
MACBRIDE, E. W. 1924. *An Introduction to the Study of Heredity*. London: Williams and Norgate.
MCCONNAUGHEY, GLORIA. 1950. "Darwin and Social Darwinism." *Osiris* 9: 394–412.

McDougall, William. 1927. "An Experiment for the Testing of the Hypothesis of Lamarck." *Brit. J. Psych.* 17: 267–304.

Mackenzie, Donald. 1976. "Eugenics in Britain." *Social Studies of Science*. 6: 499–532.

———. 1978. "Statistical Theory and Social Interests." *Social Studies of Science* 8: 35–84.

———. 1982. *Statistics in Britain, 1865–1930: The Social Construction of Scientific Knowledge*. Edinburgh: Edinburgh University Press.

McKinney, H. Lewis. 1966. "Alfred Russel Wallace and the Theory of Natural Selection." *J. Hist. Medicine* 21: 333–359.

———, ed. 1971. *Lamarck to Darwin: Contributions to Evolutionary Biology*. Lawrence, Kansas: Coronado Press.

———. 1972. *Wallace and Natural Selection*. New Haven: Yale University Press.

Macleod, Roy M. 1965. "Evolutionism and Richard Owen." *Isis* 56: 259–280.

McPherson, Thomas. 1972. *The Argument from Design*. London: Macmillan.

Maienschein, Jane. 1978. "Cell Lineage, Ancestral Reminiscence, and the Biogenetic Law." *J. Hist. Biol.* 11: 129–158.

———. 1984. "What Determines Sex: A Study of Converging Research Approaches." *Isis* 75: 457–480.

Malthus, Thomas Robert. 1914. *An Essay on the Principle of Population*. 2 vols. 7th ed., reprinted London: Everyman.

———. 1959. *Population: The First Essay*. Ann Arbor: University of Michigan Press.

Mandelbaum, Maurice. 1957. "The Scientific Background to Evolutionary Theory in Biology." *In* P. Wiener and A. Noland, eds., *Roots of Scientific Thought*. New York: Basic Books. Pp. 517–536.

———. 1958. "Darwin's Religious Views." *J. Hist. Ideas* 19: 363–378.

———. 1971. *History, Man and Reason: A Study in Nineteenth Century Thought*. Baltimore: Johns Hopkins University Press.

Manier, Edward. 1978. *The Young Darwin and His Cultural Circle: A Study of the Influences which Shaped the Language and Logic of the Theory of Natural Selection*. Dordrecht, Holland: D. Reidel.

———. 1980. "History, Philosophy and Sociology of Biology: A Family Romance." *Stud. Hist. & Phil. Sci.* 11: 1–24.

Manser, A. R. 1965. "The Concept of Evolution." *Philosophy* 40: 18–34.

Mantell, Gideon. 1831. "The Geological Age of Reptiles." *Edinburgh New Phil. J.* 11: 181–185.

Manuel, F. E. 1956. *The New World of Henri de Saint-Simon*. Cambridge, Mass.: Harvard University Press.

Marchant, James. 1916. *Alfred Russel Wallace: Letters and Reminiscences*, London: reprinted New York, Arno Press.

Marcou, Jules, 1896. *The Life, Letters and Work of Louis Agassiz*. New York: reprinted Farnborough, Gregg, 1972.

Marsh, Othniel C. 1880. *Odontornithes: A Monograph on the Extinct Toothed Birds of North America*. Washington: Report of the Geological Exploration of the Fortieth Parallel, vol. 8.

Mason, Stephen F. 1968. *A History of the Sciences*. New York: Collier.

Maupertuis, Pierre Louis Moreau de. 1768. *Oeuvres*. 4 vols. Reprinted Hildesheim: Georg Olms, 1968.

———. 1968. *The Earthly Venus*. Translated by Simon Brangier Boas. New York: Johnson Reprint Corporation.

MAYR, ERNST. 1942. *Systematics and the Origin of Species*. New York: Columbia University Press.

———. 1954. "Wallace's Line in the Light of Recent Zoogeographic Studies." *Quart. Rev. Biology* 29: 1–14. Reprinted in Mayr, *Evolution and the Diversity of Life*. Pp. 626–645.

———. 1955. "Karl Jordan's Contribution to Current Concepts in Systematics and Evolution." *Trans. Roy. Entomological Soc. London* 107: 45–66. Reprinted in Mayr, *Evolution and the Diversity of Life*. Pp. 135–143, 297–306, and 485–492.

———. 1959*a*. "Agassiz, Darwin and Evolution." *Harvard Library Bulletin* 12: 165–194. Reprinted in Mayr, *Evolution and the Diversity of Life*. Pp. 251–276.

———. 1959*b*. "Isolation as an Evolutionary Factor." *Proc. Am. Phil. Soc.* 103: 221–230. Reprinted in Mayr, *Evolution and the Diversity of Life*. Pp. 120–134.

———. 1959*c*. "Where are We?" Reprinted in Mayr, *Evolution and the Diversity of Life*. Pp. 307–328.

———. 1963. *Animal Species and Evolution*. Cambridge, Mass.: Harvard University Press.

———. 1964. Introduction to Charles Darwin, *On the Origin of Species*. Facsimile reprint, Cambridge, Mass.: Harvard University Press.

———. 1972*a*. "The Nature of the Darwinian Revolution." *Science* 176: 981–989. Reprinted in Mayr, *Evolution and the Diversity of Life*. Pp. 277–296.

———. 1972*b*. "Lamarck Revisited." *J. Hist. Biology* 5: 55–94. Reprinted in Mayr, *Evolution and the Diversity of Life*. Pp. 222–250.

———. 1973. "The Recent Historiography of Genetics." Reprinted in Mayr, *Evolution and the Diversity of Life*. Pp. 329–353.

———. 1976. *Evolution and the Diversity of Life*. Cambridge, Mass.: Harvard University Press.

———. 1977. "Darwin and Natural Selection: How Darwin may have Discovered His Highly Unconventional Theory." *Am. Scientist* 65: 321–377.

———. 1982. *The Growth of Biological Thought: Diversity, Evolution and Inheritance*. Cambridge, Mass.: Harvard University Press.

———. 1985. "Weismann and Evolution." *J. Hist. Biol.* 18: 259–322.

MAYR, ERNST, and PROVINE, WILLIAM B. 1980. *The Evolutionary Synthesis: Perspectives on the Unification of Biology*. Cambridge, Mass.: Harvard University Press.

MEAD, MARGARET, et al. 1968. *Science and the Concept of Race*. New York: Columbia University Press.

MEDAWAR, P. B. 1961. Review of Teilhard de Chardin, *The Phenomenon of Man*. *Mind* 70: 99–106.

MEDVEDEV, ZHORES. 1969. *The Rise and Fall of T. D. Lysenko*. Translated by I. Michael Lerner. New York: Columbia University Press.

MENDEL, GREGOR JOHANN. 1965. *Experiments on Plant Hybridization*. Foreword by Paul C. Mangelsdorf. Cambridge, Mass.: Harvard University Press.

MERCIER, LOUIS-SÉBASTIEN. 1770. *L'an deux mille quatre cent quarante: rêve s'il en fut jamais*. Ed. Raymond Trousson. Paris: Ducros, 1971.

MERZ, JOHN THEODORE. 1896–1903. *A History of European Thought in the Nineteenth Century*. 2 vols. Edinburgh. Reprinted New York: Dover, 4 vols.
METZGER, HELENE. 1930. *Newton, Stahl, Boerhaave et la doctrine chimique*. Paris: Alcan.
MEYER, A. W. 1935. "Some Historical Aspects of the Recapitulation Idea." *Quart. Rev. Biology* 10: 379–396.
———. 1939. *The Rise of Embryology*. Stanford: Stanford University Press.
———. 1956. *Human Generation: The Conclusions of Burdach, Dollinger and von Baer*. Stanford: Stanford University Press.
MIDGLEY, MARY. 1978. *Beast and Man: The Roots of Human Nature*. Hassocks, Sussex: Harvester Press.
MILKMAN, R., ed. 1982. *Perspectives on Evolution*. Sunderland, Mass.: Sinauer.
MILL, JOHN STUART. 1950. *Mill on Bentham and Coleridge*. London: Chatto and Windus.
———. 1957. *Theism*. Ed. Richard Taylor. New York: Bobbs Merrill.
MILLAR, RONALD. 1972. *The Piltdown Men: A Case of Archaeological Fraud*. London: Victor Gollancz.
MILLER, HUGH. 1841. *The Old Red Sandstone: Or New Walks in an Old Field*. Edinburgh. Boston, 1857 edition reprinted New York: Arno Press, 1977.
———. 1850. *Footprints of the Creator: Or the Asterolepis of Stomness*. 3d ed., Edinburgh. Edinburgh, 1861 edition reprinted Farnborough, Gregg, 1971.
MILLER, STANLEY. 1953. "A Production of Amino Acids under Possible Primitive Earth Conditions." *Science* 117: 528.
MILLHAUSER, MILTON. 1954. "The Scriptural Geologists." *Osiris* 11: 65–86.
———. 1959. *Just Before Darwin: Robert Chambers and Vestiges*. Middletown, Conn.: Wesleyan University Press.
MIVART, ST. GEORGE JACKSON. 1871. *The Genesis of Species*. London.
MONTAGU, ASHLEY. 1952. *Darwin: Competition and Cooperation*. New York: Henry Schumann.
———, ed. 1962. *Culture and the Evolution of Man*. New York: Oxford University Press.
———. 1963. *Race, Society and Humanity*. New York: Van Nostrand Reinhold.
———. 1974. *Man's Most Dangerous Myth: The Fallacy of Race*. 5th ed., Oxford and New York: Oxford University Press.
———, ed. 1980. *Sociobiology Examined*. New York: Oxford University Press.
———, ed. 1982. *Evolution and Creation*. New York: Oxford University Press.
MONTESQUIEU, CHARLES LOUIS SECONDAT, BARON DE. 1900. *The Spirit of the Laws*. Translated by Thomas Nugent. 2 vols. Revised edition, New York: Colonial Press.
———. 1965. *Considerations on the Cause of the Greatness of the Romans and Their Decline*. Translated by David Lowenthal. New York: Free Press; London: Collier MacMillan.
MONTGOMERY, WILLIAM M. 1974. "Germany." *In* Glick, ed., *The Comparative Reception of Darwinism*. Pp. 81–116.
MOORE, J. N., and SLUSHER, H. S. 1970. *Biology: A Search for Order in Complexity*. Grand Rapids, Mich.: Zonderran.

MOORE, JAMES R. 1979. *The Post-Darwinian Controversies: A Study of the Protestant Struggle to Come to Terms with Darwin in Great Britain and America, 1870–1900*. New York: Cambridge University Press.

———. 1982. "Charles Darwin Lies in Westminster Abbey." *Biological J. Linn. Soc.* 17: 97–113.

———. 1985*a*. "Herbert Spencer's Henchmen: The Evolution of Protestant Liberals in Late-Nineteenth-Century America." *In* J. Durant, ed., *Darwinism and Divinity*. Pp. 76–100.

———. 1985*b*. "Evangelicals and Evolution: Henry Drummond, Herbert Spencer, and the Naturalization of the Spiritual World." *Scottish Journal of Theology* 38: 383–417.

MOOREHEAD, ALAN. 1969. *Darwin and the Beagle*. London: Hamish Hamilton.

MORAVIA, SERGIO. 1978. "From *homme machine* to *homme sensible*: changing eighteenth century models of man's image." *J. Hist. Ideas* 39: 45–60.

MORGAN, CONWAY LLOYD. 1927. *Emergent Evolution: The Gifford Lectures Delivered at the University of St. Andrews in the Year 1922*. 2d edition, London.

MORGAN, THOMAS HUNT. 1903. *Evolution and Adaptation*. New York, reprinted 1908.

———, et al. 1915. *The Mechanism of Mendelian Heredity*. New York: reprinted with an introduction by Garland E. Allen, New York, Johnson Reprint Corporation, 1972.

———. 1916. *A Critique of the Theory of Evolution*. Princeton: Princeton University Press.

MORNET, DANIEL. 1911. *Les sciences de la nature en France au XVIIIe siècle*. Paris: reprinted New York, Franklin, 1971.

MORRIS, DESMOND. 1967. *The Naked Ape: A Zoologist's Study of the Human Animal*. London: Jonathan Cape.

MORRIS, H. M. 1974. *Scientific Creationism*. San Diego: Creation Life.

MORTON, PETER. 1984. *The Vital Science: Biology and the Literary Imagination*. London: Allen and Unwin.

MOSSE, GEORGE L. 1978. *Toward the Final Solution: A History of European Racism*. New York: Howard Fertig.

MUELLER, RONALD H. 1976. "A Chapter in the History of the Relationship between Psychology and Sociology in America." *J. Hist. Behavioural Sci.* 12: 240–253.

MULKAY, MICHAEL. 1979. *Science and the Sociology of Knowledge*. London: Allen and Unwin.

MÜLLER, FRITZ. 1869. *Facts and Arguments for Darwin*. London: reprinted Farnborough, Gregg International. 1968.

MULLER, HERMANN J. 1949. "The Darwinian and Modern Conceptions of Natural Selection." *Proc. Am. Phil. Soc.* 90: 459–70.

———. 1959. "One Hundred Years without Darwin Are Enough." *The Humanist,* 19: 139–49.

MURCHISON, RODERICK I. 1839. *The Silurian System*. . . . London.

———. 1854. *Siluria: The History of the Oldest Known Rocks containing Organic Remains*. . . . 2 vols. London. London, 1872 edition reprinted Millwood, N.Y.: Kraus Reprint Co.

MURPHY, TERENCE D. 1976. "Jean-Baptiste Robinet: The Career of a Man of Letters." *Stud. Voltaire & 18th Cent.* 150: 183–250.

MUSCHINSKE, DAVID. 1977. "The Nonwhite as Child: G. Stanley Hall on the Education of Nonwhite Peoples." *J. Hist. Behavioural Sci.* 13: 328–336.

NÄGELI, CARL VON. 1898. *A Mechanico-Physiological Theory of Organic Evolution.* Chicago.

NAVILLE, PIERRE. 1967. *D'Holbach et la philosophie scientifique au XVIIIe siècle.* New ed. Paris: Gallimard.

NEEDHAM, JOHN TURBERVILLE. 1748. "A Summary of Some Late Observations upon the Generation, Composition and Decomposition of Animal and Vegetable Substances." *Phil. Trans. Roy. Soc.* 45: 615–666.

NEEDHAM, JOSEPH. 1959. *A History of Embryology.* 2d ed. New York: Abelard-Schuman.

NELKIN, DOROTHY. 1977. *Science Textbook Controversies and the Politics of Equal Time.* Cambridge, Mass.: M.I.T. Press.

———. 1983. *The Creation Controversy: Science or Scripture in Public Schools.* New York: Norton.

NELSON, GARETH. 1978. "From Candolle to Croizat: Comments on the History of Biogeography." *J. Hist. Biology* 11: 269–305.

NELSON, GARETH, and ROSEN, DON E., eds. 1981. *Vicariance Biogeography: A Critique.* New York: Columbia University Press.

NEWELL, NORMAN D. 1982. *Creation and Evolution: Myth or Reality?* New York: Columbia University Press.

NEWTON, ISAAC. 1729. *Mathematical Principles of Natural Philosophy.* Trans. Andrew Motte, revised by Florian Cajori. Berkeley and Los Angeles: University of California Press, 1962.

———. 1730. *Opticks.* 4th ed. London: reprinted New York, Dover, 1972.

NICHOLS, CHRISTOPHER. 1974. "Darwinism and the Social Sciences." *Phil. Soc. Sci.* 4: 255–277.

NICHOLSON, A. J. 1960. "The Role of Population Dynamics in Natural Selection." *In* Sol Tax, ed., *Evolution after Darwin,* I. Chicago: University of Chicago Press. Pp. 477–522. 3 vols.

NORDENSKIÖLD, ERIK. 1946. *The History of Biology.* Reprinted New York: Tudor Publishing Co.

NORTON, B. J. 1973. "The Biometric Defense of Darwinism." *J. Hist. Biology* 6: 283–316.

———. 1975*a*. "Biology and Philosophy: The Methodological Foundations of Biometry." *J. Hist. Biology* 8: 85–93.

———. 1975*b*. "Metaphysics and Population Genetics: Karl Pearson and the Background to Fisher's Multi-Factorial Theory of Inheritance." *Annals of Science* 32: 537–553.

———. 1978. "Karl Pearson and Statistics: The Social Origins of Scientific Innovation." *Social Studies of Science* 8: 3–34.

———. 1983. "Fisher's Entrance into Evolutionary Science: The Role of Eugenics." *In* M. Grene, ed., *Dimensions of Darwinism.* Pp. 19–30.

———, and PEARSON, E. S. 1976. "A Note on the Background to, and Refereeing of, R. A. Fisher's Paper 'On the Correlation . . .'" *Notes and Records of Roy. Soc. London* 31: 151–162.

NUMBERS, RONALD L. 1977. *Creation by Natural Law: Laplace's Nebular Hypothesis in American Thought*. Seattle: University of Washington Press.

———. 1982. "Creationism in Twentieth-Century America." *Science* 218: 538–544.

NYE, R. 1976. "Heredity or Milieu: The Foundations of European Criminological Theory." *Isis* 67: 335–355.

OBERG, BARBARA BOWEN. 1976. "David Harley and the Association of Ideas." *J. Hist. Ideas* 37: 441–454.

O'BRIEN, CHARLES F. 1970. "*Eozoön canadense:* The Dawn Animal of Canada." *Isis* 61: 200–223.

———. 1971. *Sir William Dawson: A Life in Science and Religion*. Mem. Am. Phil. Soc. 84.

OGILVIE, M. B. 1975. "Robert Chambers and the Nebular Hypothesis." *Brit. J. Hist. Sci*. 7: 214–232.

OKEN, LORENZ. 1847. *Elements of Physico-Philosophy*. Translated by Alfred Tulk. London: Ray Society.

OLBY, ROBERT C. 1966. *The Origins of Mendelism*. London: Constable. 2d ed. Chicago: University of Chicago Press, 1985.

———. 1974. *The Path to the Double Helix*. Foreword by Francis Crick. Seattle: University of Washington Press.

———. 1979. "Mendel No Mendelian." *History of Science* 17: 53–72.

OLBY, ROBERT C., and GAUTRY, PETER. 1968. "Eleven References to Mendel before 1900." *Annals of Science* 24: 7–20.

OLDROYD, D. R. 1972. "Robert Hooke's Methodology of Science as Exemplified in His 'Discourse of Earthquakes.'" *Brit. J. Hist. Sci*. 6: 109–130.

———. 1980. *Darwinian Impacts: An Introduction to the Darwinian Revolution*. Milton Keynes: Open University Press.

———. 1984. "How Did Darwin Arrive at His Theory?" *History of Science* 22: 325–374.

OLDROYD, D. R., and LANGHAM, IAN, eds. 1983. *The Wider Domain of Evolutionary Thought*. Dordrecht: D. Reidel.

OPARIN, A. I. 1938. *The Origin of Life*. Translated by Sergius Morgulis. 2d ed., New York: Dover, 1953.

OPPENHEIMER, JANE. 1967. *Essays in the History of Embryology*. Cambridge, Mass.: MIT Press.

OREL, VITESLAV. 1984. *Mendel*. Oxford: Oxford University Press.

O'ROURKE, JOSEPH. 1978. "A Comparison of James Hutton's *Principles of Knowledge* and *Theory of the Earth*." *Isis* 69: 5–20.

OSBORN, HENRY FAIRFIELD. 1908. "The Four Inseparable Factors of Evolution." *Science* 27: 148–150.

———. 1912. "The Continuous Origin of Certain Unit Characters as Observed by a Paleontologist." *Am. Naturalist* 46: 185–206, 249–278.

———. 1917. *The Origin and Evolution of Life on the Theory of Action, Reaction and Interaction of Energy*. New York.

———. 1929. *The Titanotheres of Ancient Wyoming, Dakota and Nebraska*. 2 vols. Washington: U.S. Geological Survey Monograph No. 55.

———. 1931. *Cope: Master Naturalist: The Life and Writings of Edward Drinker Cope*. Princeton: Princeton University Press.

———. 1934. "Aristogenesis: The Creative Principle in the Origin of Species." *Am. Naturalist* 68: 193–235.

OSPOVAT, ALEXANDER M. 1969. "Reflections on A. G. Werner's 'Kurze Klassifica-tion.'" *In* Schneer, ed., *Toward a History of Geology*. Pp. 242–256.

OSPOVAT, DOV. 1976. "The Influence of Karl Ernst von Baer's Embryology, 1828–1859: A Reappraisal in Light of Richard Owen and William B. Carpenter's 'Paleontological Application of von Baer's Law.'" *J. Hist. Biology* 9: 1–28.

———. 1977. "Lyell's Theory of Climate." *J. Hist. Biology* 10: 317–339.

———. 1978. "Perfect Adaptation and Teleological Explanation: Approaches to the Problem of the History of Life in the Mid-Nineteenth Century." *Studies in the History of Biology* 2: 33–56.

———. 1979. "Darwin after Malthus." *J. Hist. Biology* 12: 211–230.

———. 1980. "God and Natural Selection: The Darwinian Idea of Design." *J. Hist. Biology* 13: 169–194.

———. 1981. *The Development of Darwin's Theory: Natural History, Natural Theology, and Natural Selection, 1838–59*. Cambridge and New York: Cambridge University Press.

OSTOYA, P. 1951. *Les Theories d'évolution*. Paris: Payot.

OUTRAM, DORINDA. 1984. *Georges Cuvier: Vocation, Science, and Authority in Post-Revolutionary France*. Manchester: Manchester University Press.

———. 1986. "Uncertain Legislator: Georges Cuvier's Laws of Nature and Their Intellectual Context." *J. Hist. Biol.* 19: 323–368.

OWEN, RICHARD. 1841. "Report on British Fossil Reptiles: Part 2." *Report of the British Association for the Advancement of Science: 1841*, 60–204.

———. 1846. *A History of British Fossil Mammals and Birds*. London: reprinted New York, AMS Press.

———. 1848. *On the Archetype and Homologies of the Vertebrate Skeleton*. London: reprinted New York, AMS Press.

———. 1849. *On the Nature of Limbs*. . . . London.

———. 1851. "Lyell on Life and Its Successive Development." *Quart. Rev.* 89: 412–451.

———. 1860. *Palaeontology: Or a Systematic Study of Extinct Animals and Their Geological Relations*. Edinburgh.

———. 1866–68. *The Anatomy of the Vertebrates*. 3 vols. London: reprinted New York, AMS Press, 1973.

OWEN, REV. R. *The Life of Richard Owen*. 2 vols. London: reprinted Farnborough, Gregg, 1970.

PACKARD, ALPHEUS. 1889. "The Cave Fauna of North America." *Mem. Nat. Acad. Sci.* 4: 1–156.

———. 1894. "On the Inheritance of Acquired Characters in Animals with a Complete Metamorphosis." *Proc. Am. Acad. Arts & Sci.* 29: 331–370.

———. 1901. *Lamarck, the Founder of Evolution: His Life and Work. With Translations of His Writings on Organic Evolution*. New York.

PAGE, LEROY E. 1969. "Diluvialism and Its Critics." *In* Schneer, ed., *Toward a History of Geology*. Pp. 257–271. Reprinted in C. A. Russell, ed., *Science and Religious Belief*. London: University of London Press, 1973.

PALEY, EDMUND. 1825. *An Account of the Life and Writings of William Paley*. Lon-

don: reprinted Farnborough, Gregg, 1970.
PALEY, WILLIAM. 1802. *Natural Theology: Or Evidences of the Existence and Attributes of the Deity Collected from the Appearances of Nature*. London: reprinted Farnborough, Gregg, 1970.
PARADIS, JAMES G. 1978. *T. H. Huxley: Man's Place in Nature*. Lincoln and London: University of Nebraska Press.
PARKER, G. E. 1980. *Creation: The Facts of Life*. San Diego: C.L.P. Publishers.
PARKINSON, JAMES. 1833. *Organic Remains of a Former World*. London: reprinted New York, Arno Press, 1977, 3 vols.
PASSMORE, J. 1959. "Darwin's Influence on British Metaphysics." *Victorian Studies* 3: 41–54.
PASTORE, NICHOLAS. 1949. *The Nature-Nurture Controversy*. New York: King's Crown Press.
PATTERSON, COLIN. 1980. "Cladistics." *Biologist* 27: 234–240. Reprinted in Maynard Smith, ed., *Evolution Now*. Pp. 110–120.
———. 1982. "Cladistics and Classification." *New Scientist* 94: 303–306. Reprinted in Cherfas, ed., *Darwin Up to Date*. Pp. 35–39.
PAUL, DIANE. 1984. "Eugenics and the Left." *J. Hist. Ideas* 45: 567–590.
PAULY, PHILIP J. 1982. "Samuel Butler and His Darwinian Critics." *Victorian Studs.* 25: 161–180.
PEACOCKE, A. R. 1980. *Creation and the World of Science*. Oxford: Oxford University Press.
PEARSON, KARL. 1894. "Socialism and Natural Selection." *Fortnightly Revue* n.s. 56: 1–21.
———. 1896. "Regression, Heredity and Panmixia." *Phil. Trans. Roy. Soc.* 197 A: 253–318.
———. 1898. "Mathematical Contributions to the Theory of Evolution: On the Law of Ancestral Heredity." *Proc. Roy. Soc.* 57: 386–412.
———. 1900. *The Grammar of Science*. 2d ed. London. Reprinted with the two detailed chapters on evolution omitted, London: Everyman, 1937.
———. 1914–30. *The Life Letters and Labours of Francis Galton*. 3 vols. Cambridge: Cambridge University Press.
PECKHAM, MORSE. 1959. "Darwinism and Darwinisticism." Reprinted in Peckham, *The Triumph of Romanticism: Collected Essays*. Columbia: University of South Carolina Press, 1970. Pp. 176–201.
PEEL, J. D. Y. 1971. *Herbert Spencer: The Evolution of a Sociologist*. London: Heinemann.
PEIRCE, CHARLES SANDERS. 1931–35. *The Collected Papers of Charles Sanders Peirce*. 6 vols. Cambridge, Mass.: Harvard University Press.
PERKINS, JEAN A. 1959*a*. "Diderot and La Mettrie." *Studies in Voltaire and the Eighteenth Century* 10: 49–100.
———. 1959*b*. "Voltaire and La Mettrie." *Studies in Voltaire and the Eighteenth Century* 10, 101–113.
PERSONS, STOW, ed. 1956. *Evolutionary Thought in America*. New York: George Braziller.
PETERS, R. H. 1976. "Tautology in Evolution and Ecology." *Am Naturalist* 110: 1–12.

PFEIFER, EDWARD J. 1965. "The Genesis of American Neo-Lamarckism." *Isis* 56: 156–167.

———. 1974. "United States." *In* Glick, ed., *The Comparative Reception of Darwinism*. Pp. 168–206.

PHILLIPS, JOHN. 1844. *Memoir of William Smith*. London: reprinted New York, Arno Press, 1977.

PICAVET, F. 1891. *Les idéologues: Essai sur l'histoire des idées et des théories scientifiques, philosophiques, religieuses, etc., en France depuis 1789*. Paris: Alcan.

PICKENS, D. K. 1968. *Eugenics and the Progressives*. Nashville, Tenn.: Vanderbilt University Press.

PLATE, LUDWIG. 1900. *Über Bedeutung und Tragweite des Darwin'schen Selectionsprinzip*. Leipzig.

———. 1903. *Über Bedeutung des Darwin'schen Selectionsprinzip und Probleme der Artbildung*. Leipzig.

———. 1913. *Selektionsprinzip und Probleme der Artbildung*. Leipzig and Berlin.

PLATE, ROBERT. 1964. *The Dinosaur Hunters: Othniel C. Marsh and Edward D. Cope*. New York: D. McKay.

PLAYFAIR, JOHN. 1802. *Illustrations of the Huttonian Theory of the Earth*. Edinburgh: reprinted New York, Dover, 1964.

POLIAKOV, L. 1970. *The Aryan Myth: A History of Racist and Nationalist Ideas in Europe*. New York: Basic Books.

POLLARD, J. W., ed. 1984. *Evolutionary Theory: Paths into the Future*. Chichester: John Wiley.

POLLARD, SYDNEY. 1968. *The Idea of Progress: History and Society*. Reprinted Harmondsworth, Middlesex: Penguin Books, 1971.

POPPER, KARL. 1959. *The Logic of Scientific Discovery*. London: Hutchinson.

———. 1962. *The Open Society and Its Enemies*. 2 vols. 4th ed., Princeton: Princeton University Press.

———. 1974. *The Philosophy of Karl Popper*. Edited by Paul A. Schilpp. 2 vols. La Salle, Ill.: Open Court.

———. 1978. "Natural Selection and the Emergence of Mind." *Dialectica* 32: 339–355.

PORTER, ROY. 1973. "The Industrial Revolution and the Rise of the Science of Geology." *In* M. Teich and R. M. Young, eds., *Changing Perspectives in the History of Science*. Pp. 320–343.

———. 1977. *The Making of Geology: Earth Sciences in Britain, 1660–1815*. Cambridge: Cambridge University Press.

———. 1978. "Philosophy and Politics of a Geologist: G. H. Toulmin (1754–1817)." *J. Hist. Ideas* 39: 435–450.

PORTER, ROY, and ROUSSEAU, G. S., eds. 1980. *The Ferment of Knowledge: Studies in the Historiography of Eighteenth-Century Science*. New York and London: Cambridge University Press.

POSNER, E., and SKUTIL, J. 1968. "The Great Neglect: The Fate of Mendel's Classic Paper between 1865 and 1900." *Medical History* 12: 122–136.

POULTON, EDWARD BAGNALL. 1890. *The Colors of Animals: Their Meaning and Use, Especially Considered in the Case of Insects*. New York.

———. 1908. *Essays in Evolution: 1889–1907*. Oxford.

POWELL, BADEN. 1855. *Essays on the Spirit of the Inductive Philosophy, the Unity of Worlds and the Philosophy of Creation*. London: reprinted Farnborough. Gregg International, 1969.

POWELL, J. W. 1888. "Competition as a Factor in Human Evolution." *Am. Anthropologist* 1: 297–323.

PRIEST, JOSIAH. 1843. *Slavery: As It Relates to the Negro or African Race*. Albany, N.Y.: reprinted New York, Arno, 1977.

PROVINE, WILLIAM B. 1971. *The Origins of Theoretical Population Genetics*. Chicago: University of Chicago Press.

———. 1973. "Geneticists and the Biology of Race Crossing." *Science* 182: 790–796.

———. 1978. "The Role of Mathematical Population Genetics in the Evolutionary Synthesis of the 1930s and 1940s." *Stud. Hist. Biology* 2: 167–192.

———. 1979. "Francis B. Sumner and the Evolutionary Synthesis." *Stud. Hist. Biology* 3: 211–240.

———. 1986. *Sewall Wright and Evolutionary Biology*. Chicago: University of Chicago Press.

PUNNETT, R. C. 1915. *Mimicry in Butterflies*. Cambridge: Cambridge University Press.

QUETELET, LAMBERT. 1842. *A Treatise on Man and the Development of His Faculties*. London: reprinted Delmar, N.Y., Scholars' Facsimiles and Reprints.

RACHOOTIN, STAN, and THOMSON, KEITH S. 1981. "Epigenetics, Paleontology, and Evolution." *In* G. G. E. Scudder and J. L. Reveal, eds., *Evolution Today: Proceedings of the Second International Congress of Systematic and Evolutionary Biology*. Pittsburgh: Hunt Institute for Botanical Documentation, Carnegie-Mellon University. Pp. 181–193.

RÁDL, EMMANUEL. 1930. *The History of Biological Theories*. Translated E. J. Hatfield. Oxford: Oxford University Press; London: Humphrey Milford.

RAIKOV, BORIS E. 1968. *Karl Ernst von Baer, 1792–1876: Sein Leben und sein Werke*. Leipzig: Acta Historia Leopoldina, no. 5.

RAINGER, RONALD. 1981. "The Continuation of the Morphological Tradition in American Paleontology, 1880–1910." *J. Hist. Biology* 14: 129–158.

RANDALL, J. HERMAN, JR. 1961. "The Changing Impact of Darwin on Philosophy." *J. Hist. Ideas* 22: 435–462.

RANSOM, C. J. 1976. *The Age of Velikovsky*. Glassboro, N.J.: Kronos Press.

RAPPAPORT, RHODA. 1978. "Geology and Orthodoxy: The Case of Noah's Flood in Eighteenth-Century Thought." *Brit. J. Hist. Sci.* 11: 1–18.

RAVEN, C. E. 1942. *John Ray, Naturalist: His Life and Work*. Cambridge: Cambridge University Press.

RAVIN, ARNOLD W. 1977. "The Gene as Catalyst: The Gene as Organism." *Stud. Hist. Biology*. 1: 1–45.

RAY, JOHN. 1691. *The Wisdom of God as Manifested in the Works of Creation*. London.

———. 1692. *Miscellaneous Discourses concerning the Changes of the World*. London: reprinted Hildesheim, Georg Olms, 1968.

———. 1713. *Three Physico-Theological Discourses*. 3d ed. London: reprinted New York, Arno Press, 1977.

———. 1724. *Synopsis Methodica Stirpum Britannicum*. 3d ed. London. Reprinted with Linnaeus, *Flora Anglica*, London: the Ray Society, 1973.

READER, J. 1981. *Missing Links: The Hunt for Earliest Man*. London: Collins.
REHBOCK, PHILIP F. 1975. "Huxley, Haeckel, and the Oceanographers: The Case of *Bathybius Haeckelii*." *Isis* 66: 504–533.
———. 1983. *The Philosophical Naturalists: Themes in Early Nineteenth-Century British Biology*. Madison: University of Wisconsin Press.
REIF, WOLF-ERNST. 1983. "Evolutionary Theory in German Paleontology." *In* M. Grene, ed., *Dimensions of Darwinism*. Pp. 173–204.
———. 1986. "The Search for a Macroevolutionary Theory in German Paleontology." *J. Hist. Biol.* 19: 79–130.
RENSCH, BERNHARD. 1960. *Evolution above the Species Level*. New York: Columbia University Press.
———. 1983. "The Abandonment of Lamarckian Explanations: The Case of Climatic Parallelism of Animal Characteristics." *In* M. Grene, ed., *Dimensions of Darwinism*. Pp. 31–42.
RICHARDS, EVELLEEN. 1987. "A Question of Property Rights: Richard Owen's Evolutionism Reassessed." *Brit. J. Hist. Sci.* 20: 129–172.
RICHARDS, ROBERT J. 1977. "Lloyd Morgan's Theory of Instinct: From Darwinism to Neo-Darwinism." *J. Hist. Behavioural Sci.* 13: 12–32.
———. 1981. "Instinct and Intelligence in British Natural Theology: Some Contributions to Darwin's Theory of the Evolution of Behavior." *J. Hist. Biology* 14: 193–230.
———. 1987. *Darwin and the Emergence of Evolutionary Theories of Mind and Behavior*. Chicago: University of Chicago Press.
RICHARDSON, R. ALAN. 1981. "Biogeography and the Genesis of Darwin's Ideas on Transmutation." *J. Hist. Biology* 14: 1–41.
RIDLEY, MARK. 1982*a*. "Coadaptation and the Inadequacy of Natural Selection." *Brit. J. Hist. Sci.* 15: 45–68.
———. 1982*b*. "How to Explain Organic Diversity." *New Scientist* 94: 359–361. Reprinted in Cherfas, ed., *Darwin Up to Date*. Pp. 42–43.
———. 1985. *The Problems of Evolution*. Oxford: Oxford University Press.
RITTERBUSH, PHILLIP C. 1964. *Overtures to Biology: The Speculation of the Eighteenth-Century Naturalists*. New Haven: Yale University Press.
ROBERTS, H. F. 1929. *Plant Hybridization before Mendel*. Princeton: Princeton University Press.
ROBERTS, KENNETH L. 1922. *Why Europe Leaves Home*. Indianapolis: reprinted New York, Arno, 1977.
ROBINET, J. B. 1761–1766. *De la nature*. 4 vols. Amsterdam.
ROBINSON, GLORIA. 1979. *A Prelude to Genetics: Theories of a Material Substance of Heredity, Darwin to Weismann*. Lawrence, Kan.: Coronado Press.
ROE, SHIRLEY A. 1981. *Matter, Life, and Generation: Eighteenth-Century Embryology and the Haller-Wolff Debate*. Cambridge: Cambridge University Press.
———. 1983. "John Turberville Needham and the Generation of Living Organisms." *Isis* 74: 159–184.
———. 1985. "Voltaire versus Needham: Atheism, Materialism, and the Generation of Life." *J. Hist. Ideas* 46: 65–87.
ROGER, JACQUES. 1963. *Les sciences de la vie dans la pensée française du XVIIIe siècle*. Paris: Armand Colin.
———. 1974. "La théorie de la terre au XVIIe siècle." *Rev. d'hist. des sci.* 26: 23–48.

ROGERS, JAMES ALLEN. 1972. "Darwinism and Social Darwinism." *J. Hist. Ideas* 33: 265–280.

———. 1974. "The Reception of Darwin's *Origin of Species* by Russian Scientists." *Isis* 64: 484–503.

ROLLER, DUANE H. D., ed. 1971. *Perspectives in the History of Science and Technology*. Norman: University of Oklahoma Press.

ROMANES, GEORGE JOHN. 1886. "Physiological Selection: An Additional Suggestion on the Origin of Species." *J. Linn. Soc. (Zool.)* 19: 337–411. Abstracted *Nature* 34: 314–316, 336–340, and 362–365.

———. 1888. *Mental Evolution in Man*. London: Kegan Paul, Trench Trubner.

———. 1892–97. *Darwin and after Darwin: An Exposition of the Darwinian Theory and a Discussion of Post-Darwinian Problems*. 3 vols. London.

———. 1899. *An Examination of Weismannism*. 2d ed., Chicago.

ROSE, STEVEN, KAMIN, LEON, JR., and LEWONTIN, R. C. 1984. *Not in Our Genes: Biology, Ideology, and Human Nature*. New York: Pantheon.

ROSENFIELD, LEONORA C. 1968. *From Beast-Machine to Man-Machine: Animal Soul in French Letters from Descartes to La Mettrie*. New ed., New York: Octagon Books.

ROSS, EDWARD ALSWORTH. 1927. *Standing Room Only?* New York and London: reprinted New York, Arno, 1977.

ROSSI, PAOLO. 1984. *The Dark Abyss of Time: The History of the Earth and the History of Nations from Hooke to Vico*. Chicago: University of Chicago Press.

ROSTAND, JEAN. 1932. *L'évolution des espèces: Histoire des idées transformistes*. Paris: Hachette.

RUDBERG, DANIEL. 1752. "*Peloria*." *In* Linnaeus, ed., *Amoenitates Academicae*, II: 280–298.

RUDWICK, MARTIN J. S. 1967. "A Critique of Uniformitarian Geology: A Letter from W. D. Conybeare to Charles Lyell, 1841." *Proc. Am. Phil. Soc.* 111: 272–287.

———. 1970. "The Strategy of Lyell's *Principles of Geology*." *Isis* 61: 5–33.

———. 1971. "Uniformity and Progression: Reflections on the Structure of Geological Theory in the Age of Lyell." *In* Duane H. D. Roller, ed., *Perspectives in the History of Science and Technology*. Pp. 209–227.

———. 1972. *The Meaning of Fossils: Episodes in the History of Paleontology*. 2d ed., New York: Science History Publications, 1976.

———. 1974*a*. "Poulett Scrope on the Volcanoes of Auvergne: Lyellian Time and Political Economy." *Brit. J. Hist. Sci.* 7: 205–242.

———. 1974*b*. "Darwin and Glen Roy: A 'Great Failure' in Scientific Method?" *Stud. Hist. & Phil. Sci.* 5: 97–185.

———. 1975. "Caricature as a Source for the History of Science: De la Beche's Anti-Lyellian Sketches of 1831." *Isis* 66: 534–560.

———. 1978. "Charles Lyell's Dream of a Statistical Palaeontology." *Palaeontology* 21: 225–244.

———. 1982. "Charles Darwin in London: The Integration of Public and Private Science." *Isis* 73: 186–206.

———. 1985. *The Great Devonian Controversy: The Shaping of Scientific Knowledge among Gentlemanly Specialists*. Chicago: University of Chicago Press.

RUPKE, NICOLAAS A. 1983. *The Great Chain of History: William Buckland and the English School of Geology (1814–1849)*. Oxford: Oxford University Press.
RUSE, MICHAEL. 1970. "The Revolution in Biology." *Theoria* 35: 13–22.
———. 1971*a*. "Two Biological Revolutions." *Dialectica* 25: 17–38.
———. 1971*b*. "Natural Selection in the *Origin of Species*." *Stud. Hist. & Phil. Sci.* 1: 311–351.
———. 1974. "The Darwin Industry: A Critical Evaluation." *History of Science* 7: 43–58.
———. 1975*a*. "Charles Darwin and Artificial Selection." *J. Hist. Ideas* 36: 339–350.
———. 1975*b*. "Darwin's Debt to Philosophy: An Examination of the Influence of the Philosophical Ideas of John F. W. Herschel and William Whewell on the Development of Charles Darwin's Theory of Evolution." *Stud. Hist. & Phil. Sci.* 6: 159–181.
———. 1975*c*. "Charles Darwin's Theory of Evolution: An Analysis." *J. Hist. Biology* 8: 219–241.
———. 1975*d*. "The Relationship between Science and Religion in Britain, 1830–1870." *Church History* 34: 505–522.
———. 1976. "Charles Lyell and the Philosophers of Science." *Brit J. Hist. Sci.* 9: 121–131.
———. 1977. "Karl Popper's Philosophy of Biology." *Philosophy of Science* 44: 638–661.
———. 1979*a*. *The Darwinian Revolution: Science Red in Tooth and Claw*. Chicago: University of Chicago Press.
———. 1979*b*. *Sociobiology: Sense or Nonsense?* Dordrecht: D. Reidel.
———. 1980. "Charles Darwin and Group Selection." *Annals of Science* 37: 615–630.
———. 1982. *Darwinism Defended: A Guide to the Evolution Controversies*. Reading, Mass.: Addison-Wesley.
RUSSELL, C. A., ed. 1973. *Science and Religious Belief: A Selection of Recent Historical Studies*. London: University of London Press/Open University Press.
RUSSELL, DALE A. 1982. "The Mass Extinctions of the Late Mesozoic." *Sci. American* 246: 48–55.
RUSSELL, E. S. 1916. *Form and Function: A Contribution to the History of Animal Morphology*. London: Murray, reprinted Farnborough, Gregg, 1972.
RUSSETT, CYNTHIA EAGLE. 1976. *Darwin in America: The Intellectual Response*. San Francisco: W. H. Freeman.
SAHLINS, MARSHALL. 1976. *The Use and Abuse of Biology: An Anthropological Critique of Sociobiology*. Ann Arbor: University of Michigan Press.
SAINT-SIMON, CLAUDE-HENRI DE ROUVRAY, COMTE DE. 1952. *Selected Writings*. Translated by F. M. H. Markham. Oxford: Oxford University Press.
SANDLER, IRIS. 1983. "Pierre Louis Moreau de Maupertuis—A Precursor of Mendel?" *J. Hist. Biol.* 16: 101–136.
SANTURRI, EDMUND N. 1982. "Theodicy and Social Policy in Malthus' Thought." *J. Hist. Ideas* 43: 315–330.
SAPP, JAN. 1983. "The Struggle for Authority in the Field of Heredity, 1900–1932." *J. Hist. Biol.* 16: 311–342.

———. 1987. *Beyond the Gene: Cytoplasmic Inheritance and the Struggle for Authority in Genetics*. Oxford: Oxford University Press.

SARJEANT, WILLIAM. 1980. *Geologists and the History of Geology: An International Bibliography from the Origins to 1978*. London: Macmillan.

SAVIOZ, RAYMOND. 1948. *La philosophie de Charles Bonnet de Genève*. Paris: Vrin.

SCHELLING, F. W. J. 1942. *The Ages of the World*. Translated by F. Bolman, Jr. New York.

SCHILLER, JOSEPH, ed. 1971*a*. *Collque international "Lamarck" tenue au Muséum national d'histoire naturelle*. Paris: Blanchard.

———. 1971*b*. "L'échelle des êtres et la série chez Lamarck." *In* Schiller, ed., *Colloque "Lamarck."* Pp. 87–103.

———. 1974. "Queries, Answers, and Unsolved Problems in Eighteenth-Century Biology." *Hist. of Science* 12: 184–199.

———. 1978. *La notion d'organisation dans l'histoire de la biologie*. Paris: Maloine.

SCHNEER, CECIL J., ed. 1969. *Toward a History of Geology*. Cambridge, Mass.: MIT Press.

SCHNEIDER, H. W. 1946. *A History of American Philosophy*. New York: Columbia University Press.

SCHOPF, J. W. 1978. "The Evolution of the Earliest Cells." *Sci. American* (September): 110–138.

SCHUCHERT, CHARLES, and LEVENE, CLARA MAE. 1940. *O. C. Marsh: Pioneer in Paleontology*. New Haven: Yale University Press.

SCHULTZ, ALFRED P. 1908. *Race or Mongrel: A Brief History of the Ancient Races of the Earth*. Boston: reprinted New York, Arno, 1977.

SCHWARTZ, JOEL S. 1974. "Charles Darwin's Debt to Malthus and Edward Blyth." *J. Hist. Biology* 7: 301–318.

———. 1984. "Darwin, Wallace, and the Descent of Man." *J. Hist. Biol.* 17: 271–289.

SCHWEBER, SYLVAN S. 1977. "The Origin of the *Origin* Revisited." *J. Hist. Biology* 10: 229–316.

———. 1979. "Essay Review: The Young Charles Darwin." *J. Hist. Biology* 13: 175–192.

———. 1980. "Darwin and the Political Economists: Divergence of Character." *J. Hist. Biology* 13: 195–289.

SCOPES, JOHN THOMAS. 1967. *Center of the Storm*. New York: Holt, Rinehart and Winston.

SCOTT, CLIFFORD H. 1976. *Lester Frank Ward*. Boston: Twayne Publishers.

SCROPE, GEORGE POULETT. 1827. *Memoir on the Geology of Central France. . . .* London.

———. 1858. *The Geology and Extinct Volcanoes of Central France*. 2d ed. London: reprinted New York, Arno Press, 1977.

SEARLE, G. R. 1976. *Eugenics and Politics in Britain: 1900–1914*. Leiden: Noordhoff International Publishing.

———. 1979. "Eugenics and Politics in Britain in the 1930s." *Annals of Science* 36: 159–169.

SECORD, JAMES A. 1981. "Nature's Fancy: Charles Darwin and the Breeding of Pigeons." *Isis* 72: 163–186.

———. 1986. *Controversy in Victorian Geology: The Cambrian-Silurian Debate*. Princeton: Princeton University Press.

SEDGWICK, ADAM. 1845. "Vestiges of the Natural History of Creation." *Edinburgh Review* 82: 1–85.

SELLARS, ROY WOOD. 1922. *Evolutionary Naturalism*. Chicago: reprinted New York, Russell & Russell, 1969.

SEMMEL, BERNARD. 1960. *Imperialism and Social Reform: English Socio-Imperial Thought, 1895–1914*. London: Allen and Unwin.

SETTLE, M. L. 1972. *The Scopes Trial*. New York: Franklin Watts.

SHAPIN, STEVEN. 1982. "History of Science and Its Sociological Reconstructions." *History of Science* 20: 157–211.

SHAW, GEORGE BERNARD. 1970–74. *The Bodley Head Bernard Shaw*. 7 vols. London.

SHEETS-JOHNSTONE, MAXINE. 1982. "Why Lamarck Did Not Discover the Principle of Natural Selection." *J. Hist. Biology* 15: 443–65.

SHINE, IAN B., and WROBEL, SYLVIA. 1976. *Thomas Hunt Morgan: Pioneer of Genetics*. Lexington: University of Kentucky Press.

SHKLOVSKII, I. S., and SAGAN, CARL. 1966. *Intelligent Life in the Universe*. New York: Holden and Day, reprinted London: Picador, 1977.

SHOR, ELIZABETH NOBLE. 1974. *The Fossil Feud between E. D. Cope and O. C. Marsh*. Hicksville, N.Y.: Exposition Press.

SHUTE, EVAN. 1961. *Flaws in the Theory of Evolution*. Philadelphia: Presbyterian and Reformed Publishing Co.

SIMPSON, GEORGE GAYLORD. 1944. *Tempo and Mode in Evolution*. New York: Columbia University Press.

———. 1949. *The Meaning of Evolution*. New Haven: Yale University Press.

———. 1953*a*. "The Baldwin Effect." *Evolution* 7: 110–117.

———. 1953*b*. *The Major Features of Evolution*. New York: Columbia University Press.

———. 1963. *This View of Life; the World of an Evolutionist*. New York: Harcourt, Brace and World.

———. 1973. "The Concept of Progress in Organic Evolution." *Social Research* 41: 28–51.

SLOAN, PHILLIP R. 1972. "John Locke, John Ray and the Problem of the Natural System." *J. Hist. Biology* 5: 1–53.

———. 1976. "The Buffon-Linnaeus Controversy." *Isis* 67: 356–375.

———. 1979. "Buffon, German Biology, and the Historical Interpretation of Species." *Brit. J. Hist. Sci.* 12: 109–153.

———. 1985. "Darwin's Invertebrate Program, 1826–1836." *In* D. Kohn, ed., *The Darwinian Heritage*. Pp. 71–120.

———. 1986. "Darwin, Vital Matter, and the Transformism of Species." *J. Hist. Biol.* 19: 369–445.

SMILES, SAMUEL. 1859. *Self-Help*. London.

SMITH, ADAM. 1910. *The Wealth of Nations*. 2 vols. Reprinted London: Everyman.

SMITH, SIR GRAFTON ELLIOT. 1924. *The Evolution of Man: Essays*. Oxford: Oxford University Press.

SMITH, J. MAYNARD. 1975. *The Theory of Evolution*. 3d. ed. Harmondsworth:

Penguin Books.
———. 1982. *Evolution and the Theory of Games*. Cambridge: Cambridge University Press.
———, ed. 1982. *Evolution Now: A Century after Darwin*. London: *Nature*, in association with Macmillan.
SMITH, J. PERCY. 1965. *The Unrepentant Pilgrim: A Study of the Development of George Bernard Shaw*. Boston: Houghton Mifflin.
SMITH, R. 1972. "Alfred Russel Wallace: Philosophy of Nature and Man." *Brit. J. Hist. Sci.* 6: 177–199.
SMITH, SYDNEY. 1960. "The Origin of the *Origin*." *Advancement of Science* 16: 391–401.
SMITH, WILLIAM. 1815. *A Memoir to the Map and Delineation of the Strata of England. . . .* London.
SNYDER, LOUIS L. 1962. *The Idea of Racialism: Its Meaning and History*. Princeton: Princeton University Press.
SPALLANZANI, LAZZARO. 1769. *Nouvelles recherches sur les découvertes microscopiques, et la génération des corps organizées*. 2 vols. London and Paris.
SPENCER, HERBERT. 1851. *Social Statics: Or the Conditions Essential to Human Happiness Specified, and One of Them Adopted*. London: reprinted Farnborough, Gregg International.
———. 1855. *Principles of Psychology*. London: reprinted Farnborough, Gregg International. 1881 ed. reprinted (2 vols.) Boston: Longwood Press.
———. 1862. *First Principles of a New Philosophy*. London.
———. 1864. *Principles of Biology*. 2 vols. London.
———. 1883. *Essays Scientific, Political and Speculative*. 3 vols. London.
———. 1893. "The Inadequacy of Natural Selection." *Contemporary Review* 63: 153–166, 439–456.
———. 1904. *An Autobiography*. 2 vols. New York.
———. 1969. *The Man versus the State*. Edited by Donald Macrae. Harmondsworth: Penguin Books.
SPENGLER, OSWALD. 1926–28. *The Decline of the West*. 2 vols., reprinted New York: Alfred A. Knopf, 1939.
SPINOZA, BENEDICT DE. 1887. *The Chief Works of Benedict de Spinoza*. 2 vols. Translated by R. H. M. Elwes. London: Bohn's Philosophical Library.
STAFLEU, F. 1971. *Linnaeus and the Linnaeans: The Spreading of Their Ideas in Systematic Botany*. Utrecht: Oosthoek.
STANLEY, STEVEN M. 1981. *The New Evolutionary Timetable: Fossils, Genes, and the Origin of Species*. New York: Basic Books.
STANTON, WILLIAM. 1960. *The Leopard's Spots: Scientific Attitudes toward Race in America, 1815–1859*. Chicago: Phoenix Books.
STARK, WERNER. 1961. *Montesquieu: Pioneer of the Sociology of Knowledge*. Toronto: University of Toronto Press.
STAUFFER, ROBERT C. 1960. "Ecology in the Long Manuscript Version of Darwin's *Origin of Species* and Linnaeus' Oeconomy of Nature." *Proc. Am. Phil. Soc.* 104: 235–241.
STAUM, MARTIN S. 1974. "Cabanis and the Science of Man." *J. Hist. Behavioural Sci.* 10: 135–143.

———. 1980. *Cabanis: Enlightenment and Medical Philosophy in the French Revolution*. Princeton: Princeton University Press.

STEBBINS, ROBERT E. 1974. "France," *In* Glick, ed., *The Comparative Reception of Darwinism*. Pp. 117–167.

STEELE, E. J. 1979. *Somatic Selection and Adaptive Evolution: On the Inheritance of Acquired Characters*. Toronto: Williams and Wallace International. Reprinted Chicago: University of Chicago Press, 1982.

———. 1981. "Lamarck and Immunity: A Conflict Resolved." *New Scientist* 90: 360–361.

STENO, NICHOLAS. 1916. *The Prodromus of Nicholas Steno's Dissertation concerning a Solid Body Enclosed by Process of Nature within a Solid*. Translated by J. G. Winter. University of Michigan Humanistic Studies, vol. 1, pt. 2. Reprinted New York: Hafner Publishing Co., 1968.

STEPAN, NANCY. 1982. *The Idea of Race in Science: Great Britain, 1800–1960*. London: Macmillan.

STEPHENS, LESTER G. 1976. "Joseph LeConte on Evolution, Education and the Structure of Knowledge." *J. Hist. Behavioural Sci.* 12: 103–119.

———. 1978. "Joseph LeConte's Evolutionary Idealism: A Lamarckian View of Cultural History." *J. Hist. Ideas* 39: 465–480.

———. 1982. *Joseph LeConte: Gentle Prophet of Evolution*. Baton Rouge: Louisiana State University Press.

STERLING, KEIR B., ed. 1974. *Selected Works in Nineteenth-Century North American Paleontology*. New York: Arno.

———. 1975. *Contributions to the History of American Natural History*. New York: Arno.

STERN, CURT, and SHERWOOD, E. R., eds. 1966. *The Origin of Genetics: A Mendel Sourcebook*. San Francisco: W. H. Freeman.

STEVENS, ROBERT. 1978. *Charles Darwin*. Boston: Twayne.

STOCKING, GEORGE W. 1962. "Lamarckianism in American Social Science." *J. Hist. Ideas* 23: 239–256.

———. 1968. *Race, Culture and Evolution*. New York: Free Press.

———. 1987. *Victorian Anthropology*. New York: Free Press.

STOMPS, T. G. 1955. "The Rediscovery of Mendel's Work by Hugo De Vries." *J. Heredity* 45: 293–344.

STONE, IRVING. 1981. *The Origin*. London: Cassell.

STUBBE, HANS. 1972. *History of Genetics: From Prehistoric Times to the Rediscovery of Mendel's Laws*. Translated by T. H. Waters. Cambridge, Mass.: MIT Press.

STURTEVANT, A. H. 1965. *A History of Genetics*. New York: Harper & Row.

SULLOWAY, FRANK J. 1979. "Geographic Isolation in Darwin's Thinking: The Vicissitudes of a Crucial Idea." *Stud. Hist. Biology* 3: 23–65.

———. 1982*a*. "Darwin and His Finches: The Evolution of a Legend." *J. Hist. Biology* 15: 1–54.

———. 1982*b*. "Darwin's Conversion: The *Beagle* Voyage and Its Aftermath." *J. Hist. Biology* 15: 325–396.

SUMNER, F. B. 1924. "The Stability of Subspecific Characters under Changed Conditions of Environment." *Am. Naturalist* 58: 481–505.

SWEET, J. M., and WATERSTON, C. 1967. "Robert Jameson's Approach to the Wernerian Theory of the Earth, 1796." *Annals of Science* 23: 81–95.

SWINBURNE, R. G. 1965. "Galton's Law-Formulation and Development." *Annals of Science* 21: 15–31.

SWINTON, WILLIAM E. 1970. *The Dinosaurs*. 2d ed. London: Allen and Unwin.

SYMONDSON, ANTHONY, ed. 1970. *The Victorian Crisis of Faith*. London: SPCK.

TALBOTT, STEPHEN L., ed. 1976. *Velikovsky Reconsidered*. New York: Warner Books.

TAYLOR, GORDON RATTRAY. 1983. *The Great Evolution Mystery*. London: Secker and Warburg.

TAYLOR, KENNETH L. 1969. "Nicholas Desmarest and Geology in the Eighteenth Century." *In* Schneer, ed., *Toward a History of Geology*. Pp. 339–356.

TEICH, M., and YOUNG, R. M., eds. 1973. *Changing Perspectives in the History of Science*. London: Heinemann.

TEILHARD DE CHARDIN, PIERRE. 1959. *The Phenomenon of Man*. Introduced by Julian Huxley. London: Collins.

TEMKIN, OWSEI. 1950. "German Concepts of Ontogeny and Development around 1800." *Bull. Hist. Medicine* 24: 227–246.

———. 1959. "The Idea of Descent in Post-Romantic German Biology." *In* Glass et al., eds., *Forerunners of Darwin*. Pp. 323–355.

TENNENBAUM, J. 1956. *Race and Reich: The Story of an Epoch*. Boston: Twayne Publishers.

TENNYSON, ALFRED LORD. 1973. *In Memoriam*. Edited by Robert H. Ross. New York: A Norton Critical Edition.

THEUNISSEN, B. 1986. "The Relevance of Cuvier's *Lois zoologiques* for His Paleontological Work." *Annals of Science* 43: 543–556.

THODAY, J. M. 1962. "Naturalist Selection and Biological Progress." *In* S. A. Barnett, ed., *A Century of Darwin*. Pp. 313–333.

TORREY, NORMAN L. 1930. *Voltaire and the English Deists*. New Haven: Yale University Press.

TOULMIN, STEPHEN, and GOODFIELD, JUNE. 1965. *The Discovery of Time*. Chicago: University of Chicago Press.

TREMBLEY, ABRAHAM. 1973. *Memoirs on the Natural History of the Freshwater Polyp*. Translated by R. F. Ewer. New York: Johnson Reprints.

TURGOT, ANNE ROBERT JACQUES. 1973. *Turgot on Progress, Sociology and Economics*. Translated by Robert L. Meek. Cambridge: Cambridge University Press.

TURNER, FRANK MILLER. 1974. *Between Science and Religion: The Reaction to Scientific Naturalism in Late Victorian England*. New Haven and London: Yale University Press.

———. 1978. "The Victorian Conflict between Science and Religion: A Professional Dimension." *Isis* 69: 356–376.

TURRILL, WILLIAM BERTRAM. 1963. *Joseph Dalton Hooker: Botanist, Explorer and Administrator*. London: Scientific Book Guild.

TYNDALL, JOHN. 1902. *Fragments of Science*. 2 vols. New ed., New York.

VAN DOREN, CHARLES L. 1967. *The Idea of Progress*. New York: F. A. Praeger.

VARTANIAN, ARAM. 1949. "From Deist to Atheist." *Diderot Studies* 1: 46–61.

———. 1950. "Trembley's Polyp, La Mettrie and Eighteenth-Century French

Materialism." *J. Hist. Ideas* 11: 259–286.
———. 1953. *Diderot and Descartes: A Study of Scientific Naturalism in the Enlightenment*. Princeton: Princeton University Press.
———. 1968. "The Enigma of Diderot's *Eléments de physiologie*." *Diderot Studies* 10: 285–301.
VELIKOVSKY, IMMANUEL. 1950. *Worlds in Collision*. Garden City, N.Y.: Doubleday.
———. 1952. *Ages in Chaos*. Garden City, N.Y. Doubleday.
———. 1955. *Earth in Upheaval*. Garden City, N.Y.: Doubleday.
VERNIÈRE, PAUL. 1954. *Spinoza et la pensée française avant la revolution*. 2 vols. Paris: Presses Universitaires de France.
VICO, GIANBATTISTA. 1948. *The New Science of Gianbattista Vico*. Translated by Thomas Goddard Bergin and Max Harold Fisch. Ithaca, N.Y.: Cornell University Press.
VOLTAIRE, FRANCOIS MARIE AROUET DE. 1876. *Oeuvres complétes*. 13 vols. Paris.
———. 1947. *Candide and Other Writings*. Edited by H. M. Block. New York: Modern Library.
———. 1961. *The Age of Louis XIV*. Translated by Martin P. Pollach. London: Everyman.
———. 1965. *The Philosophy of History*. Translated by Thomas Kierman. New York: Philosophical Library.
VORZIMMER, PETER J. 1963. "Charles Darwin and Blending Inheritance." *Isis* 54: 371–390.
———. 1965. "Darwin's Ecology and Its Influence upon His Theory." *Isis* 56: 148–155.
———. 1968. "Darwin and Mendel: The Historical Connection." *Isis* 59: 72–82.
———. 1969*a*. "Darwin, Malthus and the Theory of Natural Selection." *J. Hist. Ideas* 30: 527–542.
———. 1969*b*. "Charles Darwin's 'Questions on the Breeding of Animals.'" *J. Hist. Biology* 2: 269–281.
———. 1970. *Charles Darwin, the Years of Controversy: The* Origin of Species *and Its Critics, 1859–82*. Philadelphia: Temple University Press.
———. 1975. "An Early Darwin Manuscript: The 'Outline and Draft of 1839.'" *J. Hist. Biology* 8: 191–217.
———. 1977. "The Darwin Reading Notebooks (1838–1860)." *J. Hist. Biology* 10: 107–152.
WADDINGTON, C. H. 1957. *The Strategy of the Gene*. London: Allen and Unwin.
———. 1975. *The Evolution of an Evolutionist*. Edinburgh: Edinburgh University Press.
WADE, IRA O. 1971. *The Intellectual Origins of the French Enlightenment*. Princeton: Princeton University Press.
WAERDEN, B. L. VAN DER. 1968. "Mendel's Experiments." *Centaurus* 12: 275–288.
WAGNER, MORITZ. 1873. *The Darwinian Theory and the Law of the Migration of Organisms*. Translated by J. L. Laird. London.
———. 1889. *Die Enstehung der Arten durch raumliche Sonderung*. Basel.
WALLACE, ALFRED RUSSEL. 1855. "On the Law which Has Regulated the Introduction of New Species." *Annals & Magazine of Natural History* 26: 184–196. Reprinted in Wallace, *Contributions to the Theory of Natural Selection*. Also in Wallace, *Natural Selection and Tropical Nature*.

———. 1869. *The Malay Archipelago: The Land of the Orang-Utan and the Bird of Paradise: A Narrative of Travel with Studies of Man and Nature*. London. Rev. ed. 1890, reprinted New York: Dover, 1962.

———. 1870. *Contributions to the Theory of Natural Selection*. London: reprinted New York, AMS Press, 1973.

———. 1876. *The Geographical Distribution of Animals*. 2 vols. London: reprinted New York, Hafner, 1962.

———. 1880. *Island Life: Or the Phenomena and Causes of Insular Faunas: Including a Discussion and an Attempted Solution to the Problem of Geological Climates*. London.

———. 1889. *Darwinism: An Exposition of the Theory of Natural Selection*. London: reprinted New York, AMS Press, 1975.

———. 1891. *Natural Selection and Tropical Nature*. London: reprinted Farnborough, Gregg International, 1969.

———. 1905. *My Life: A Record of Events and Opinions*. London: reprinted Farnborough, Gregg International, 1969.

WARD, LESTER. 1883. *Dynamic Sociology: Or Applied Social Science as Based upon Statistical Sociology and the Less Complex Sciences*. Reprinted with an introduction by David W. Noble, New York: Johnson Reprint Corporation, 1968.

———. 1906. *Applied Sociology: A Treatise on the Conscious Improvement of Society by Society*. Reprinted New York: Johnson Reprint Corporation, 1968.

WASSERMAN, GERHARD D. 1978. "Testability and the Role of Natural Selection within Theories of Population Genetics and Evolution." *Brit. J. Phil. Sci.* 29: 223–242.

———. 1982. "On the Nature of the Theory of Evolution." *Philosophy of Sci.* 48: 416–437.

WATCHTOWER BIBLE AND TRACT SOCIETY. 1967. *Did Man Get Here by Evolution or by Creation?* New York.

WEGENER, ALFRED. 1966. *The Origin of Continents and Oceans*. 4th German ed., 1929. Translation by J. Biram. New York: Dover.

WEINER, J. S. 1955. *The Piltdown Hoax*. Oxford: Oxford University Press.

WEINSTEIN, A. 1977. "How Unknown Was Mendel's Paper?" *J. Hist. Biology* 10: 341–364.

WEISMANN, AUGUST. 1882. *Studies in the Theory of Descent*. Translated by Raphael Meldola. London: reprinted (2 vols.) New York, AMS Press.

———. 1891–92. *Essays upon Heredity and Kindred Biological Problems*. Edited by E. B. Poulton et al. 2 vols. Oxford.

———. 1893*a*. *The Germ Plasm: A Theory of Heredity*. Translated by W. Newton Parker and Harriet Ronfeldt. London.

———. 1893*b*. "The All-Sufficiency of Natural Selection." *Contemporary Review* 64: 309–338, 596–610.

———. 1896. *On Germinal Selection*. Chicago. Also printed in *The Monist* 6: 250–293.

———. 1904. *The Evolution Theory*. 2 vols. Translated by J. Arthur Thomson and M. R. Thomson. London.

WEISS, SHEILA FAITH. 1986. "Wilhelm Schallmayer and the Logic of German Eugenics." *Isis* 77: 33–46.

WELDON, W. F. R. 1894–95. "An Attempt to Measure the Death Rate due to the Selective Destruction of *Carcinas moenas* with Respect to Particular Dimensions." *Proc. Roy. Soc.* 57: 360–379.

———. 1898. President's Address, Zoological Section. *Report of the British Association for the Advancement of Science,* 887–902.

———. 1901. "A First Study of Natural Selection in *Clausilia laminata.*" *Biometrika* 1: 109–124.

WELLS, GEORGE A. 1967. "Goethe and Evolution." *J. Hist. Ideas* 28: 537–550.

WELLS, HERBERT GEORGE. 1920. *The Outline of History: Being a Plain History of Life and Mankind.* Rev. ed. London.

WELLS, H. G., HUXLEY, JULIAN, and WELLS, G. P. 1930. *The Science of Life.* New York: Doran.

WELLS, KENTWOOD D. 1971. "Sir William Lawrence (1783–1867): A Study of Pre-Darwinian Ideas on Heredity and Variation." *J. Hist. Biology* 4: 319–362.

———. 1973*a*. "The Historical Context of Natural Selection: The Case of Patrick Matthew." *J. Hist. Biology* 6: 225–258.

———. 1973*b*. "William Charles Wells and the Races of Man." *Isis* 64: 215–225.

WERNER, ABRAHAM GOTTLOB. 1971. *Short Classification and Description of the Various Rocks.* Translated by Alexander M. Ospovat. New York: Hafner Publishing Co.

WERSKEY, GARY. 1978. *The Visible College: The Collective Biography of British Scientific Socialists of the 1930s.* New York: Holt, Rinehart and Winston. London: Allan Lane.

WESTFALL, RICHARD S. 1958. *Science and Religion in Seventeenth-Century England.* New Haven: Yale University Press.

WHEWELL, WILLIAM. 1847*a*. *History of the Inductive Sciences.* 3 vols. New ed. London: reprinted Hildesheim, Georg Olms.

———. 1847*b*. *Philosophy of the Inductive Sciences.* 2 vols. New ed., London: reprinted with an introduction by J. W. Herivel, New York, Johnson Reprint Corporation.

WHISTON, WILLIAM. 1696. *A New Theory of the Earth from its Original to the Consummation of All Things.* London: reprinted New York, Arno Press, 1977.

WHITCOMB, J. C., and MORRIS, H. M. 1961. *The Genesis Flood.* Nutley, N.J.: Presbyterian and Reformed Publishing Co.

WHITE, ANDREW DICKSON. 1896. 2 vols. *A History of the Warfare of Science with Theology.* Reprinted New York: Dover, 1969.

WHITE, PETER. 1976. *The Past Is Human.* New York: Taplinger.

WHITEHEAD, ALFRED NORTH. 1925. *Science and the Modern World.* New York: Macmillan.

———. 1929. *Process and Reality: An Essay in Cosmology.* Cambridge: Cambridge University Press.

WHITMAN, CHARLES OTIS. 1894. "Bonnet's Theory of Evolution" and "The Palingenesia and the Germ Theory of Bonnet." *Biological Lectures, Woods Hole,* 225–240, 241–272.

WIENER, PHILIP P. 1949. *Evolution and the Founders of Pragmatism.* Cambridge, Mass.: Harvard University Press.

WILKIE, J. S. 1955. "Galton's Contribution to the Theory of Evolution, with Special

Reference to His Use of Models and Metaphors." *Annals of Science* 11: 194–205.
———. 1956. "The Idea of Evolution in the Writings of Buffon." *Annals of Science* 12: 48–62, 212–247, 255–266.
———. 1962. "Some Reasons for the Rediscovery and Appreciation of Mendel's Work in the First Years of the Present Century." *Brit. J. Hist. Sci.* 1: 5–17.
WILLEY, BASIL. 1940. *The Eighteenth-Century Background: The Idea of Nature in the Thought of the Period*. London: Chatto and Windus.
———. 1949. *Nineteenth-Century Studies: Coleridge to Matthew Arnold*. London: Chatto and Windus.
———. 1956. *More Nineteenth-Century Studies: A Group of Honest Doubters*. London: Chatto and Windus.
———. 1960. *Darwin and Butler: Two Versions of Evolution*. London: Chatto and Windus.
WILLIAMS, G. C. 1966. *Adaptation and Natural Selection*. Princeton: Princeton University Press.
———, ed. 1971. *Group Selection*. Chicago and New York: Aldine Atherston.
WILLIAMS-ELLIS, AMABEL. 1966. *Darwin's Moon: A Biography of Alfred Russel Wallace*. London and Glasgow: Blackie.
WILSON, DAVID B. 1977. "Victorian Science and Religion." *History of Science* 15: 52–67.
WILSON, EDWARD O. 1975. *Sociobiology: The New Synthesis*. Cambridge, Mass.: Harvard University Press.
———. 1978. *On Human Nature*. Cambridge, Mass.: Harvard University Press.
WILSON, LEONARD G. 1967. "The Origins of Charles Lyell's Uniformitarianism." *Geol. Soc. Am., Special Paper* 89: 35–62.
———. 1969. "The Intellectual Background in Charles Lyell's *Principles of Geology*, 1830–33." *In* Schneer, ed., *Toward a History of Geology*. Pp. 426–443.
———. 1972. *Charles Lyell: The Years to 1841: The Revolution in Geology*. New Haven: Yale University Press.
———. 1980. "Geology on the Eve of Charles Lyell's First Visit to America, 1841." *Proc. Am. Phil. Soc.* 124: 168–202.
WINSLOW, JOHN. 1971. "Darwin's Victorian Malady: Evidence for Its Medically Induced Origins." *Mem. Am. Phil. Soc.* 88.
WINSOR, MARY P. 1976. *Starfish, Jellyfish and the Order of Life*. New Haven: Yale University Press.
———. 1979. "Louis Agassiz and the Species Question." *Stud. Hist. Biology* 3: 89–117.
WOHL, R. 1960. "Buffon and the Project for a New Science." *Isis* 51: 186–199.
WOOD, ROBERT MUIR. 1985. *The Dark Side of the Earth*. London: Allen and Unwin.
WOODCOCK, GEORGE. 1969. *Henry Walter Bates: Naturalist of the Amazons*. London: Faber and Faber.
WOODWARD, JOHN. 1695. *An Essay toward a Natural History of the Earth and Terrestrial Bodyes*. London: reprinted New York, Arno Press, 1977.
WRIGHT, SEWALL. 1930. "The Genetical Theory of Natural Selection: A Review." *J. Heredity* 21: 349–356.
———. 1931. "Evolution in Mendelian Populations." *Genetics* 16: 97–159.

———. 1966. "Mendel's Ratios." *In* Stern and Sherwood, eds., *The Origins of Genetics*. Pp. 173–175.

———. 1968–78. *Evolution and the Genetics of Populations*. 4 vols. Chicago: University of Chicago Press.

———. 1986. *Evolution: Selected Papers*. Chicago: University of Chicago Press.

WYLLIE, IRVINE. 1959. "Social Darwinism and the Businessman." *Proc. Am. Phil. Soc*. 103: 629–635.

WYNNE-EDWARDS, V. C. 1962. *Animal Dispersion and the Relation to Social Behavior*. Edinburgh: Oliver and Boyd.

YEO, RICHARD. 1984. "Science and Intellectual Authority in Mid-Nineteenth-Century Britain: Robert Chambers and *Vestiges of the Natural History of Creation*." *Victorian Studs*. 28: 5–31.

YOLTON, JOHN W. 1983. *Thinking Matter: Materialism in Eighteenth-Century Britain*. Minneapolis: University of Minnesota Press.

YOUNG, ROBERT M. 1967. "Animal Soul." *In* Paul Edwards, ed., *The Encyclopedia of Philosophy*. 8 vols. New York and London: Collier Macmillan. 1: 122–127.

———. 1969. "Malthus and the Evolutionists: The Common Context of Biological and Social Theory." *Past and Present* 43: 109–145.

———. 1970*a*. *Mind, Brain and Adaptation in the Nineteenth Century: Cerebral Localization and Its Biological Context from Gall to Ferrier*. Oxford: Clarendon Press.

———. 1970*b*. "The Impact of Darwin on Conventional Thought." *In* Symondson, ed., *The Victorian Crisis of Faith*. Pp. 13–36.

———. 1971*a*. "Darwin's Metaphor: Does Nature Select?" *Monist* 55: 442–503.

———. 1971*b*. "Evolutionary Biology and Ideology: Then and Now." *Science Studies* 1: 177–206.

———. 1973. "The Historiographical and Ideological Context of the Nineteenth-Century Debate on Man's Place in Nature." *In* M. Teich and R. M. Young, eds., *Changing Perspectives in the History of Science*. Pp. 344–438.

———. 1985. *Darwin's Metaphor: Nature's Place in Victorian Culture*. Cambridge: Cambridge University Press.

YULE, G. UDNEY. 1902. "Mendel's Laws and Their Probable Relations to Intraracial Heredity." *New Phytologist* 1: 193–207, 222–238.

ZILSEL, EDGAR. 1945. "The Genesis of the Idea of Scientific Progress." *J. Hist. Ideas* 6: 346.

ZIRKLE, CONWAY. 1946. "The Early History of the Idea of the Inheritance of Acquired Characters and Pangenesis." *Trans. Am. Phil. Soc*. 35: 91–151.

———. 1949. *Death of a Science in Russia: The Fate of Genetics as Described in Pravda and Elsewhere*. Philadelphia: University of Pennsylvania Press.

———. 1951. "Gregor Mendel and His Precursors." *Isis* 42: 97–104.

———. 1959*a*. "Species before Darwin." *Proc. Am. Phil. Soc*. 103: 634–644.

———. 1959*b*. *Evolution, Marxian Biology and the Social Scene*. Philadelphia: University of Pennsylvania Press.

———. 1964. "Some Oddities in the Delayed Discovery of Mendelism." *J. Heredity* 55: 65–72.

———. 1968. "The Role of Liberty Hyde Bailey and Hugo De Vries in the Rediscovery of Mendelism." *J. Hist. Biology* 1: 205–218.

ZITTEL, KARL VON. 1901. *History of Geology and Paleontology*. Translated by M. M. Ogilvie-Gordon. London: reprinted Weinheim, J. Cramer, 1962.

ZMARZLIK, GUNTER. 1972. "Social Darwinism in Germany." *In* H. Holborn, ed., *Republic to Reich: The Making of the Nazi Revolution*. New York: Pantheon. Pp. 435–474.

Index

Designer: U.C. Press Staff
Compositor: Prestige Typography
Text: 10/12 Caledonia
Display: Helvetica Roman
Printer: Maple-Vail
Binder: Maple-Vail